# RADIO AMATEURS GUIDE TO THE IONOSPHERE

Leo F. McNamara

**KRIEGER PUBLISHING COMPANY**
MALABAR, FLORIDA
1994

Original Edition 1994
(Based on THE IONOSPHERE)

Printed and Published by
**KRIEGER PUBLISHING COMPANY
KRIEGER DRIVE
MALABAR, FL 32950**

Copyright © 1991 (Chapters 1–9) by Krieger Publishing Company
Copyright © 1994 (Chapter 10 and Appendixes) by Krieger Publishing Company (special edition for radio amateurs)

All rights reserved. No part of this book may be reproduced in any form or by any means, electronic or mechanical, including information storage and retrieval systems without permission in writing from the publisher.
*No liability is assumed with respect to the use of the information contained herein.*
Printed in the United States of America.

> FROM A DECLARATION OF PRINCIPLES JOINTLY ADOPTED BY A COMMITTEE OF THE AMERICAN BAR ASSOCIATION AND A COMMITTEE OF PUBLISHERS:
> This publication is designed to provide accurate and authoritative information in regard to the subject matter covered. It is sold with the understanding that the publisher is not engaged in rendering legal, accounting, or other professional service. If legal advice or other expert assistance is required, the services of a competent professional person should be sought.

**Library of Congress Cataloging-in-Publication Data**
McNamara, L. F.
   Radio amateurs guide to the ionosphere / Leo F. McNamara.
     p.  cm.
   ISBN 0-89464-804-7 (alk. paper)
   1. Ionospheric radio wave propagation.   I. Title.
   QC973.4.I6M35   1994
   621.384'11—dc20
                                                 92-32988
                                                      CIP

10 9 8 7 6 5 4 3 2

To Clarrie McCue, for his support in the difficult times

# Contents

*Preface* .................................................................................................. ix

## 1. THE IONOSPHERE IN PERSPECTIVE .................................................. 1

1.1 Introduction ......................................................................................... 1
1.2 Users of the Ionosphere ........................................................................ 2
1.3 About This Book .................................................................................. 3

## 2. THE QUIET SUN ..................................................................................... 5

2.1 Introduction ......................................................................................... 5
2.2 The Sun in White Light ....................................................................... 5
2.3 The Sun in $H_\alpha$ Light .............................................................................. 6
2.4 The Sun at EUV and X-Ray Wavelengths ........................................... 8
2.5 The Sun at Radio Wavelengths ............................................................ 10
2.6 The Solar Cycle .................................................................................... 11
2.7 Prediction of Future Solar Cycles ........................................................ 11
2.8 References ............................................................................................ 14
2.9 Problems .............................................................................................. 14

## 3. THE IONOSPHERE .................................................................................. 17

3.1 Introduction ......................................................................................... 17
3.2 Formation of the Ionosphere ............................................................... 17
3.3 Observing the Ionosphere ................................................................... 20
3.4 Variations of the Ionosphere ............................................................... 22
3.5 Sporadic E ........................................................................................... 28
3.6 The Equatorial Ionosphere .................................................................. 30
3.7 The Polar Ionosphere .......................................................................... 31
3.8 The Earth's Magnetic Field ................................................................. 31
3.9 Irregularities in the F Region .............................................................. 33
3.10 Travelling Ionospheric Disturbances ................................................... 34
3.11 Thermospheric Winds ......................................................................... 37
3.12 References ............................................................................................ 37
3.13 Problems .............................................................................................. 38

v

## 4. HF RADIO PROPAGATION ........ 39

    4.1 Introduction ........ 39
    4.2 Polarization ........ 40
    4.3 The Maximum Usable Frequency (MUF) ........ 42
    4.4 The Skip Zone ........ 43
    4.5 Propagation Modes ........ 44
    4.6 Multipath Interference ........ 45
    4.7 Choosing the Correct Antenna ........ 45
    4.8 Absorption ........ 47
    4.9 The Lowest Usable Frequency ........ 48
    4.10 Variability of the MUF ........ 49
    4.11 References ........ 49
    4.12 Problems ........ 49

## 5. IONOGRAMS AND THEIR INTERPRETATION ........ 51

    5.1 Introduction ........ 51
    5.2 Vertical Incidence (VI) Ionograms ........ 51
    5.3 Oblique Incidence (OI) Ionograms ........ 57
    5.4 Backscatter Ionograms ........ 61
    5.5 Real Height Analysis of Ionograms ........ 66
    5.6 Automatic Scaling of Ionograms ........ 72
    5.7 References ........ 72
    5.8 Problems ........ 73

## 6. PREDICTIONS FOR HF COMMUNICATIONS ........ 75

    6.1 Introduction ........ 75
    6.2 Models of the Ionosphere ........ 75
    6.3 Geometry of the Circuit ........ 76
    6.4 Calculation of the MUF and LUF ........ 77
    6.5 GRAFEX Frequency Predictions ........ 78
    6.6 Uses of HF Predictions ........ 83
    6.7 Errors in Predictions ........ 84
    6.8 Field Strength of the Received Signal ........ 86
    6.9 "Now-casting" the Ionosphere ........ 89
    6.10 Sources of HF Predictions and Solar-Terrestrial Forecasts ........ 89
    6.11 References ........ 89
    6.12 Problems ........ 90

## 7. COMMUNICATION PROBLEMS UNDER NORMAL CONDITIONS ........ 93

    7.1 Introduction ........ 93
    7.2 Fading ........ 93
    7.3 Sporadic E ........ 97
    7.4 Problems at Low Latitudes ........ 97
    7.5 Problems at High Latitudes ........ 98
    7.6 Polarization Mismatch ........ 100
    7.7 Noise and Interference ........ 101
    7.8 References ........ 102
    7.9 Problems ........ 103

## 8. DISTURBANCES TO NORMAL COMMUNICATIONS — 105

- 8.1 Introduction — 105
- 8.2 Solar Flares and Their Effects — 106
- 8.3 Coronal Holes and HSSWS — 114
- 8.4 Sudden Disappearing Filaments — 116
- 8.5 Geomagnetic Effects — 117
- 8.6 Bartels Charts — 117
- 8.7 UT and Seasonal Control of Geomagnetic Disturbances — 121
- 8.8 References — 122
- 8.9 Problems — 123

## 9. UNUSUAL HF PROPAGATION MODES — 125

- 9.1 Introduction — 125
- 9.2 Mid-Latitude $E_s$ Modes — 125
- 9.3 Low-Latitude $E_s$ Modes — 125
- 9.4 Transequatorial Propagation — 126
- 9.5 Round-the-World Propagation — 128
- 9.6 Non-Great-Circle Propagation — 129
- 9.7 6-Meter Observations of TEP — 130
- 9.8 References — 134
- 9.9 Problems — 134

## 10. PROPAGATION PREDICTION PROGRAMS — 135

- 10.1 Introduction — 135
- 10.2 Predictions of Usable Frequencies — 135
- 10.3 Field Strength Predictions — 141
- 10.4 Ionospheric Indices — 149
- 10.5 Programs Available — 151
- 10.6 Limitations of the Programs — 152
- 10.7 Conclusions — 153
- 10.8 References — 153

## APPENDIX A  FORMULAS FOR S/N AND FIELD STRENGTH — 155

## APPENDIX B  ANSWERS TO PROBLEMS — 157

*Bibliography* — 163

*Index* — 165

# Preface

There has long been a gap in the literature available to the serious radio amateur, with esoteric texts at one end and technically inadequate tracts at the other. This book bridges that gap, and provides a solid foundation for learning more about the fascinating world of HF communications. It covers the ionosphere and HF radio communications, as well as things that go wrong with both, simply and clearly, with just enough physics and mathematics to suit each reader. There are no attempts to bamboozle, the text is technically straightforward, and there are signposts to follow for more advanced treatments of the subject matter.

With the advent of the personal computer, and an explosion in the number of HF propagation prediction programs available, it has become possible for radio amateurs to really understand what is happening on their circuits and thus be able to optimize their communications. However these programs are just someone else's canned knowledge, and using them does not guarantee an increased understanding of the things that go wrong with HF communications and make the field such a challenging one. Nor does it guarantee success in efforts to establish an HF link. What that requires is this book, one of the better programs, and a desire to succeed.

The first nine chapters of this book have already appeared in *The Ionosphere: Communications, Surveillance and Direction Finding*. *The Ionosphere* has been well received, but reviewers from the radio amateur community have expressed concern that the material in Chapters 10 through 16 is of little interest to many of their readers. *Radio Amateurs Guide to the Ionosphere* therefore contains the first nine chapters of *The Ionosphere* untouched, since these chapters deal with the ionosphere and HF propagation. Missing are the chapter on Real Time Channel Evaluation, the four chapters on the ionospheric aspects of HF Direction Finding, the chapter on Over the Horizon Backscatter Systems, and the chapter on problems to test the keenest reader. The present Chapter 10 on Propagation Prediction Programs is entirely new, as are the appendixes. *The Ionosphere* is still available for readers who wish to follow up on the other topics.

Earlier versions of Chapters 1 to 9 (excluding 5) were serialized in the mid 1980's in Roger [VK2ZTB] Harrison's *Australian Electronics Monthly*, with the series title "Radio Communicators Guide to the Ionosphere." This material drew heavily on a Users' Training Course provided by my then employer, the Australian Government IPS Radio and Space Services. Many of the present illustrations, together with much moral support, were provided freely by World Data Center A in Boulder, Colorado; the Space Environment Services Center, also in Boulder; the University of Massachusetts Lowell Center for Atmospheric Research; and IPS Radio and Space Services in Sydney.

The new Chapter 10 discusses some of the programs available for use on personal computers, and uses one in particular (Advanced Stand Alone Prediction System, or ASAPS) to illustrate features of HF propagation which are best dealt with by example. ASAPS is first introduced in Chapter 6. PC-based propagation prediction programs are appearing all the time. One of the latest available is SPARC (System Prediction Analysis Radio Communication), which is issued by the Institute of Applied Geophysics in Moscow.

Radio amateurs have played an important role in advancing the study and use of the ionospheric propagation medium, since the very inception of radio. Before World War II amateurs reported strange bursts of noise on the upper HF bands, and these observations led to the creation of the whole new field of solar radio astronomy. The role of amateurs in the early studies of transequatorial propagation and other unusual propagation modes is also recognized as a key one, as can be seen from Chapter 9. The continuing activity of amateurs in the HF band will help to ensure that we will always be learning more and more about this fascinating subject.

Most writers associated with HF have a little soapbox they like to stand upon, and the present author is no exception. HF is cheap, mobile, robust, and all that, but it is not always that simple. It has come and gone over the years, alternately falling out of favor, and then being hailed as a savior. In the mid 1980's, it took just one satellite to be deliberately destroyed for the powers that be to recognize the truth of the ancient adage about eggs in baskets. In the early 1990's, a well-known international dispute took place rather inconveniently at a location where transionospheric satellite communications had to cope with the signals passing through the equatorial anomaly. Then peace broke out, and everything changed again!

HF is never going to go away, any more than the humble letter has gone away in the modern era of personal computers, fax machines, and cellular telephones. To be sure, the move with "wireless" communications is towards higher frequencies which are less and less affected by the ionosphere, and it is beginning to look like HF means *higher* frequencies. But there is still a class of communicator who must get messages through, but who does not have sufficient priority to snare a satellite link, especially in troubled times. There are also those users who want their message to get through immediately, and not be trapped in the uncertain world of communications networks. It is in areas such as these that you will find HF still going strong.

This book is dedicated to the memory of Clarrie McCue. Clarrie became my first boss at IPS in Sydney when I decided that I should finally face the real world and get a proper job. Clarrie always put people first, and he taught all of us at IPS some good lessons about the ionosphere and life in general. One lesson that I never forgot, once having learned it, was not to bring up the topic of his beloved football any time after lunch if I wanted to be home in time for tea!

# Chapter 1

# The Ionosphere in Perspective

## 1.1 Introduction

The first suggestion that an electrically conducting layer, which we now call the Ionosphere, exists in the Earth's atmosphere came from physicists studying the Earth's magnetic field. The small daily variations of the magnetic field had been known since the eighteenth century, and C. F. Gauss speculated in 1839 that their origin might lie in atmospheric electric currents. In 1860, Lord Kelvin also speculated on the existence of a conducting layer, in connection with the phenomenon of atmospheric electricity.

The conducting layer theory soon became topical in quite another way. In 1901, soon after the invention of radio, Guglielmo Marconi succeeded in transmitting signals from England to North America over the bulge of the intervening Atlantic Ocean. This achievement was difficult to explain, for it was known that radio signals were a form of electromagnetic wave motion like light, the only difference being that the wavelength was much longer. It was also known that electromagnetic waves travel nearly, but not quite, in straight lines and the inevitable small amount of bending, which was produced by the process known as diffraction, was calculated to be much too small to account for the travel of radio waves around the Earth.

In 1902, three men independently furnished, albeit in somewhat vague terms, the correct explanation of Marconi's demonstration, suggesting a permanent electrically conducting layer high in the atmosphere. These were O. Heaviside in Great Britain ("a somewhat wayward scientific genius"), A. K. Kennelly at Harvard, and K. Nagaoka in Japan. In an article in the Encyclopedia Britannica, Heaviside wrote that "There may possibly be a sufficiently conducting layer in the upper air. If so, the waves will, so to speak, catch on it, more or less. Then the guidance will be by the sea on the one side and the upper layer on the other." A similar idea was proposed by Kennelly, although he was doubtful whether such a layer would alone account for all the reports of long distance reception. In his original paper Kennelly seems to have foreseen the future of ionospheric physics: "As soon as the long distance wireless waves come under the sway of accurate measurements, we may hope to find, from the observed attenuations, data for computing the electrical conditions of the upper atmosphere."

The theoretical idea of a conducting layer was given more substance about a decade later when the radio engineer W. H. Eccles suggested that the number of ions, or electrically charged atoms, increased with the height of the atmosphere above the Earth. If this were so, he postulated, then radio waves passing through these increasingly ionized regions would "surely be bent to follow the curvature of the Earth."

In the meantime, J. E. Taylor (1903) and J. A. Fleming (1906), among others, had suggested that the conducting layer was produced by the ionizing action of ultraviolet light from the sun on the upper atmosphere. This implied the solar control of radio propagation, which was confirmed as soon as commercial communication links were set up across the Atlantic. The measured strengths of the signals were found to vary in a regular way throughout the day, season, and solar cycle, and to be disturbed when the Earth's magnetic field was disturbed. Experiments which indirectly measured the height of the ionized reflecting layer were conducted over more than a decade, between 1910 and 1925.

In his 1910 textbook, *Principles of Wireless Telegraphy*, G. W. Pierce proposed that waves reflected from the ionized layer might interfere with the direct or ground wave. Pierce states: "This interference, if it should exist, would intensify waves of some wavelength, so that by changing the wavelength through a range corresponding to a half period it should be possible to turn the interference to advantage. No such effects have been found . . . ." However, in 1912, Lee De Forest described how a 3,260 meter (90 kHz) "main wave" from Los Angeles would fade out at San Francisco (560 km north), while the 3,100 m "compensation wave" remained at full strength. At the same time, the main wave as observed at Phoenix, Arizona (400 km to the east), maintained normal amplitude. De Forest proposed that the new phenomenon must be caused by wave interference effects, since nothing else could explain such changes in amplitude at frequencies separated by less than 5%. He then ventured to suggest that the evidence to which Pierce alluded had been found. The two men corresponded on the subject, and even calculated a reflection height of 99 km, a very good estimate as it turned out.

De Forest's work was done while he was with the Federal Telegraph Company. What we today call frequency selective fading (interference between the ground and sky waves) must have become apparent as soon as the chain of Federal stations was placed in service, although the effect occurred primarily in the evening hours and was not seen by day. However the Federal engineers, while recognizing the unusual nature and scientific implications of the fading, gave first priority to solution of the problem. This resulted in the quick-shift-telegraph-compensation key.

Final experimental proof of the existence of a reflecting layer came in 1924 with the "frequency change" experiments of E. V. Appleton and M. Barnett, using the BBC transmitter at Bournemouth. The idea was basically simple. The signals broadcast from Bournemouth at a slowly varied wavelength would be picked up after coming by two different routes, one along the ground and the other after reflection from the hypothetical reflecting layer. When the difference in lengths of these two routes was a whole number of wavelengths, they would combine to produce a strong signal. When the path difference was equal to an odd number of half wavelengths, the received signals would tend to cancel each other out and the resulting signal would be weak.

A simple equation relating the mean wavelength, the small continuous change in wavelength by the transmitter, and the number of "fadings" would enable the difference in length of the two routes to be estimated. Since the distance between the transmitter and receiver was known, simple triangulation would allow the height of the reflecting layer to be calculated.

Appleton reported that "natural fading" occurred while the fringes were being produced by the comparatively slow wavelength change. But when the results were examined there was no doubt that there were more fades while the wavelength was changing than there were when it was constant, and from the number of these extra fades or fringes, it was deduced that the reflection had been from a height of about 100 km. The first direct measurements of the reflecting layer were made by the Americans G. Breit and M. A. Tuve in 1925, using a pulse-sounding technique, a forerunner of modern radar.

Meanwhile, Appleton fully realized the possible implications of his experiments and proceeded to concentrate on the upper atmosphere to the exclusion of all else. The name "ionosphere" was proposed by the Australian R. A. Watson-Watt in 1926, although it did not enter the literature until a few years later. Ionospheric physics became an experimental subject in its own right, its development being heavily influenced by Appleton. It is interesting to note that it was in the same year, a quarter of a century after Marconi's successful communications between Cornwall and Newfoundland, that a well-known Post Office electrical engineer, was deploring that "a good deal of time, money and energy is being wasted on the pursuit of this academic myth, of a useful ionized layer."

In his own words, Appleton was kept "as busy as a bee" and had long strings of questions to answer. He had proved that the ionized layer did indeed exist, but what were its properties? How was it formed? How did it vary with height, time of day, season, and so on?

Appleton was not alone, of course, and by the time of the Second World War, which gave a strong impetus to radar and radio communications technology, the overall structure and variations of the ionosphere had already been well described, but remained, in general, poorly understood. The post-war decades, especially with the advent of scientific rockets and satellites, have seen an enormous increase in our understanding of the ionosphere and its place in the whole complex solar terrestrial environment. The December 1974 special issue of the *Journal of Atmospheric and Terrestrial Physics* is a very good source for anyone interested in the early history of the ionosphere.

These years have also seen great advances in our understanding of how the ionosphere affects, as well as supports, radio wave propagation, with advances in communications technology going hand in hand with advances in our knowledge of the ionosphere and all its peculiarities. We have come a long way since the days of the "academic myth of a useful ionized layer."

## 1.2 Users of the Ionosphere

The traditional user of the ionosphere for the first six decades after Marconi's successful demonstration of "wireless" communications has been the HF communicator. It was not that the communicator ever wanted to rely on such an uncooperative medium. Given the choice, most professional communicators would prefer to use another medium, but in some circumstances HF communications are all that is available. HF communications win out over the more reliable methods of real-time communications, the cables and the satellites, only when its peculiar advantages are overwhelming. These advantages include cheapness, mobility, and robustness. HF communications find important applications in the less-developed countries, which have not yet built up the expensive infrastructure for other methods of communication. The local geography can also make it impossible to use line of sight techniques such as satellite communications and even VHF communications. It is also a reasonably straightforward procedure to disrupt a communications link by disabling a satellite or cutting a cable. On the other hand, it would take a deadly nuclear war to disrupt HF communications via the ionosphere.

In the late 1950's, R. F. Treharne, Z. R. Jeffrey and C. G. McCue developed an operational system for determining the location of an HF transmitter which *relied* on the signals being reflected by the ionosphere. This technique, known as single station location (SSL), demands an even greater knowledge of the ionosphere than does HF communications. An SSL system is a direction finding (DF) system which measures the elevation angle of the incoming radio waves, as well as its azimuth. If we know the ionosphere well enough, radio waves can be traced in simulation from the SSL site back towards the transmitter—where the rays hit the ground is where the transmitter should be. HF communications are adequately described by the average behavior for a month at a fixed hour, provided this behavior can be predicted some time ahead with acceptable accuracy. SSL, on the other hand, requires a much more detailed knowledge of the ionosphere at the time the SSL observations are made. Practical applications of the SSL technique include search and rescue, frequency regulation, law enforcement in general, and military intelligence.

The level of ionospheric knowledge required by the user increased dramatically with the introduction of operational over the horizon HF backscatter radars for frequency management and surveillance. The returning signals for this radar have been reflected by the ionosphere on the way out to the reflection area, and on the way back. Over the horizon (OTH) radars are similar to SSL systems in the way that they must take the ionosphere into account. The main difference is that an SSL system measures the elevation and azimuth of the incoming wavefront, whereas an OTH backscatter (BS) system measures the time delay for signals at different frequencies. OTH BS radars are used for frequency management of no-acknowledge HF links, to determine the location of aircraft and ships, and for the remote sensing of the state of the sea.

In all three applications, the more we know about the ionosphere the more likely our deductions are to be correct. Simply ignoring the ionosphere, or looking for an "engineering" solution which does not get to the physical cause of a particular problem, can only lead to less than ideal solutions.

## 1.3 About This Book

Knowledge of the ionosphere means different things to different people. The users, who actually use the systems to communicate at HF, to determine the location of an HF transmitter, to determine the location of a remote aircraft or ship, or to monitor the state of the sea for meteorological and other reasons, will need to know a lot about their systems, but only enough about the ionosphere so they can use those systems effectively and with the highest obtainable accuracy. At the other end of the scale, there are the ionospheric physicists, who want to know all about the ionosphere, especially the more interesting things. They usually do this by experiment and analysis of the experimental data and are often only vaguely concerned with the limitations which the ionosphere will place on any operational system.

This book is designed to bridge the gap between the two extremes, bringing some basic ionospheric physics to the user, and some flavor of the problems faced in practice to the physicist. The physicist works by measurement, analysis, theories, and models. The user concentrates on measurements, and uses the models developed by the physicist to relate the measurements to the particular problem to be solved. Between the two extremes, there is a spectrum of workers and ideas. The more the attention paid to all aspects of this spectrum, the more effective the operational solutions will be.

The subject matter covered is very much a function of my own professional experience, and there is an infinite amount of material not even alluded to. Some of this is covered well in other books. The books which I happen to refer to most are those by Davies (1965, 1969, 1990) and Hargreaves (1979). The conference proceedings which do not gather dust on my shelves are those of the Ionospheric Effects Symposia (every three years from 1975 to 1990), and the International Solar Terrestrial Workshops (Boulder, 1979; Meudon, 1984; Leura, 1989). My copy of CCIR Volume 6 "Propagation in Ionized Media" is also rather battered.

The book consists of three main sections, covering HF communications (Chapters 4 to 10), SSL (Chapters 11 to 14), and OTH BS (Chapter 15). Chapter 2 is a brief introduction to the sun, since it is the sun which is the overwhelming driving force in the ionosphere, while Chapter 3 is a very brief overview of the ionosphere. The level of complexity of the book increases in step with the level of complexity of the subject matter. Because of the nature of the text, many of the questions set at the end of each chapter are review questions, and the reader should be able to answer them with little difficulty. The problems posed in Chapter 16 are a different matter. Solutions to these problems are available to bona fide professors from the author.

# Chapter 2

# The Quiet Sun

## 2.1 Introduction

Before discussing the ionosphere and our uses of it, it is important to consider some of the features of the sun and see how they directly or indirectly influence the ionosphere. The sun can be studied at two levels, the "quiet" sun and the "disturbed" sun. In this chapter we shall be concerned with the quiet sun, which means that we shall be concerned with what happens most of the time. When the sun is even mildly disturbed, HF communications and other applications of the ionosphere can be completely disrupted, but we shall defer discussion of these problems to Chapter 8.

The sun is just an average star, with a radius of $7 \times 10^5$ km, over 100 times that of the Earth, and a mass of $2 \times 10^{30}$ kg, over 300,000 times that of the Earth. When we "look" at the sun, we see different things, depending on the wavelength of the radiation that we decide to use. (*No one ever looks directly at the sun; the sun's image is either projected onto a screen, viewed using very expensive filters, or studied by photographic means.*) In practice, the sun is studied at all possible wavelengths, ranging from X rays to radio waves, enabling a comprehensive picture of the sun to be built up. Figure 2.1 shows pictures of the sun in white light (left) and in the red $H_\alpha$ line of hydrogen (right), and illustrates the different features which are seen at different wavelengths. Figure 2.2 is a cutaway sketch of the sun, as it appears at different wavelengths. We shall look more closely at the various features in the following sections.

## 2.2 The Sun in White Light

If we use a telescope and project an image of the sun on to a sheet of paper, without using any filters, we get what is called a **white light image** of the sun. Basically, it will be a bright disc which will shimmer as we watch it. This shimmering is caused by the movement of the air between the sun and the telescope. When there is virtually no shimmer, we say that the "seeing" is good and solar

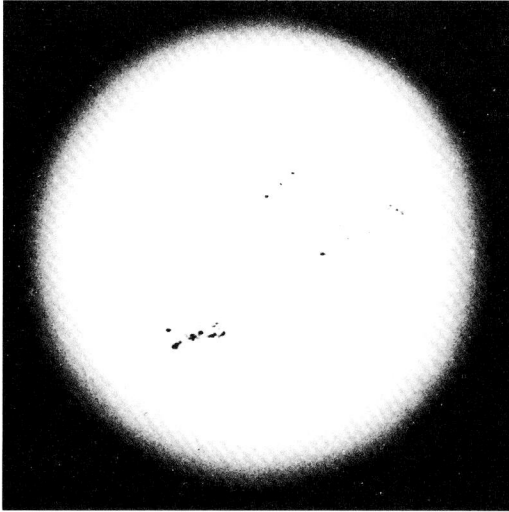
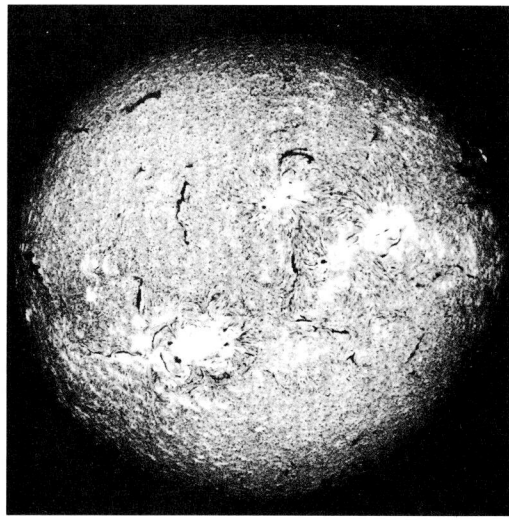

**Figure 2.1** Simultaneous photographs of the sun in white light (left) and in the red hydrogen $H_\alpha$ line. The white light photograph shows the *photosphere,* while the $H_\alpha$ photograph shows the overlying *chromosphere.* The low-lying sunspots are not always visible in $H_\alpha$ photographs. *Courtesy CSIRO Division of Physics.*

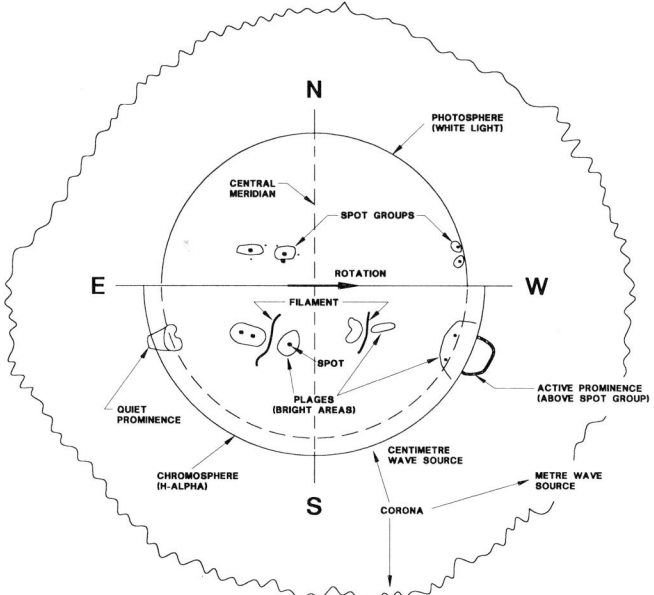

**Figure 2.2** Cutaway sketch of the sun, showing its appearance at different wavelengths. When we look at the sun in ordinary white light (upper half of sketch), we see the photosphere and any sunspots that happen to be present. When we look at a wavelength of 6,563 Å, we see some of the light emitted by hydrogen, in this case called the $H_\alpha$ light. At this wavelength, the chromosphere and its various features are visible—prominences, filaments, plages, and sunspots again. If we use EUV light or X rays, we can see the corona, which appears as a tenuous halo around the chromosphere. If we monitor the sun with radio telescopes, we detect radiation at a wavelength which depends on the altitude of the emitting region. *Courtesy IPS Radio and Space Services.*

observers go to great lengths and usually to very remote locations to ensure good "seeing" conditions. The solar observatories in Australia, for example, are near Narrabri (New South Wales) and Exmouth (Western Australia), while those in other countries include Mount Wilson (California), Kitt Peak (New Mexico), the Canary Islands, and Palehua (Hawaii).

The part of the sun that we "see" in white light is called the **photosphere** and has a temperature of about 6,000° C. It is only about 500 km thick, compared to the $1.4 \times 10^5$ km diameter of the sun, which is why the edge of the sun seems so sharp. The photosphere itself is of no direct concern to us here. We are more interested in the small dark patches that we can usually see on our white light image. These are called **sunspots** and have been recorded for thousands of years, Chinese observations of sunspots seen with the naked eye being available back to the first century B.C. However it was not until 1610, just after the invention of the telescope, that Galileo showed that the sunspots were actually on the surface of the sun. This caused great consternation at the time because the sun was supposed to be a perfect heavenly body and it had been shown to have warts! Galileo found that the

spots moved from east to west across the face of the sun, taking about 13 days to move the full width of the sun. The spots then disappeared for another 13 days or so, before reappearing again at the east limb of the sun. The movement of the spots indicated that the sun rotates, with a **solar rotation period** of approximately 27 days. The axis of rotation of the sun is found to be approximately north-south in the sky, parallel to the axis of rotation of the Earth.

As we shall see later, the number of sunspots on the face of the sun is a good indicator of the general level of the effect that the sun is having on the ionosphere. We would also like to know what spots are on the side of the sun facing away from the Earth, since we would then know what the sun has in store for us. However until there is a satellite launched into a suitable orbit around the sun, we will have to do without this useful information.

Sunspots normally occur in **sunspot groups** which may contain several clearly discernible spots, but sunspots may also occur by themselves. Figure 2.3 shows what the face of the sun looked like on 12 March 1989, which was a rather interesting time to be making SSL observations (see Chapter 11). Sunspots never occur near the north and south poles of the sun, but tend to cluster within about 30° of the solar equator. On the Earth, this would correspond roughly to being confined within the tropics. Sunspots look dark because they are cooler than the surrounding photosphere, but the temperatures within them are typically 4,000° C. The diameter of a typical sunspot is greater than the diameter of the Earth.

In studies of the ionosphere and HF propagation, we define a number called the **sunspot number** by counting the number of spot groups and individual spots and then forming the weighted sum

Sunspot Number = 10 × Number of sunspot groups + Number of individual spots.

This is not another case of "add on the number you first thought of"—what we are saying is that a sunspot group is 10 times as important as an individual spot. The sunspot number can be zero (completely spot-less), eleven (one sunspot, which is also regarded as one sunspot group) or more. Sunspot numbers in excess of 250 have been observed.

## 2.3 The Sun in $H_\alpha$ Light

$H_\alpha$ is one of the many spectral lines emitted by hydrogen. With a wavelength of 6,563 angstroms (Å) or $6.563 \times 10^{-7}$ m, it lies in the red part of the visible spectrum. We can see the sun in $H_\alpha$ by fitting a special filter in front of the eyepiece of a telescope. If we fitted a special

# The Quiet Sun

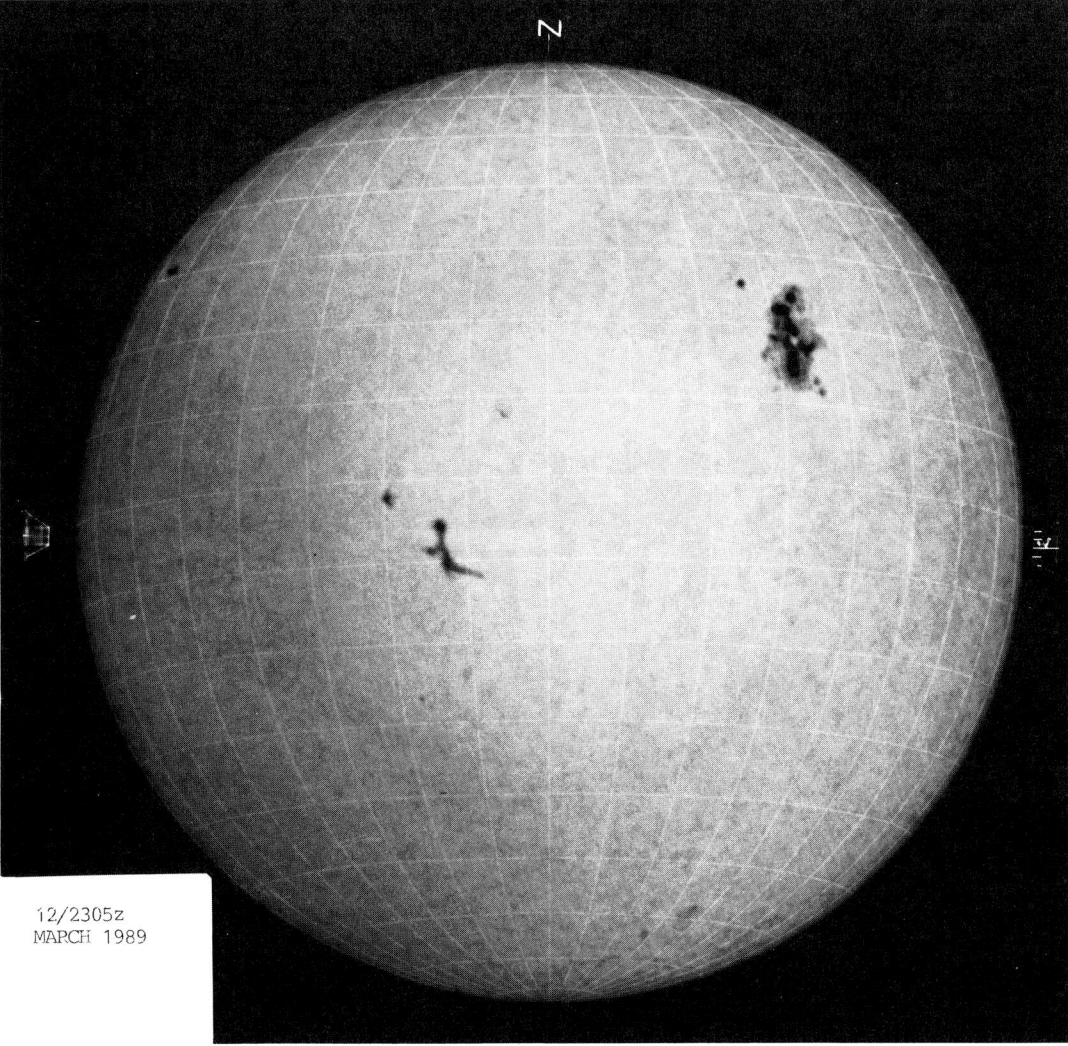

**Figure 2.3** White light photograph of the sun as it appeared on 12 March 1989, near the peak of solar cycle 22. *Courtesy IPS Radio and Space Services.*

yellow filter, we could see what the sun looks like in sodium light. The common yellow street lights use sodium vapor which is heated until it emits its characteristic yellow light. Actually what we would be seeing is what the *sodium* in the sun is doing.

When we look at the sun in $H_\alpha$, we cannot see as far into the sun as we do in white light. In other words, we see higher layers of the sun's atmosphere, and the sun appears larger. The layer above the photosphere which we can see only if we restrict ourselves to looking at a single wavelength or color, is called the **chromosphere.** "Chromos" is the Greek word for color.

The chromosphere lies directly above the photosphere, and is about 3,000 km thick. The temperature of the chromosphere rises from about 4,500° C at the top of the photosphere to nearly a million degrees Celsius at the top of the chromosphere. Such high temperatures are beyond our normal comprehension, but they can be considered simply as indicators that the charged particles (ions) which make up the chromosphere are moving at extremely high speeds.

The chromosphere exhibits a wide range of very detailed structure and is a beautiful sight to those lucky enough to see it through a telescope at a solar observatory. Figure 2.4 shows some of the main features which can be seen: plages, sunspots, prominences, filaments, and fibrils.

**Plages** (from the French word for "beach") are large, irregularly shaped bright areas, usually, but not always, associated with sunspots. Sunspots do not show up very well in $H_\alpha$ since they are lower in the sun's atmosphere than the chromosphere and are thus often hidden by the overlying chromosphere. The sunspots that we see in $H_\alpha$ are usually only the large ones. Plages are important to us because they emit copious amounts of EUV light, which we will find in Chapter 3 to be responsible for the

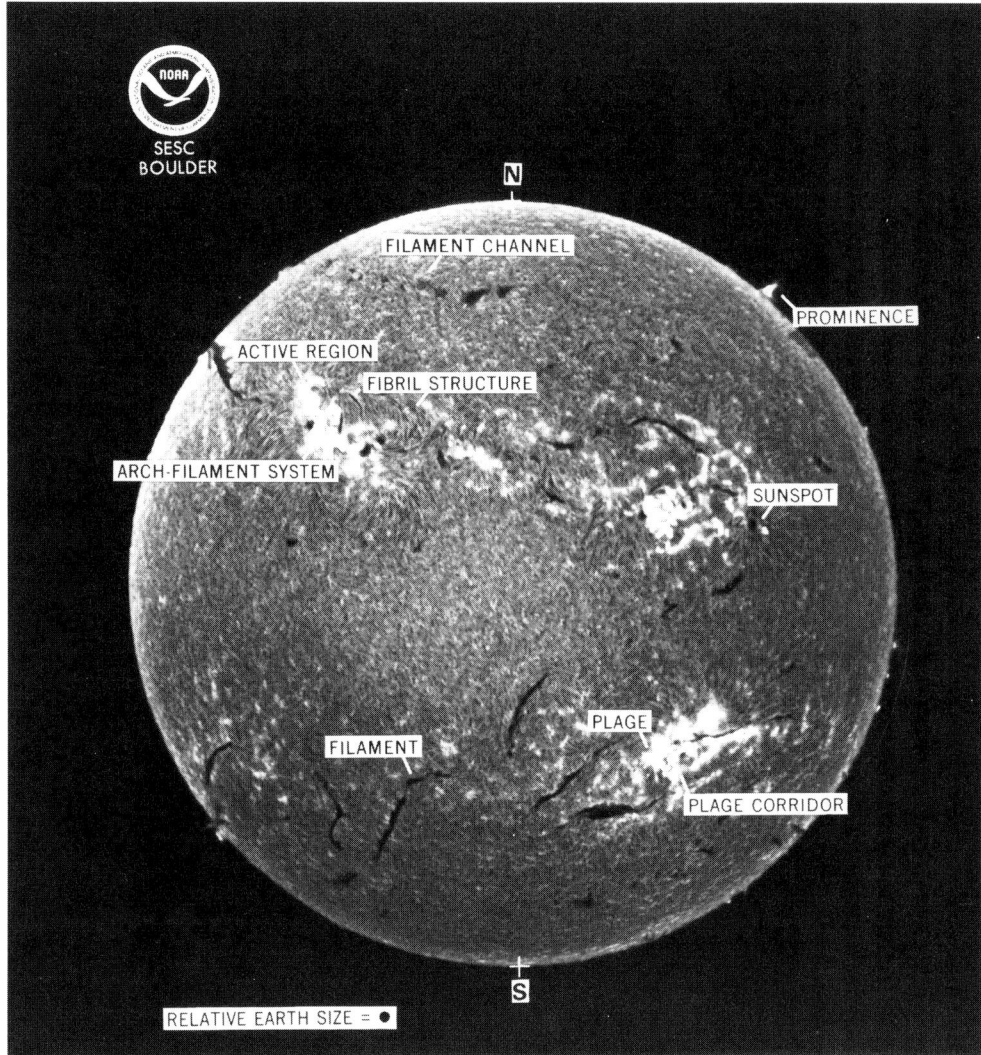

**Figure 2.4** A routine photograph of the sun at the $H_\alpha$ wavelength of 6,563 Å, with various features labelled.

formation of the ionosphere and thus the support of HF radio propagation.

Plages also go by other names, depending on the wavelength of the light used to observe them. If we use white light, they are called **faculae.** The regions containing plages and sunspots are known as **active regions** because they are continually changing. It is these regions which are of most importance to users of the ionosphere.

Prominences and filaments are the same thing seen from different perspectives. A **prominence** is a large cloud of relatively cool gas which is suspended above the surface of the sun by magnetic fields which restrain it from falling down. When this cloud of gas is seen on the edge of the sun, against the dark background of space, it appears bright and is known as a prominence. When viewed against the face of the sun itself, the cloud appears dark because it is relatively cool, and is known as a **filament.**

Filaments can reach lengths of $3 \times 10^8$ km and heights of $10^5$ km above the photosphere. They can be very stable, lasting for months, but may suddenly erupt and send a cloud of solar material out into space. If this cloud hits the Earth, it can cause changes to the Earth's magnetic field, to the ionosphere, and possibly to HF communications (see Chapter 8).

The background chromosphere between the features mentioned above shows a great deal of fine detail which is called the **fibril** structure because of its fibrous appearance. Around active regions, this structure is often ordered into large swirling patterns, apparently by magnetic fields. The fibrils are difficult to discern, except in high resolution photographs such as Figure 2.5.

## 2.4 The Sun at EUV and X-Ray Wavelengths

With the advent of scientific satellites and space stations such as Skylab, we are finally able to get telescopes

# The Quiet Sun

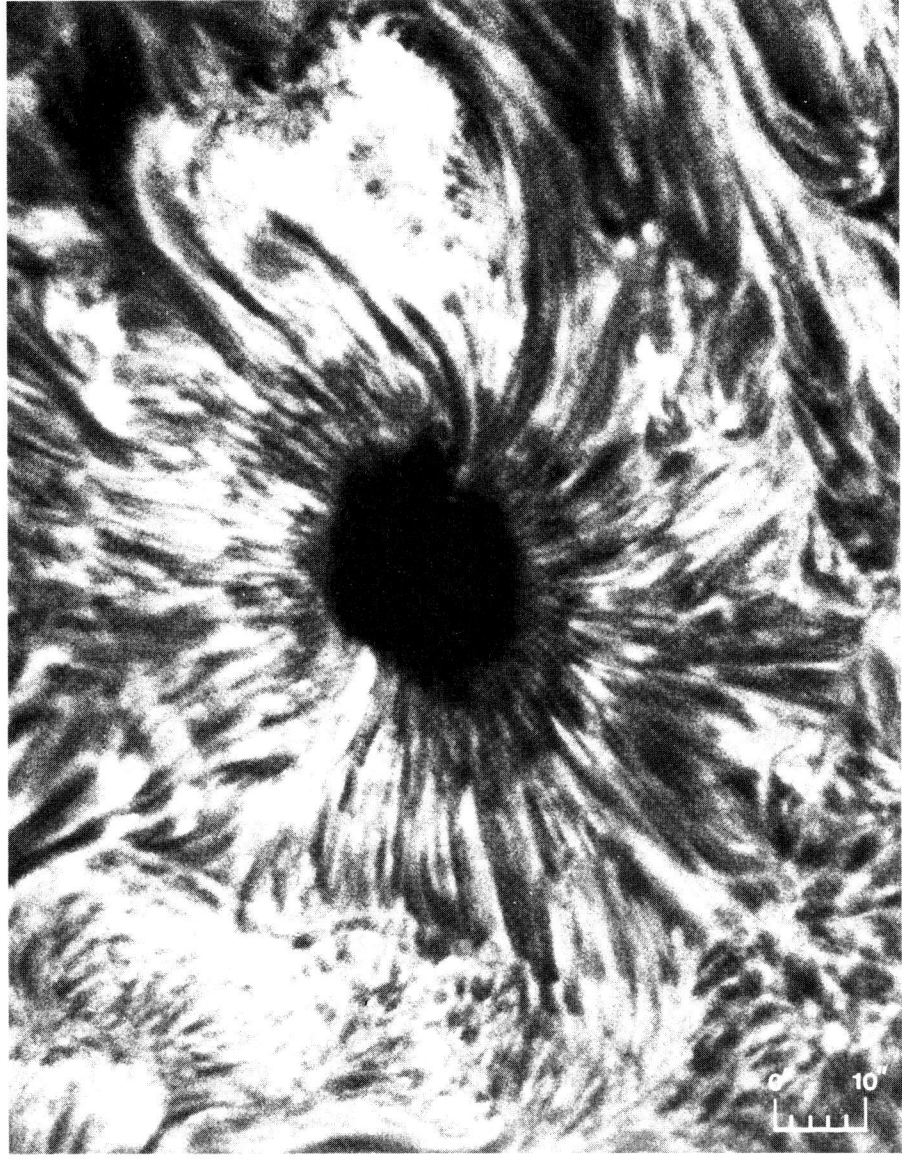

**Figure 2.5** A large sunspot photographed with a 30 cm telescope through an $H_\alpha$ filter with a pass band of 0.5Å. The spot is surrounded by a pattern of bright and dark streaks which map variations in the strength and direction of its magnetic field, which funnels outwards from the dark umbra. *Courtesy CSIRO Division of Physics.*

above the Earth's atmosphere and see what the sun looks like at very short wavelengths in the electromagnetic spectrum. The atmosphere absorbs extreme ultraviolet (EUV) radiation and X rays, forming the ionosphere in so doing and protecting mankind from annihilation. However grateful we may be for this benefit, it does mean that we cannot observe the sun in EUV or X rays from the surface of the Earth.

When we look at the sun in these very short wavelengths we see what is known as the **corona,** which is a tenuous halo or crown overlying the chromosphere. The optical brightness of the corona is only one millionth of that of the photosphere (roughly comparable with the full moon) and is less than that of the light scattered in a clear blue sky. Consequently the corona can be seen in visible light only when the light from the photosphere is removed, as in an eclipse of the sun. Figure 2.6 shows what the corona looked like during the eclipse of 12 November 1966.

The corona can be readily seen, however, at very short wavelengths. The temperature of the corona is very high, about two million degrees Celsius, and consequently it emits copious amounts of energy in the EUV and X-ray wavelengths. The cooler photosphere does not emit much energy at these wavelengths, which means that the corona appears relatively bright. Figure 2.7 shows what

**Figure 2.6** The solar corona photographed from the ground during the solar eclipse of 12 November 1966. A special filter compensated for the rapid decrease of intensity with radial distance from the sun, allowing distant features to be studied, as well as the brighter features close to the limb of the moon. A large coronal hole is clearly visible near the south pole, where there is apparently no corona. The white spot on the left is the planet Venus. *Courtesy High Altitude Observatory.*

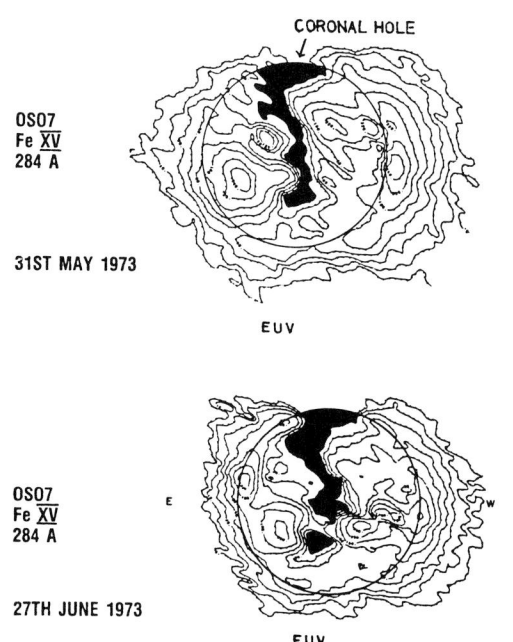

**Figure 2.7** The coronal hole observed by Skylab during its mission in early 1973, in extreme ultraviolet light. The black areas denote areas of low emission at these wavelengths and correspond to magnetically "open" regions in the corona which are the source of high speed solar wind streams. The lower illustration corresponds to a time 27 days (one solar rotation) after the upper one.

the corona looked like during the Skylab mission in 1973, in EUV (284 Å) light.

An interesting and important feature of the corona is the relatively cool and therefore dark areas which stretch equatorwards from either of the poles of the sun. These are known as **coronal holes.** The coronal hole observed by Skylab is shown in Figure 2.7. The time interval between the observations is 27 days, which means that we are seeing the same coronal hole for successive rotations of the sun. The eclipse photograph in Figure 2.6 also shows evidence of a large coronal hole covering the south polar region. Coronal holes are not important in themselves, but are important to the user of the ionosphere because they are the source of streams of charged particles which affect the ionosphere as they sweep over the Earth [1]. We shall return to them in Chapter 8.

The corona is very dynamic and contributes to the general outflow of material from the sun into interplanetary space in what is known as the **solar wind.** This "wind" carries several million tons of solar material away from the sun per second—a sobering thought for a man weighing a tenth of a ton at most. However the sun is not about to disappear. At the present rate, it would take 150 billion years for the sun to lose just 1% of its total mass. The solar wind flows at a speed of about 400 km/sec (roughly 900,000 mi/h), so that it takes about 5 days for individual charged particles to travel from the sun to the Earth. We cannot feel the solar wind at the surface of the Earth because there are only a few wind particles per cubic centimeter, which is a density far lower than any vacuum yet achieved on Earth. However the particles, being electrically charged, do affect the Earth's magnetic field and the ionosphere. It is the solar wind which pushes comet tails so that they always point away from the sun.

## 2.5 The Sun at Radio Wavelengths

Reversing the trend towards shorter wavelengths, we can also look at the sun at longer wavelengths, in particular at radio wavelengths. To do this, we use a radio telescope, which is essentially just a sophisticated radio receiver coupled to an extremely good antenna pointed at the sun. We can use frequencies between about 20 MHz (15 m wavelength) and 20 GHz (1.5 cm wavelength). The lower limit is set by the ionosphere, which will not allow lower frequencies to penetrate through to the ground, while the upper limit is set by practical considerations such as attenuation by water vapor or rain in the Earth's atmosphere. Radio telescopes work through cloud cover, especially at the low frequency end, and consequently offer important advantages over optical telescopes.

Many radio telescopes work on a single frequency, the output being a plot of signal amplitude versus time. The way this amplitude changes with time, and with fre-

# The Quiet Sun

quency, gives us important information about what is going on in the sun. It is also possible to tune some telescopes through a wide frequency range, measuring how the amplitude changes with both frequency and time. Both these types of radio telescope, the fixed frequency and swept frequency types, look at the sun as a whole and give no information about which active region may be giving rise to the changes. A third type of radio telescope, called a radio interferometer, which uses a large array of receiving antennas, can actually map the distribution of radio emission at a single frequency over the surface of the sun. However interferometers are very expensive and consequently rather rare.

One of the favorite frequencies at which the sun and the rest of the universe is studied is 2800 MHz or 2.8 GHz. This frequency corresponds to a wavelength of 10.7 cm, and the energy is emitted by neutral hydrogen in an atomic fine-structure transition. By studying the sun at all possible wavelengths in the electromagnetic spectrum, we are able to build up a comprehensive picture of what the sun does and how it affects the Earth. Once we have built up enough experience, we can forecast some time ahead what a particular event on the sun will do to the Earth, and in particular in the present context, what it will do to the ionosphere.

## 2.6 The Solar Cycle

If we were to observe the sun every day and calculate the sunspot number, we would find that not only would it vary as the sun rotates, but that it would also vary from zero to around 100 every 11 years or so. Of course we do not have to make these observations ourselves because scientists have been making them for hundreds of years, ever since Galileo first turned his telescope on the sun, and will continue to make them for many years to come.

Figure 2.8 shows how the sunspot number, smoothed to eliminate sudden changes from month to month, has varied from the year 1610 up to 1975. The sunspot number clearly goes up and down every 11 years or so, i.e., it has a cycle of about 11 years. Virtually everything associated with the sun of relevance to the ionosphere occurs with an approximately 11 year cycle, hence the name solar cycle. It can be seen from the figure that all solar cycles have not had the same number of sunspots at solar cycle maximum i.e., at a time when the number has its maximum value for the cycle. The 1957 maximum was the highest ever observed. Sometimes the maximum does not get very far above the minimum—between 1645 and 1715 there seem to have been virtually no sunspots at all, and certainly no evidence for a solar cycle. This period is known as the Maunder minimum, after Walter Maunder who, in the 1890s, drew attention to this inconvenient fact which had been ignored for the 200 years after the original observations.

Work by J. A. Eddy has confirmed the reality of the Maunder minimum, as well as indicating the existence of an earlier minimum of a similar nature between 1460 and 1550 [2]. It is quite feasible that we could currently be headed for another such minimum within the next few decades. Figures 2.9 and 2.10 illustrate the last few solar cycles, as they appeared in the smoothed sunspot number (Figure 2.9) and 10.7 cm flux (Figure 2.10). It can be seen that the flux and sunspot number are quite highly correlated. This means that we can often use the flux as an indicator of the general level of solar activity if we cannot get hold of the sunspot number. The 10.7 cm flux is relatively easy to observe, and is measured routinely several times a day at Ottawa, Canada.

## 2.7 Prediction of Future Solar Cycles

At the time of writing, solar activity has reached the highest expected values for cycle 22, and the question of how large the next maximum will be has already been asked. Why would we want to know now what the size of the year 2000 maximum (cycle 23) will be? The answer is that there are some very large and very expensive terrestrial and space programs which will take many years to

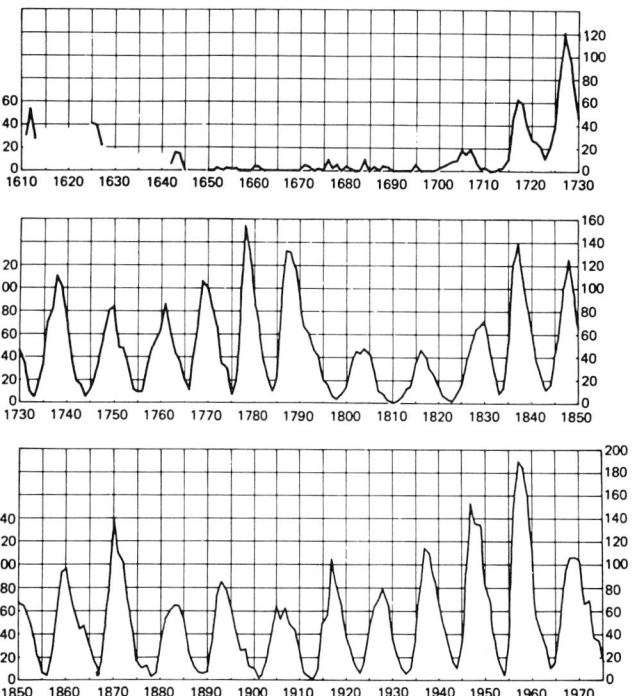

**Figure 2.8** The annual mean sunspot number (i.e., the average for a year) from 1610 to 1975. The number reaches a maximum every 11 years or so, but the maxima are not equal. The 1957 maximum was the largest ever recorded. The current cycle is one of the highest ever recorded. *After Jursa (1985).*

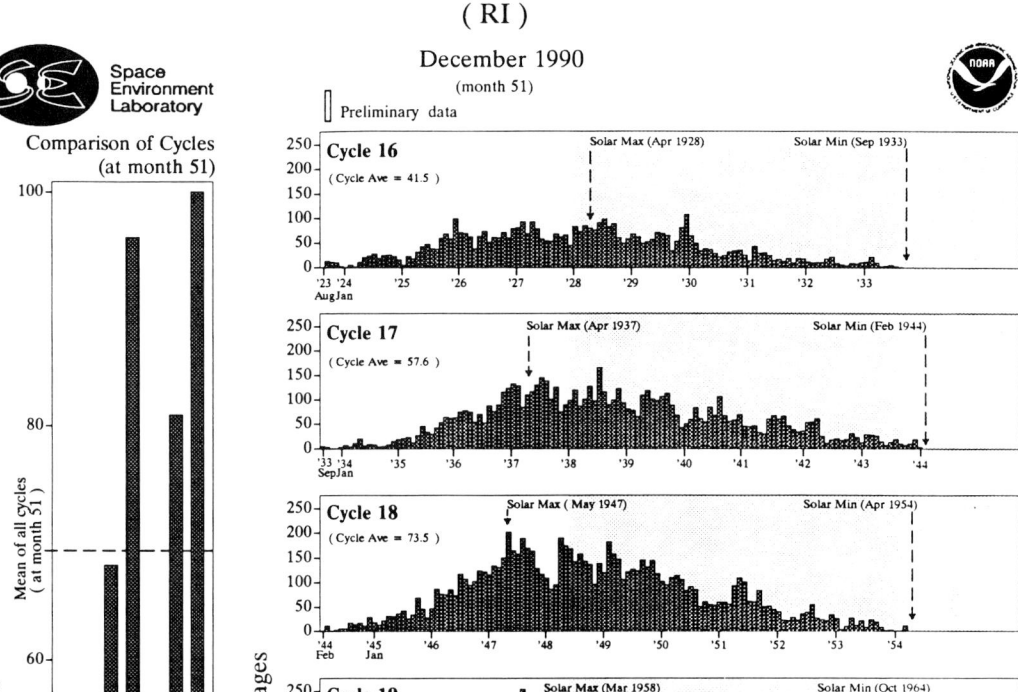

**Figure 2.9** Monthly averaged sunspot numbers from 1923 to 1990. The left-hand plot shows the integrated total sunspot number up to the end of month 51 for each cycle from cycle 16 onwards.

plan and implement, and which will be expected to last for at least a decade, thus covering a full solar cycle.

A good example of the need to predict the general level of solar activity some years in advance was given rather dramatically by the unplanned, and certainly unwanted, demise of Skylab over Australia in July 1979. One feature of high solar activity is that the higher the activity, the hotter the atmosphere of the Earth becomes, and the more it expands out into space. This means that a satellite revolving around the Earth encounters more resistance from the atmosphere that it passes through, known as **satellite drag,** and the satellite orbit decays to lower altitudes.

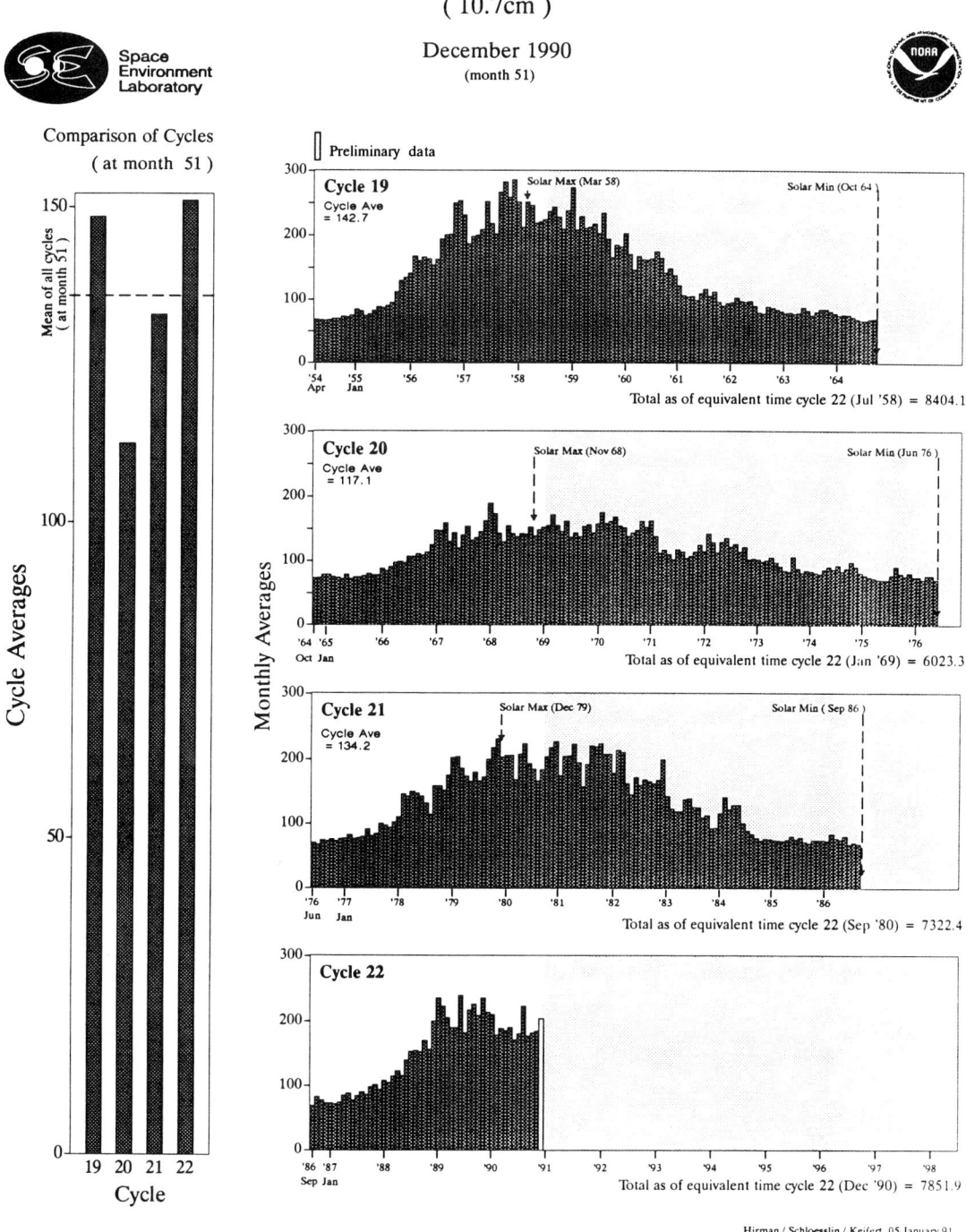

**Figure 2.10** Monthly averaged 10.7 cm solar flux from 1954 to 1990. The left-hand plot shows the integrated total flux up to the end of month 51 for each cycle from cycle 16 onwards.

The prediction of the size of cycle 21 (1976–1986) made by the Marshall Space Flight Center in 1977 was a low value of 72, which would be reached in January 1981. Assuming that this prediction was correct, it followed that NASA had several years of grace to mount a rescue mission for Skylab and boost it into a higher orbit. Unfortunately for NASA, cycle 21 passed through R = 72 in 1979, several years ahead of the predicted time, the rescue mission (using the Shuttle) was not mounted in time, and Skylab came tumbling down.

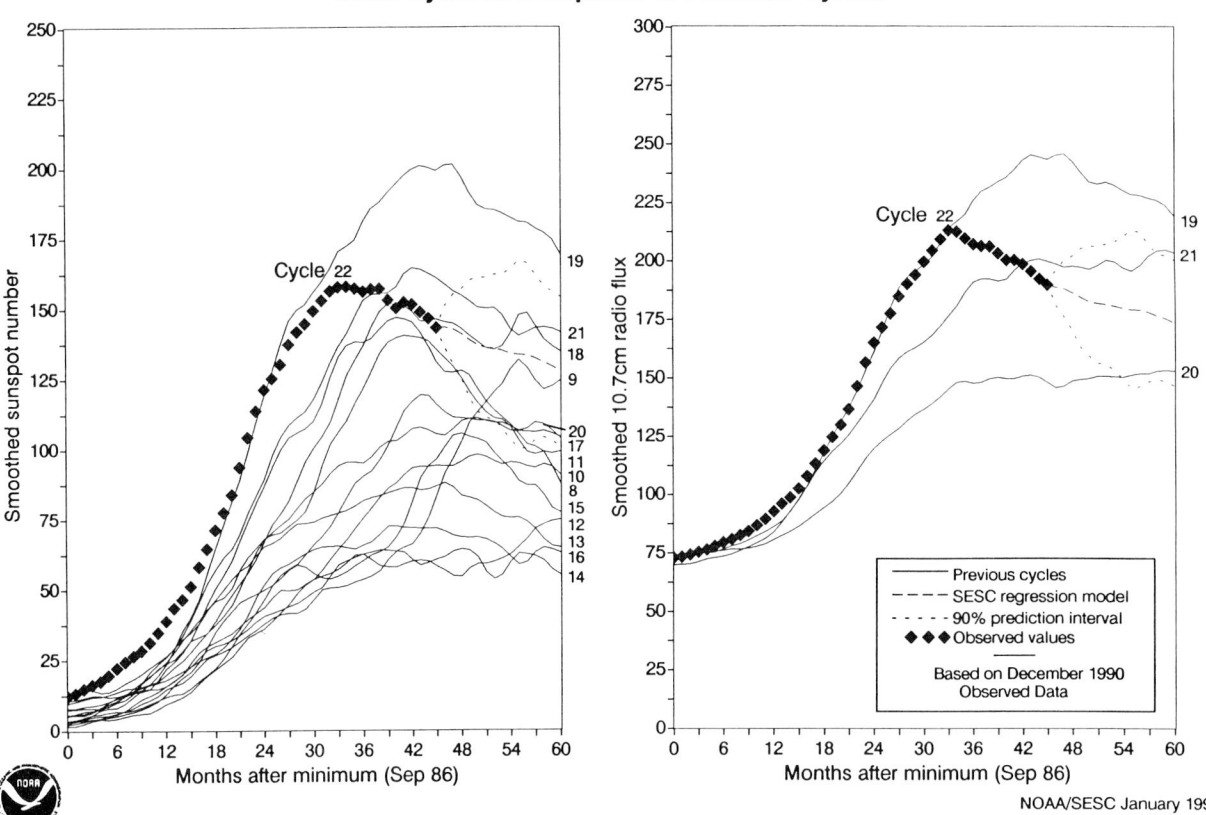

**Figure 2.11** The rise of solar cycle 22 compared to previous cycles. The rise for the first 18 months after solar minimum was the fastest ever observed.

The average cycle, averaged over cycles 1 to 22, reaches a peak of around 110 and has a period of 11 years—see Table 2.1. Figure 2.11 shows the sunspot number recorded during the past 14 cycles, up to the present peak. Cycle 22 appears to be the second largest observed in modern times. Figures 2.9 and 2.10, as well as several in Chapter 8, illustrate how cycle 22 compares with the previous cycles in terms of the different features which can be observed on the sun. Most people feel that it is not possible to make accurate predictions of the next maximum until after the next minimum is reached, in the mid-nineties [3, 4].

## 2.8 References

1. Zirker, J. B. (Ed.). *Coronal Holes and High Speed Wind Streams*. Colorado Associated University Press, Boulder, 1977.
2. Eddy, J. A. (Ed.). *The New Solar Physics*. Westview Press, Boulder, 1978.
3. Thompson, R. J. *The Rise of Solar Cycle 22*. Solar Physics 11, p. 279, 1989.
4. Thompson, R. J. *Geomagnetic Precursors of the Solar Cycle*. Solar-Terrestrial Physics Workshop, Leura, 1989.

## 2.9 Problems

1. Explain what you would see when you "look" at the sun in (a) white light (b) $H_\alpha$ light (c) EUV and X ray wavelengths, and (d) radio wavelengths. Are there any advantages to be gained by looking at the sun at all these wavelengths?

2. What difficulties are encountered when observing the sun? Are observations at some wavelengths harder to make than others?

3. What evidence exists that there are magnetic fields on the sun?

4. Given a white-light picture of the sun, how would you go about calculating the sunspot number?

5. What do we mean by the "solar cycle," and what is its typical period? Is this period a stable one?

6. When was the last solar maximum, and how did it compare to others which have been observed?

7. Billions of dollars can be lost because of an incorrect long-term forecast of the level of activity of the next solar cycle. Explain why this is so.

**Table 2.1** Statistics of the solar cycles 1 to 22, covering the period 1755 to 1990. The average size of the peak of the cycle maximum is 111.7, and the average period is 11.02 years. On average, a cycle peak is reached 4.29 years after solar minimum, and a minimum is reached 6.73 years after maximum. *Courtesy Space Environment Services Center.*

# Solar Cycles 1 – 22

( Min = smallest smooth # before Max )

( Smooth # = 13 month average )

| Cycle | Start (Solar min) Year Month | Solar Max Year Month | End (Month before min) Year Month | Max SSN | Length Years Months | Rise to Max Years Months | Max to End Years Months |
|---|---|---|---|---|---|---|---|
| 1 | 1755 Mar | 1761 Jun | 1766 May | 86.5 | 11.25  135 | 6.25  75 | 5.00  60 |
| 2 | 1766 Jun | 1769 Sep | 1775 May | 115.8 | 9.00  108 | 3.25  39 | 5.75  69 |
| 3 | 1775 Jun | 1778 May | 1784 Aug | 158.5 | 9.25  111 | 2.92  35 | 6.33  76 |
| 4 | 1784 Sep | 1788 Feb | 1798 Apr | 141.2 | 13.67  164 | 3.42  41 | 10.25  123 |
| 5 | 1798 May | 1805 Feb | 1810 Jul | 49.2 | 12.25  147 | 6.75  81 | 5.50  66 |
| 6 | 1810 Aug | 1816 Apr | 1823 Apr | 48.7 | 12.75  153 | 5.67  68 | 7.08  85 |
| 7 | 1823 May | 1829 Nov | 1833 Oct | 71.7 | 10.50  126 | 6.50  78 | 4.00  48 |
| 8 | 1833 Nov | 1837 Mar | 1843 Jun | 146.9 | 9.67  116 | 3.33  40 | 6.33  76 |
| 9 | 1843 Jul | 1848 Feb | 1855 Nov | 131.6 | 12.42  149 | 4.58  55 | 7.83  94 |
| 10 | 1855 Dec | 1860 Feb | 1867 Feb | 97.9 | 11.25  135 | 4.17  50 | 7.08  85 |
| 11 | 1867 Mar | 1870 Aug | 1878 Nov | 140.5 | 11.75  141 | 3.42  41 | 8.33  100 |
| 12 | 1878 Dec | 1883 Dec | 1890 Feb | 74.6 | 11.25  135 | 5.00  60 | 6.25  75 |
| 13 | 1890 Mar | 1894 Jan | 1901 Dec | 87.9 | 11.83  142 | 3.83  46 | 8.00  96 |
| 14 | 1902 Jan | 1906 Feb | 1913 Jul | 64.2 | 11.58  139 | 4.08  49 | 7.50  90 |
| 15 | 1913 Aug | 1917 Aug | 1923 Jul | 105.4 | 10.00  120 | 4.00  48 | 6.00  72 |
| 16 | 1923 Aug | 1928 Apr | 1933 Aug | 78.1 | 10.08  121 | 4.67  56 | 5.42  65 |
| 17 | 1933 Sep | 1937 Apr | 1944 Jan | 119.2 | 10.42  125 | 3.58  43 | 6.83  82 |
| 18 | 1944 Feb | 1947 May | 1954 Mar | 151.8 | 10.17  122 | 3.25  39 | 6.92  83 |
| 19 | 1954 Apr | 1958 Mar | 1964 Sep | 201.3 | 10.50  126 | 3.92  47 | 6.58  79 |
| 20 | 1964 Oct | 1968 Nov | 1976 May | 110.6 | 11.67  140 | 4.08  49 | 7.58  91 |
| 21 | 1976 Jun | 1979 Dec | 1986 Aug | 164.5 | 10.25  123 | 3.50  42 | 6.75  81 |
| 22 | 1986 Sep | | | | | | |
| Avg | | | | 111.7 | 11.02  132.3 | 4.29  51.5 | 6.73  80.8 |

# Chapter 3

# The Ionosphere

## 3.1 Introduction

We saw in Chapter 1 that the ionosphere is not just an academic myth, but a very useful ionized layer. It is the ionosphere which reflects HF radio waves, making possible all of the applications to which the ionosphere has been put over the last six decades. From the point of view of a radio wave, the ionosphere is a thick shell of free electrons, embedded in the Earth's neutral atmosphere, which envelops the Earth at altitudes from about 90 to 600 km, as illustrated in Figure 3.1. To the plasma physicist, the ionosphere is a "lightly ionized" plasma to experiment with and to understand. To the practical users, it is a propagation medium which allows them to receive HF radio signals from locations anywhere else in the world, enabling them to communicate, mount surveillance, or locate the source of the signals.

## 3.2 Formation of the Ionosphere

The ionosphere is formed when extreme ultraviolet (EUV) light from the sun strips electrons from the neutral atoms of the Earth's atmosphere. The more familiar ultraviolet light has a shorter wavelength than visible light, and is more energetic. Extreme ultraviolet light is even more energetic. When a bundle of EUV light (called a photon) hits a neutral atom such as an oxygen atom, its energy is transferred to an electron in the neutral atom which can then escape from the atom and dart freely around. The neutral atom thereby becomes positively charged (because it has lost a negatively charged electron) and is known as a positive ion. The process in which the photon strips an electron from a neutral atom, thus creating a positively charged ion, is known as **photoionization,** and is illustrated in Figure 3.2. That part of the atmosphere in which the ions are formed is called the ionosphere. Actually this is not a terribly good name from our point of view because it is the free electrons which reflect radio waves. The ions are over 20,000 times as heavy as the electrons, and are just too massive to respond to the rapid oscillations of a radio wave.

**Recombination** is the reverse of photoionization, with negatively charged electrons and positively charged ions combining together again to produce neutral atoms. This is the main process by which electrons are "lost" in the higher parts of the ionosphere. In the lower levels of the ionosphere, electrons are lost by the process of **attachment,** in which they are "attached" to neutral atoms which thus become negatively charged ions. Like the positively charged ions, the negative ions are much heavier than electrons and do not respond to the electromagnetic oscillations of radio waves.

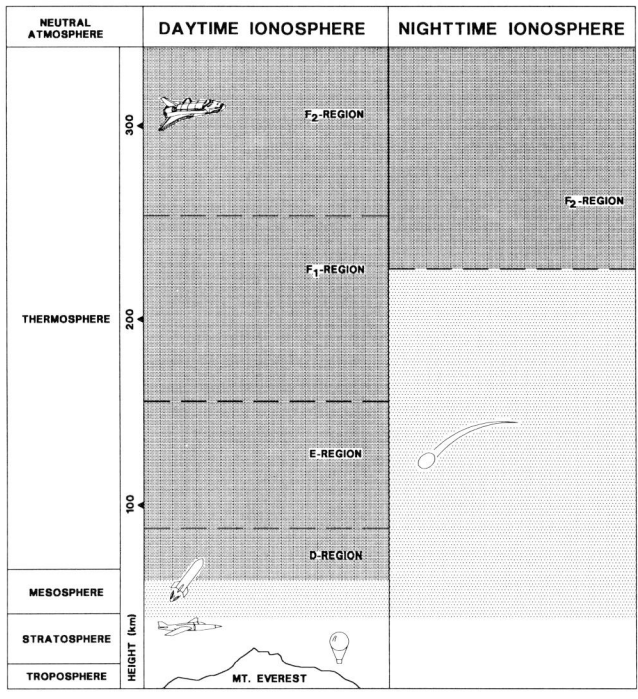

**Figure 3.1** Sketch of the various regions of the ionosphere as it appears during the day and during the night. The altitude regimes of the neutral atmosphere are shown for comparison. Four relatively distinct regions, the D, E, $F_1$, and $F_2$ regions, exist during a typical day, whereas only the $F_2$ region exists during the night. The shading represents the electron concentration, which exhibits local maxima in each of the four regions.

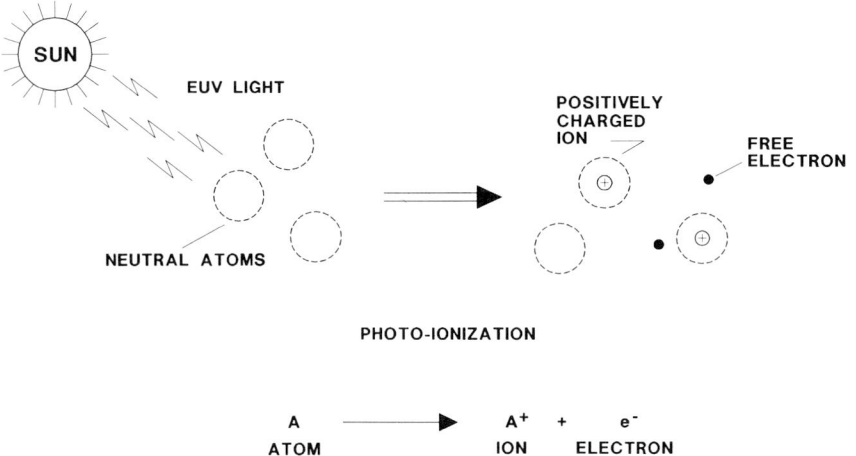

**Figure 3.2** Photoionization of a neutral atom, A, by extreme ultraviolet light from the sun, yielding a positively charged ion, $A^+$, and a free electron, $e^-$.

There are two types of recombination, radiative and dissociative, which are illustrated in Figure 3.3. In **radiative recombination,** the electrons combine directly with positively charged ions, converting them into neutral atoms and losing their own freedom. This process is not as important as **dissociative recombination** which occurs by a two-stage process and is much more efficient. In the first stage, positive ions, $X^+$ (formed by photoionization), interact with the numerous neutral molecules, $A_2$ (such as oxygen and nitrogen), replacing one of the atoms in the molecule:

$$X^+ + A_2 \rightarrow AX^+ + A. \quad (3.1)$$

In the second stage, electrons combine with the positively charged molecule $AX^+$, giving two neutral atoms and again losing their freedom:

$$AX^+ + e^- \rightarrow A + X. \quad (3.2)$$

An important thing to note is that the rate at which the electron density (we actually mean electron *concentration,* but it seems that there is tacit agreement to be sloppy in this particular case) is decreased by attachment and dissociative recombination will depend on how many neutral molecules are present at the altitude considered.

Recombination and attachment are always taking place, at all levels of the ionosphere. Photoionization, on the other hand, can occur only during the day when the sun is above the horizon. The net density of electrons in the ionosphere is the result of the imbalance between the two processes. A useful analogy is attempting to fill a leaky bucket with water. The amount of water in the bucket once equilibrium has been reached represents a balance between water coming in (photoionization) and water leaving through the holes (recombination). The electron density is greatest in the middle of the day when photoionization is at its greatest. When photoionization ceases at sunset, recombination eats away at the free electrons unimpeded and the density of electrons drops steadily as the night wears on. Recombination is not completely effective throughout the whole ionosphere, some free electrons surviving until dawn when their numbers are rapidly replenished by the rising sun. We know that recombination is not completely effective because we can still use the ionosphere at night.

The chances of some electrons remaining free, and not recombining with positively charged ions, are also increased by the fact that after sunset the ionosphere rises (or drifts) vertically to higher altitudes. The density of the neutral atmosphere decreases rapidly with height, so there are fewer neutral atoms around at the greater altitudes. Consequently, recombination is less effective at these higher altitudes.

The structure of the ionosphere at any particular location is not simple. The strength or intensity of the EUV light from the sun is not constant at all wavelengths but is much stronger at some particular wavelengths which correspond to the type of atom (for example, hydrogen) which is emitting it. The neutral atmosphere is also complex, with a wide range of atoms and molecules such as oxygen, nitrogen, and nitric oxide which can be photoionized. The situation is further complicated because the density of atoms which can be photoionized decreases as the altitude increases (recall the problems that climbers have on Mt. Everest), while the intensity of the EUV light which does the photoionizing decreases towards lower altitudes because the light has been partially absorbed (attenuated) on its way down through the upper levels of the atmosphere. The net result of these opposing effects, as illustrated in Figure 3.4, is to produce a layer

# The Ionosphere

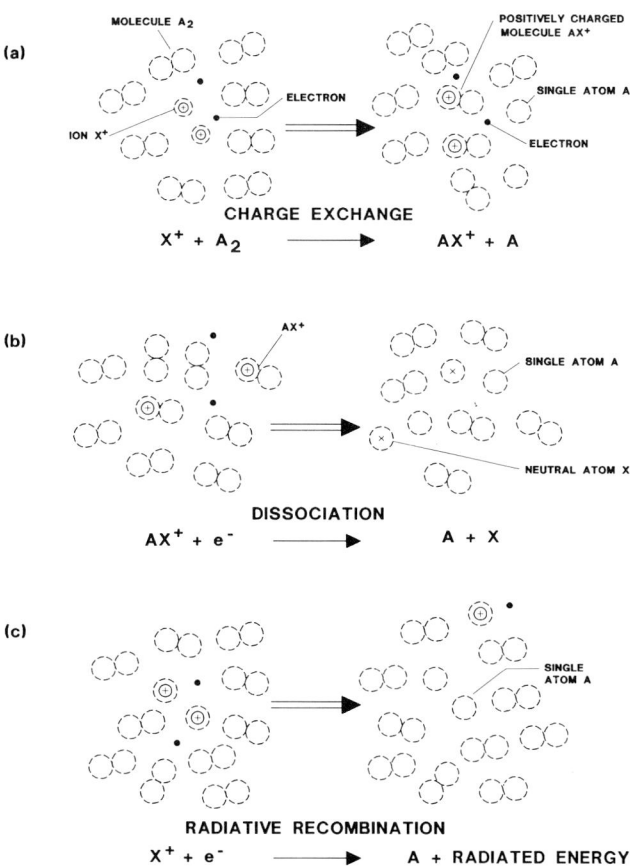

**Figure 3.3** Recombination of an electron, e⁻, with a positively charged ion, by dissociative recombination (panels a and b) and radiative recombination (panel c). In dissociative recombination, a positively charged ion ($X^+$) attaches itself to a neutral molecule ($A_2$), replacing one of the neutral atoms in the molecule, and giving rise to the positively charged molecule $AX^+$. In due course, an electron splits $AX^+$ apart (in other words, dissociates $AX^+$), yielding two neutral atoms A and X. The previously free electron thus becomes part of the atom X and is lost from the ionosphere. In radiative recombination, the free electron combines directly with any positively charged ion, giving a neutral atom, A, and radiating away any excess energy.

of electrons with a maximum electron density at some particular altitude and lower electron densities above and below this altitude. In fact, because the intensity of the EUV radiation varies with wavelength and because the neutral atmosphere contains as many different types of atoms and molecules as it does, the ionosphere may contain up to *four* different layers at different altitudes. The various ion production and recombination processes at work in the ionosphere are summarized in Table 3.1 [1].

Historically, the first layer of the ionosphere discovered was at around 100 km altitude and was called the E layer, with the "E" denoting electric field. The D layer covers the altitude range of about 50 to 90 km and the F layer covers the range of about 150 to 600 km. During the daytime in summer, the F layer splits into two separate layers known as the $F_1$ and $F_2$ layers, giving four layers D, E, $F_1$, and $F_2$. At night, recombination (and attachment) wins out over photoionization and the D, E, and $F_1$ layers almost completely disappear. The $F_2$ layer, on the other hand, survives throughout the night, albeit in a somewhat depleted fashion. This is one reason why it is the most important layer as far as HF propagation is concerned.

Figure 3.5 shows a typical daytime variation of the electron density with height. The ionosphere "starts" off at around 50 km altitude, with an electron density of only about 10 per cubic centimeter. At an altitude of about 80 km which marks the approximate boundary between the D and E layers, the density has increased to about 100 (or $10^2$—note the logarithmic scale) per cubic centimeter. The peak of the E layer lies at about 100 km and has an electron density of around $5.10^4$ per cm³. The $F_1$ region covers the altitude range of about 150 to 200 km, above which is the $F_2$ layer. The peak electron density in the $F_2$ layer is around $10^6$ per cm³. Above the $F_2$ peak, the density decreases approximately exponentially until it merges with the solar wind and loses its identity. The corresponding densities of the neutral atmosphere at the altitudes of the D and $F_2$ peaks are approximately $10^{14}$ per cm³ and $10^8$ per cm³, telling us that the ionosphere is a very wispy thing indeed in relation to the neutral atmosphere. This means that the ionosphere will be blown to and fro whenever the atmosphere chooses to move, for example when there are winds blowing. However it is not a wispy thing as far as radio waves are concerned.

The electron density profile, i.e., the way in which the electron density varies with height, N(h), undergoes dramatic changes, especially from day to night. Figure 3.6 illustrates how the profile changes from that for a typical summer day (as in Figure 3.5). At night, the D and E regions disappear almost completely, and the $F_1$ layer ceases to be a separate part of the F layer, so we are left with just an F layer. There is no $F_1$ layer during winter.

Talking in terms of electron densities is not going to get us very far because we do not have any real feeling for what an electron density means. What does a density of $10^6$ per cm³ or, even worse, $10^{12}$ per m³ mean? What does it tell us about the frequencies we can use for HF communications?

We can make the relation between the ionosphere and our use of it somewhat more comprehensible by introducing the terms **plasma frequency** and **critical frequency**. If we consider a slab of plasma in which the light electrons and heavy positive ions are displaced from each other somehow, when they are "let go," they will oscillate to and fro in simple harmonic motion at an angular frequency, $\omega$, given by $\omega^2 = N e^2 / (\epsilon_0 m)$, where N is the electron density, e is the charge on an electron, m is the mass of an electron, and $\epsilon_0$ is the permittivity of free

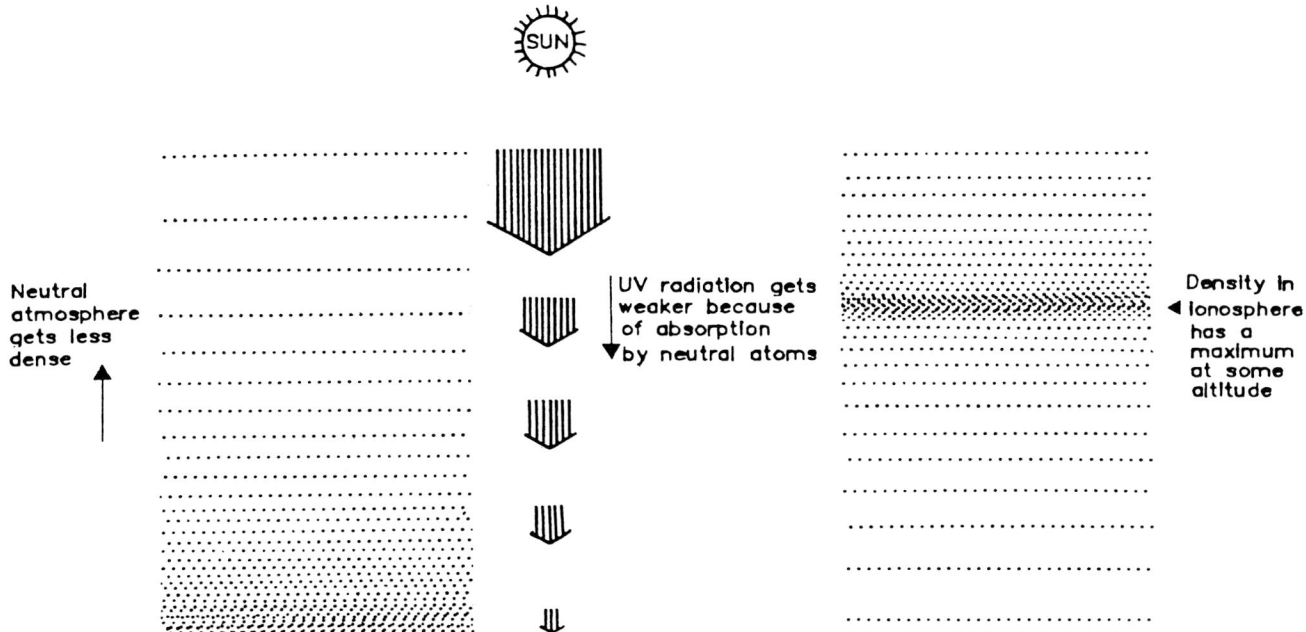

**Figure 3.4** The fact that the electrons in the ionosphere form a layer at some altitude, with the density decreasing as we move away from that height, is the result of two opposing phenomena—the density of the neutral atmosphere decreases as altitude increases, while the amount of EUV light increases as altitude increases. Thus, as we go to lower altitudes, even though there is more matter which could potentially be photoionized by the EUV light, there is less EUV light to do the job because it has been progressively absorbed as it penetrates to lower altitudes.

space [2]. The angular frequency $\omega$ is called the **angular plasma frequency**. If we change to the plasma frequency $f_N = 2\pi / \omega$, and substitute the values for the atomic constants, we find that

$$f_N^2 = 80.5 \, N,$$

where $f_N$ is the plasma frequency in hertz, and N is the number of electrons per m$^3$ (MKS units). The critical frequency of a layer, $f_c$, is related to the maximum electron density in that layer, $N_m$, by the formula

$$f_c \approx 9 \times 10^{-6} \, N_m^{1/2}, \qquad (3.3)$$

where $f_c$ is now in MHz, and $N_m$ is in electrons per m$^3$. There is a critical frequency for each of the E, $F_1$, and $F_2$ layers, denoted by $f_oE$, $f_oF_1$, and $f_oF_2$. For example, if the maximum electron density has the typical $F_2$ layer value of $10^{12}$ electrons per m$^3$, the corresponding critical frequency $f_oF_2$ is 9 MHz. The usefulness of the concept of critical frequency is that the critical frequency of a layer is equal to the maximum frequency which can be reflected from it at vertical incidence. (Note that we have ignored the effect of the Earth's magnetic field here.) Waves at higher frequencies will simply pass right through the layer. The concepts of plasma frequency and critical frequency are found to be so useful in practice that very few people think or work in terms of electron density.

## 3.3 Observing the Ionosphere

Much of our current knowledge of the ionosphere comes from radio probing techniques, although some comes from in situ probe measurements made using artificial satellites. The field has been comprehensively reviewed by Hunsucker [3]. Our main interest here is the use of ground-based swept frequency HF techniques. There are, for example, also ground-based VHF techniques, transionospheric propagation techniques, and ionospheric modification techniques.

If we transmit HF radio energy vertically upwards, and time how long it takes for a burst of energy to return to the Earth, we find that the time delay is a function of frequency. If we picture the reflection process as simple reflection from a mirror at the appropriate height, with the rays travelling to and from the mirror, we can convert the time delay into the "virtual" height of the mirror, $h' = c\,t/2$. An **ionogram** is then the plot of virtual height against operating frequency. Ionograms may also be obtained when the transmitter and receiver are separated by long distances, giving an **oblique ionogram**. When the transmitter and receiver are close to each other and the signals being received have been scattered back towards the transmitter by ground backscatter, we get a **backscatter ionogram**. The equipment used to generate an ionogram is known as an **ionosonde**. We return to a discussion of ionograms in Chapter 5.

**Table 3.1** The ionospheric production and loss processes and experimental values of the effective recombination rate, $\alpha_{eff}$, for the principal ionospheric layers. *After Chamberlain, J. W. Theory of Planetary Atmospheres, Academic Press, New York, 1978.*

| Region | Nominal height of layer peak (km) | $N_e^{(max)}$ (cm$^{-3}$) | $\alpha_{eff}$ (cm$^3$/sec) | Ion production | Recombination |
|---|---|---|---|---|---|
| D | 90<br>Lower following solar flare | $1.5 \times 10^4$ (noon); absent at night | $3 \times 10^{-8}$ | Ionization by solar x-rays, or Lyman alpha ionization of NO. Enhanced ionization following solar flares due to x-ray ionization of all species. Electron attachment to O and O$_2$ forms negative ions; ratio of negative ions to electrons increases with depth and at night. | Electrons form negative ions which are destroyed by photodetachment (daytime only), associative detachment (O + O$^-$ → O$_2$ + e), and mutual neutralization (O$^-$ + A$^+$ → O + A). |
| E | 110 | $1.5 \times 10^5$ (noon); $< 1 \times 10^4$ (night) | $10^{-8}$ | Ionization of O$_2$ may occur directly by absorption in the first ionization continuum ($h\nu > 12.0$ eV). Coronal x-rays also contribute, ionizing O, O$_2$, and N$_2$. Nighttime E and sporadic E (thin patches of extra ionization) are due to electron and meteor bombardment. Some sporadic E radio reflections may be due to turbulence in normal E layer. | Dissociative recombination<br>O$_2^+$ + e → O + O<br>and NO$^+$ + e → N + O. |
| F1 | 200 | $2.5 \times 10^5$ (noon); absent at night | $7 \times 10^{-9}$ | Ionization of O by Lyman "continuum" or by emission lines of He. This ionization probably accompanied by N$_2$ ionization, which disappears rapidly after sunset. | O$^+$ ions readily transfer charge to NO and perhaps to O$_2$. Most of the ionization is thus in molecular form and disappears by dissociative recombination. |
| F2 | 300<br>Height and electron density highly variable. Large daily, seasonal, and sunspot-cycle variations are combined with general erratic behavior. | $10^6$ (noon); $10^5$ (midnight) | $10^{-10} - 10^{-9}$<br>Variable; probably decreases with increasing height | Ionization of O by same process producing F1; F2 formed because $\alpha_{eff}$ decreases with increasing height; F2 region produces little attenuation of radiation. Additional ionization processes may contribute in F2 that are attenuated in F1. | Recombination of molecular ions as in F1; but limiting process here is charge transfer, giving an attachment-like recombination law. |

## 3.4 Variations of the Ionosphere

We have already seen that the ionosphere varies with height, ranging from the D layer at 50 to 90 km up to the F layer at 150 to 600 km. The fact that the ionosphere is created by the sun immediately suggests that it will vary with time of day, season, and position on the surface of the Earth. In practice its variations are very similar to those of the air temperature which weather forecasters predict. Generally speaking, the electron density in the ionosphere is greatest in summer, in the middle of the day, and near the equator. This simple picture is not quite true as we shall see, but it will suffice for the present. The ionosphere also varies significantly with solar activity, as the amount of EUV radiation from the sun waxes and wanes every 11 years or so, as described in Chapter 2. We shall consider the variations of the ionosphere in some detail because the frequencies available for HF communications and other uses of the ionosphere have the same variations. The five main variations are illustrated in Figure 3.7.

### 3.4.1 Diurnal Variation

The word "diurnal" simply means "throughout the day." The diurnal variation of the critical frequencies of the D, E, and $F_1$ layers is very simple. These layers are not there at night, and during the day the critical frequencies depend almost exclusively on the zenith angle of the sun.

The **zenith angle** of the sun (or any other object in the sky) is the angle between the line from the observer to the position directly overhead (called the zenith) and a line from the observer to the sun—see Figure 3.8. If the sun is vertically overhead, the zenith angle is zero, whereas at sunrise and sunset it is around 90° because the line from the observer to the sun is more or less horizontal. At midday on September 23 and March 21 (the equinoxes), the sun will be vertically overhead of an observer at the equator and the zenith angle will be zero. At midday on

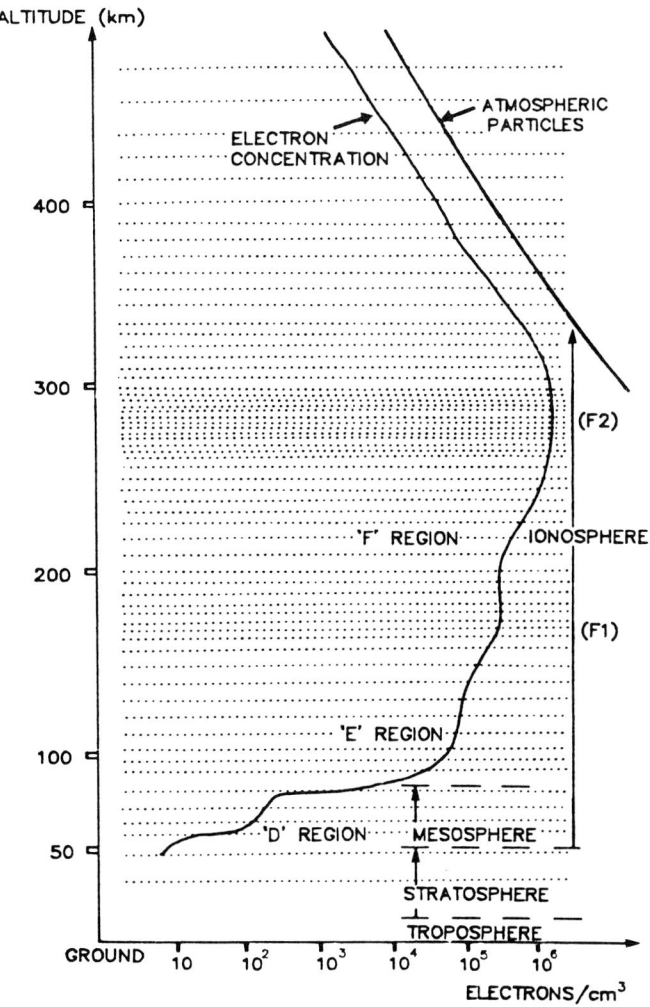

**Figure 3.5** Typical daytime variation with altitude of the electron density of the ionosphere, illustrating the four possible layers. The ionosphere covers the height range from about 50 km to about 1,000 km. (In fact there is no clearly defined upper limit, the electron density just decreasing continuously as the altitude increases.) The density of the neutral atmosphere is also shown. Note that at no time does the density of electrons exceed about 1% of the density of the neutral atmosphere.

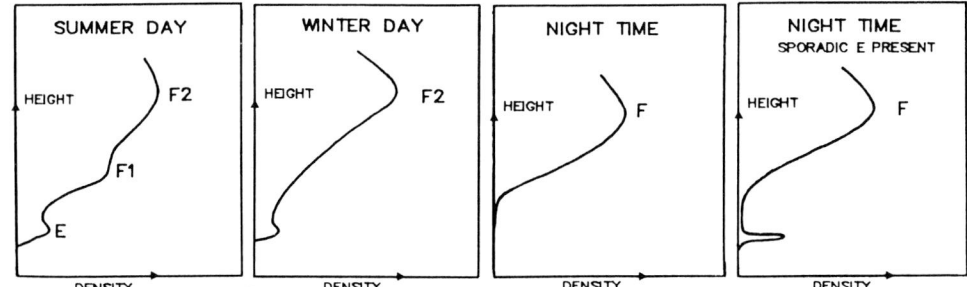

**Figure 3.6** Sketches of the variation with altitude of the electron density of the ionosphere under different conditions. The electron density profiles shown illustrate the variation of the ionosphere with season and from day to night. The fourth panel illustrates a sporadic E layer described in Section 3.5.

# The Ionosphere

# VARIATIONS OF THE IONOSPHERE

1. **DIURNAL** (THROUGHOUT THE DAY)

2. **SEASONAL** (THROUGHOUT THE YEAR)

3. **LOCATION** (GEOGRAPHIC & GEOMAGNETIC)

4. **SOLAR ACTIVITY** (SOLAR CYCLE & DISTURBANCES)

5. **HEIGHT** (DIFFERENT LAYERS)

**Figure 3.7** The five main variations of the ionosphere which must be taken into account in order to predict HF propagation conditions successfully.

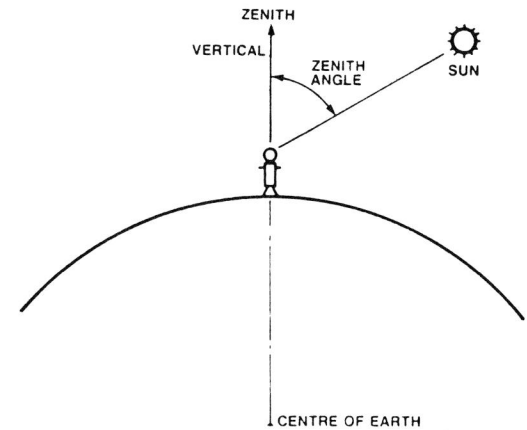

**Figure 3.8** The zenith angle of the sun is the angle between the line to the point directly overhead of the observer (the zenith) and the line to the sun.

December 21, the sun will be vertically overhead at the Tropic of Capricorn, while at midday on June 21 it will be vertically overhead at the Tropic of Cancer. The zenith angle can be readily derived for any location and universal time, working in terms of the solar declination and the equation of time [4].

Figure 3.9 shows the zenith angle for June as a function of local time. Local noon occurs when the sun reaches its highest point in the sky (smallest zenith angle for the day). Suppose, for example, that we are considering a location at 30° N. At this latitude, the sun will be at a zenith angle of 100° at 0400. The sun will rise at about 0500, when the zenith angle passes through 90° (provided we ignore atmospheric refraction and the finite size of the sun). The minimum zenith angle will be around 8°, and will by definition occur at local noon. Sunset will occur at 1900. Note that in June the sun never rises at latitudes southward of 70° S, and never sets above 70° N.

If the zenith angle is called "Z" and the level of solar activity (Chapter 2) is described by the sunspot number "R", then the critical frequencies for the E and $F_1$ layers, $f_oE$ and $f_oF_1$, are given approximately by

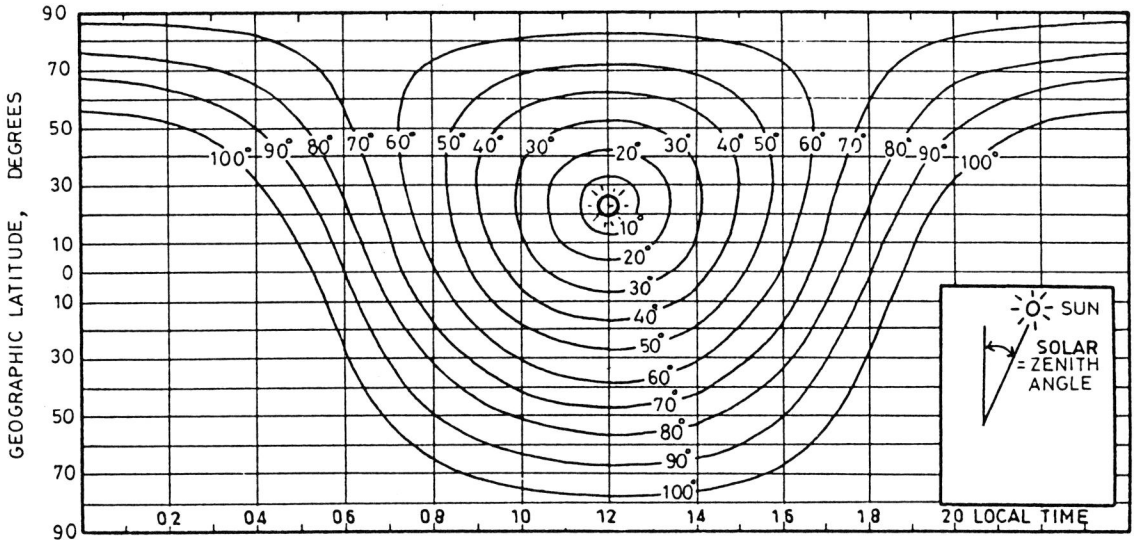

**Figure 3.9** The variation of the solar zenith angle for June, as a function of latitude and local time. At a given latitude, sunrise and sunset, for example, occur when the zenith angle passes through 90°. At local midday in June, the sun in directly overhead at the Tropic of Cancer. In June, the sun never rises at high southern latitudes (winter), and never sets at high northern latitudes (summer).

$$f_oE = 0.9[(180 + 1.44R)\cos Z]^{1/4} \text{ MHz} \quad (3.4)$$

and

$$f_oF_1 = (4.3 + 0.01R)\cos^{0.2}Z \text{ MHz.} \quad (3.5)$$

These are empirical equations which fit the observations of $f_oE$ and $f_oF_1$ fairly well. In practice, somewhat more complex formulas are used.

The power of 1/4 outside the square brackets in equation 3.4 tells us that the E layer behaves very much like a Chapman layer, which is the theoretical shape describing how the electron density will vary with height under given conditions, provided certain assumptions are valid [5, 6]. These assumptions are that (a) the temperature is independent of height, (b) there is only one species of atom or molecule being ionized, and (c) the ionizing radiation is monochromatic (i.e., is all at the one wavelength), and are usually not valid! However they are often valid enough to ensure that the real N(h) profile can be described as Chapman-like. The exponent 0.2 for the $F_1$ layer shows us that the $F_1$ layer is not particularly Chapman-like, and as we shall see, the $F_2$ layer is even worse.

Figure 3.10 shows (among other things) the observed diurnal variation of $f_oE$ and $f_oF_1$ for Canberra, Australia, for two seasons and two levels of solar activity. While easy to calculate, $f_oE$ and $f_oF_1$ are not as important for HF communications as $f_oF_2$, the critical frequency for the $F_2$ layer. Unfortunately, $f_oF_2$ is not easy to calculate, as we shall see later. However $f_oF_2$ also displays diurnal, seasonal, latitudinal, and solar cycle variations, and it is these which are of interest in this chapter. These variations are illustrated in Figure 3.10, along with those of $f_oE$ and $f_oF_1$, and in Figure 3.11.

The diurnal variations of $f_oE$ and $f_oF_1$ are more or less what we would expect, the critical frequencies reaching their greatest values at noon. The $F_1$ layer shows up as a separate layer only during the day, for 3 to 4 hours either side of local noon. The E layer does not completely vanish at night, usually staying at around $f_oE = 0.6$ MHz. However such low critical frequencies are difficult to observe and have little consequence for practical uses of the ionosphere, so we shall assume that $f_oE$ drops to zero at night.

The diurnal variation of $f_oF_2$ is often rather complicated. It reaches its lowest value just before dawn, recombination having eaten away at the electrons all night. Then the sun comes up, and $f_oF_2$ rises rapidly as photoionization starts creating a supply of free electrons again. The $F_2$ layer differs from the E and $F_1$ layers in that it survives the night. This, together with the fact that the critical frequencies are highest in the $F_2$ layer, makes the $F_2$ layer the most important layer as far as HF communications are concerned.

### 3.4.2 Seasonal Variation

The ionosphere varies throughout the year, partly because the solar zenith angle has a seasonal as well as diurnal variation, but also because of changes in the neutral atmosphere from which the ionosphere is created. In the winter, the zenith angle at noon is always greater than the corresponding angle in summer. We would therefore expect the critical frequencies of each of the layers to be greater in summer than in winter. This is found to be the case for the D, E, and $F_1$ layers, but not for the $F_2$ layer at mid-latitudes.

Figures 3.10 and 3.11 show how the critical frequencies $f_oE$, $f_oF_1$, and $f_oF_2$ vary with season. January is midsummer at Canberra and mid-winter at Manila, while June is mid-winter at Canberra and mid-summer at Manila. Note that Manila is a low-latitude station, whereas Canberra is a mid-latitude station. $f_oF_2$ is clearly much greater during the day at Canberra in winter than in summer, which is contrary to what simple ideas would predict. This unexpected difference is known as the **mid-latitude seasonal anomaly** and has its roots in seasonal changes in the relative concentrations of atomic and molecular species [7].

### 3.4.3 Latitudinal Variations

As with the seasonal variation, part of the variation of the ionosphere with position on the Earth, particularly latitude, is due to the variation with solar zenith angle. Once we get out of the tropical zone between the Tropics of Capricorn and Cancer, the solar zenith angle can never be zero, and for a given time of day it increases as we go towards the poles. However even when this effect is taken into account, the ionosphere is found to have considerable variation with latitude. The extreme cases of the equatorial and polar ionospheres are found to bear little resemblance to each other, as we shall see in Section 3.6 and Chapter 8. Matching panels of Figures 3.10 and 3.11 may be compared to determine the variations of the ionosphere from mid to low latitudes.

### 3.4.4 Variations From Day to Day

Just as the air temperature varies from day to day, so do the critical frequencies of the ionosphere. Our knowledge of the ionosphere and the observations we are able to make of it are not yet good enough to allow us to understand why it varies from day to day, except in a general sense in which we attribute the changes to changes in the flux of EUV radiation from active regions on the sun, to changes in the neutral winds blowing in the atmosphere, and to changes in the electric currents flowing in the ionosphere. Because of this, and because in practical uses of the ionosphere it is usually not necessary to worry

# The Ionosphere

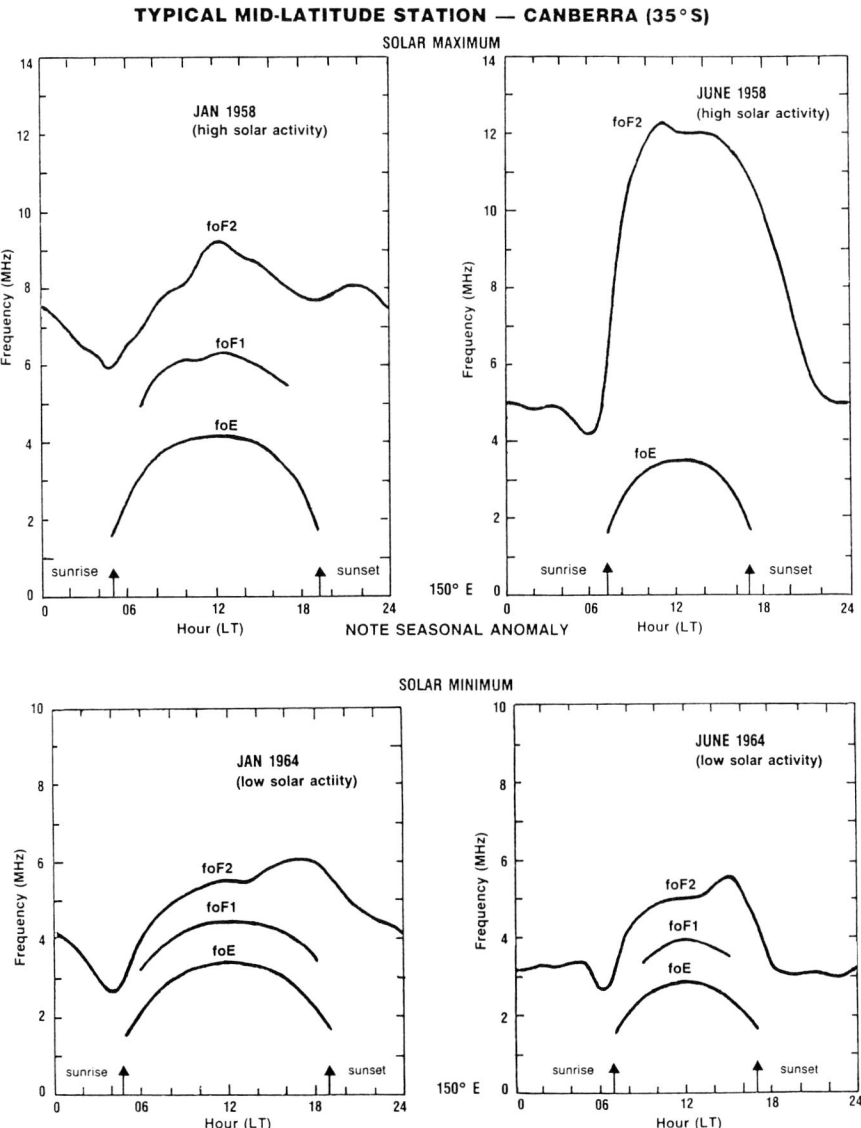

**Figure 3.10** Variations throughout the day of the critical frequencies of the E, $F_1$, and $F_2$ layers, $f_oE$, $f_oF_1$, and $f_oF_2$, for Canberra for summer (January) and winter (June) and for two levels of solar activity (high, 1958; low, 1964). Note that there is no $F_1$ layer for winter at high levels of solar activity, and that the winter values of $f_oF_2$ during the day at high levels of activity exceed the summer values. The latter phenomenon is known as the mid-latitude seasonal anomaly. Critical frequencies are significantly higher at higher levels of solar activity.

too much about such details, we can ignore the day-to-day variations to a large extent and work in terms of the average behavior of the ionosphere for the month, at each of the 24 hours of the day.

In a 31-day month, there will be 31 observations of $f_oF_2$, say, at a particular hour. A very good representative value of these 31 observations is the value which is exceeded on 15 days, and which itself exceeds the values observed on the other 15 days. This is called the **median** value. We would also like some estimate of the range of observed values about the median. If the median is 10 MHz and the individual values range from 5 to 20 MHz, the median does not tell us very much since the range covers most of the HF band. If, however, the range is from 8 to 12 MHz, we can never be more than 2 MHz out if we take 10 MHz as the value of $f_oF_2$ for every day of the month at the hour considered. In practice, we represent the range of values by the third lowest and third highest values. The third lowest value is exceeded on 28 days or approximately 90% of the month, while the third highest value exceeds the other observations for 90% of the month. The value exceeded for 90% of the month is called the **lower decile** (lowest 10%), while the value exceeded for 10% of the month is called the **upper decile.** In the same sense, the median is the fifth decile (50%).

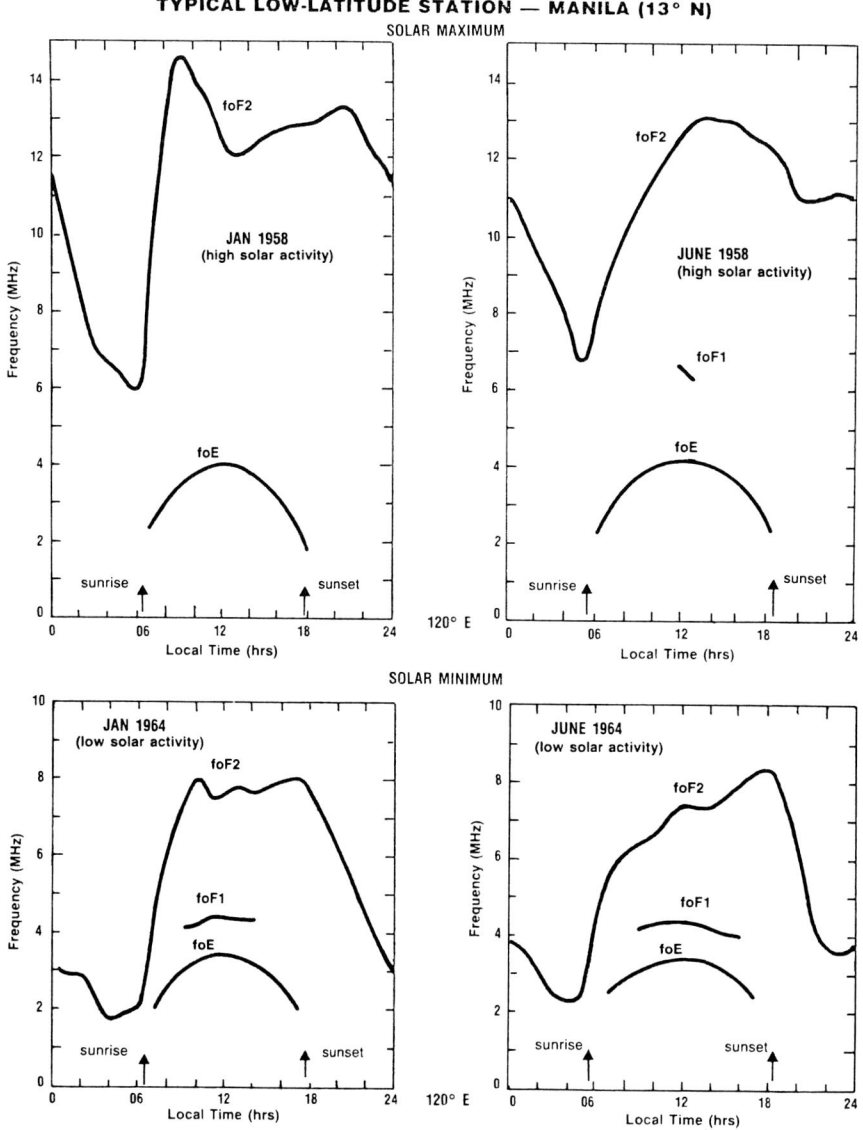

**Figure 3.11** Variations throughout the day of the critical frequencies of the E, $F_1$, and $F_2$, layers, $f_oE$, $f_oF_1$, and $f_oF_2$, for Manila for summer (June) and winter (January), and for two levels of solar activity (high, 1958; low, 1964). Note that the $F_1$ layer is not important at this low-latitude station.

Figure 3.12 illustrates the complete 24 × 31 observations of $f_oF_2$ for Canberra in December 1980, together with the values of the median and deciles at each hour. The figure also shows how the median and decile values are deduced for one (0200 LT) of the 24 sets. The diurnal and seasonal variations illustrated earlier in Figures 3.10 and 3.11 in fact correspond to *median* values of the critical frequencies. On some days, $f_oF_2$ is much lower than on the remaining days. These are called **disturbed** days and happen when the sun becomes disturbed, as we shall see in Chapter 8.

The use of this statistical description of the ionosphere reduces from 31 to 3 the number of parameters necessary to describe the behavior of the ionosphere at a given hour of a given month at a given location. This 10-fold decrease is a very useful reduction in the enormous amount of data required to describe the ionosphere and all its variations.

### 3.4.5 Variations With Solar Activity

As with the day-to-day variations, we are not really able to describe in detail how the ionosphere varies with changes in solar activity, because there are still too many unknown things going on. However we can be fairly successful if we just use some simple indicators of what the sun is doing. The sunspot number which we met in Chapter 2 is one of the parameters most widely used to

# The Ionosphere

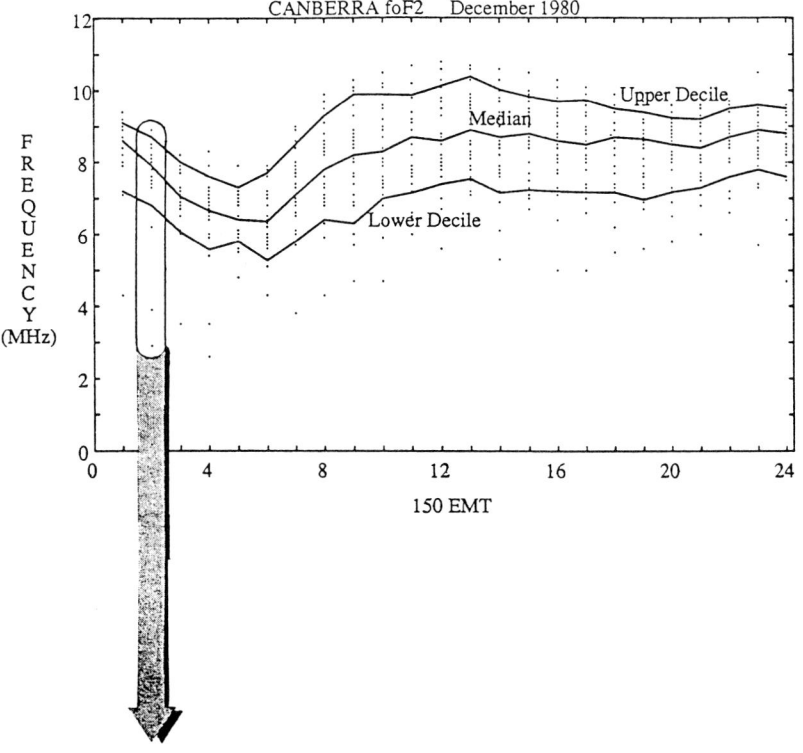

foF2 at 0200 LT as recorded

| DAY | 1 | 2 | 3 | 4 | 5 | 6 | 7 | 8 | 9 | 10 | 11 | 12 | 13 | 14 | 15 | 16 |
|---|---|---|---|---|---|---|---|---|---|---|---|---|---|---|---|---|
| foF2 | 8.5 | 6.2 | 8.7 | 7.8 | 7.6 | 8.7 | 8.1 | 8.4 | 7.9 | 8.1 | 8.4 | 7.9 | 6.8 | 7.8 | 7.6 | 7.8 |
| DAY | 17 | 18 | 19 | 20 | 21 | 22 | 23 | 24 | 25 | 26 | 27 | 28 | 29 | 30 | 31 | |
| foF2 | 7.5 | 7.3 | 8.7 | 2.9 | 3.9 | 7.9 | 7.9 | 8.2 | 8.9 | 8.9 | 8.0 | 7.4 | 8.4 | 7.4 | 7.8 | |

foF2 at 0200 LT ordered by magnitude

| foF2 | 2.9 | 3.9 | 6.2 | 6.8 | 7.3 | 7.4 | 7.4 | 7.5 | 7.6 | 7.6 | 7.8 | 7.8 | 7.8 | 7.8 | 7.9 | 7.9 |
|---|---|---|---|---|---|---|---|---|---|---|---|---|---|---|---|---|
| foF2 | 7.9 | 7.9 | 8.0 | 8.1 | 8.1 | 8.2 | 8.4 | 8.4 | 8.4 | 8.5 | 8.7 | 8.7 | 8.7 | 8.9 | 8.9 | |

**Figure 3.12** The variability of $f_oF_2$ at a typical mid-latitude station, Canberra, for December, 1980. There are 31 observations at each hour. The middle curve is the median curve—at each hour, 15 points lie above this line while 15 points lie below it (or on the line). The upper curve is the upper decile curve, and there are 3 points (10% of 31) lying above the curve, corresponding to the highest 3 observed values of $f_oF_2$. The lower curve is the lower decile curve, and there are 3 points lying below it. These three values of $f_oF_2$ were recorded on disturbed days which are described in Chapter 8. The tables illustrate how the median and decile values are calculated for the 0200 LT data. *Courtesy IPS Radio and Space Services.*

describe the behavior of the sun as it affects the ionosphere. It is found that the monthly median values of the critical frequencies of the ionosphere for a particular month are linearly related to the monthly average value of the sunspot number, smoothed or averaged over 12 months, $R_{12}$. In other words, if we draw a graph of $f_oF_2$, say, against $R_{12}$, the points will lie close to a straight line.

Figure 3.13 shows the 35 values of the monthly median $f_oF_2$ for noon in December (winter) and June (summer) at Washington, D.C., from 1934 to 1968 versus the sunspot number for each of the 35 months. The straight lines are the ones which best fit the data points. It can be seen that the lines are very good fits, except for very high values of sunspot number. Figures such as Figure 3.13 can be drawn up for all stations for which data are available (about 150), for each month (12), and for each hour (24). This gives a total of about 40,000 graphs. Each of these graphs can be thought of as a calibration curve describing how the ionosphere responds to the sun. For example, if the smoothed monthly sunspot number is expected to be 50 in some December in the future, Figure 3.13 tells us

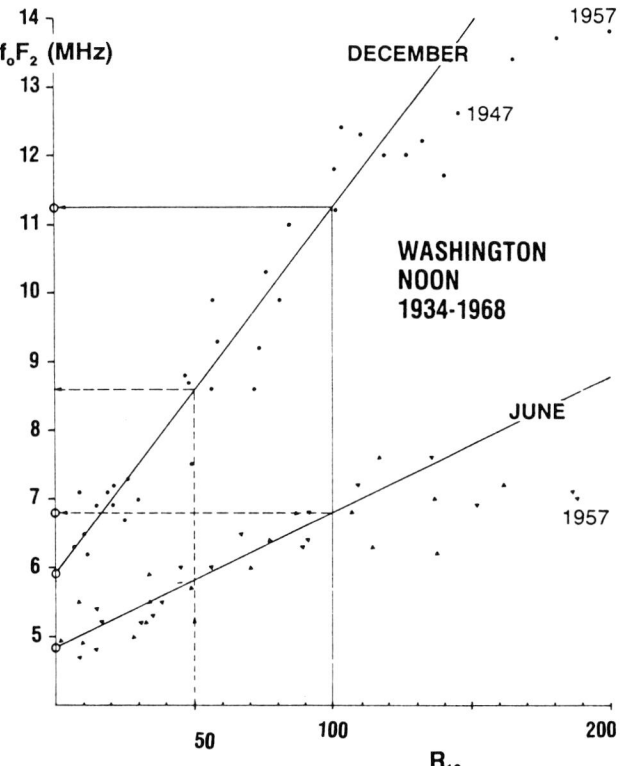

**Figure 3.13** The calibration graph of monthly median values of $f_oF_2$ versus the monthly sunspot number smoothed over twelve months, $R_{12}$, for Washington, D. C., for noon in December (winter) and June (summer). There are 35 data points, one for each December or June since the station opened in 1934, up until 1968. The straight lines are the best-fit straight lines. The values of $f_oF_2$ at $R_{12} = 0$ (4.8 and 5.9 MHz in this case) and $R_{12} = 100$ (6.8 and 11.2 MHz) are used to describe the variations of $f_oF_2$ with solar activity for noon in June and December respectively at Washington.

that at noon in that December at Washington the observed value of $f_oF_2$ will be 8.6 MHz, with a little uncertainty because the data points do not all lie exactly on the straight line.

One advantage of a straight line is that it is completely defined by specifying two points which lie on it. It is therefore common practice to represent each of the calibration curves by the two points $R = 0$ (sunspot minimum) and $R = 100$ (a fairly high sunspot number and level of solar activity). We can therefore fully describe the solar-cycle variation of the ionosphere at any location, month, and hour by just two data points and the straight line joining them.

It may seem strange at first sight to find that the behavior of the ionosphere is so closely related to the sunspot number. After all, as we saw in Chapter 2, sunspots are relatively cool areas of the sun, and it is hardly likely that they would do any more to the ionosphere than the hotter areas. What is actually important are the plage regions surrounding the sunspots in the active regions, since these are sources of increased EUV radiation which causes increased ionization of the Earth's atmosphere. The special value of the sunspots is that they are much easier to observe than plage regions, while at the same time being very good indicators of how much of the sun is covered by plage areas. They have also been observed on a systematic basis for over 300 years, giving us some idea of what to expect in the future.

We can in practice do better than the sunspot number when drawing the calibration curves. In many cases, straight lines do not fit the data very well, especially for very high levels of solar activity, as we have seen. In general, it is better to replace the sunspot number, which is a measure of what some vaguely relevant feature of the sun is doing, by an **ionospheric index**, which is a measure of what the ionosphere itself is doing [8, 9]. After all, it is the ionosphere in which we are interested. This is done by averaging the behavior of the ionosphere over a group of stations, canceling out any variations which occur at only individual stations, and leaving the variations which are common to all the stations. We will not pursue this point further because the description of the procedures for calculating an ionospheric index is rather tedious. Suffice it to say that virtually any ionospheric index, no matter what goes into making it, is a better indicator than the sunspot number of what the ionosphere is doing or will be doing. However the differences are not too great, and we can get along quite nicely for illustrative purposes by using the sunspot number. We shall denote the ionospheric index by the letter T.

The effects of different levels of solar activity on the ionosphere can be seen from the differences between the upper and lower panels of Figures 3.10 and 3.11. As a general rule, the critical frequencies of all ionospheric layers will be greater at higher levels of solar activity. The exceptions to this rule are described in Chapter 8, which describes disturbed conditions. The higher critical frequencies arise because of higher EUV flux levels which occur because there are more active regions on the surface of the sun at high levels of solar activity. This, after all, is the definition of high solar activity.

## 3.5 Sporadic E

We shall encounter sporadic E many times in this book, both as a help and as a hindrance. As its name implies, sporadic E is a reflecting layer in the ionosphere which comes and goes sporadically at E-region heights. At mid-latitudes, sporadic E ($E_s$ for short) layers are made up of clouds of electrons a few kilometers thick and a few hundred kilometers across, and occur at altitudes between approximately 90 and 130 km. To a radio wave, they often look like rather good quality mirrors. The most important aspects of $E_s$ as far as HF propagation is concerned appear to be the maximum electron density or critical frequency, $f_oE_s$, and how this varies with time of day and season. Attempts to predict these features of $E_s$ have to date been fairly unsuccessful, but we are able to make general statements which are useful under some conditions. The occurrence properties of $E_s$ at different latitudes, described in terms of the probability of $f_oE_s$ exceeding 5 MHz, are summarized in Figure 3.14.

Mid-latitude $E_s$ is essentially a daytime summer phenomenon, occurring most often at these times and with the highest critical frequencies. Midday values of $f_oE_s$ during the summer typically reach about 10 MHz. At least some mid-latitude $E_s$ is thought to be caused by high altitude winds blowing in opposite directions and compressing the very fine debris of meteors into a narrow sheet. There are still many unanswered questions about mid-latitude $E_s$ [10].

Low-latitude $E_s$ is essentially a daytime phenomenon, with little seasonal variation, and the critical frequencies are higher than at mid-latitudes. Near the geomagnetic equator (see Section 3.8) the critical frequency $f_oE_s$ exceeds 5 MHz for 90% of the time during daylight hours. Equatorial $E_s$ is due to a plasma instability caused by the high electron drift velocity. It is very patchy and transparent, and vertically incident signals are little

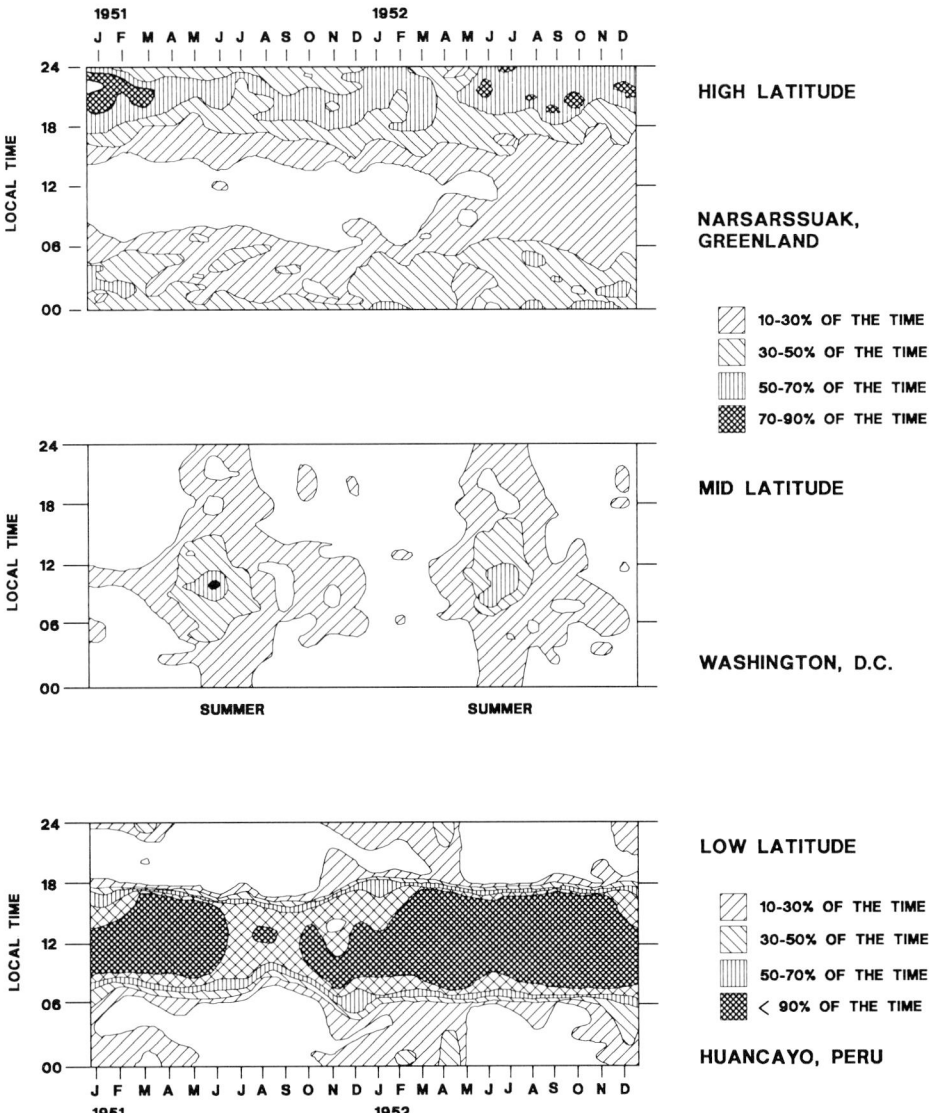

**Figure 3.14** Contour plots of the time for which the critical frequency of the sporadic E layer, $f_oE_s$, exceeds 5 MHz. In the equatorial zone, $f_oE_s$ exceeds 5 MHz over 90% of the time during daylight hours. Mid-latitude $E_s$, on the other hand, has its highest critical frequencies during the day in summer. At high latitudes, high values of $f_oE_s$ occur mainly at night and do not depend very much on season of the year. Note that Washington and Narsassuaq are northern hemisphere stations, for which summer occurs in June/July. *After E. K. Smith,* World-wide Occurrence of Sporadic E. *NBS Circular 582, Washington, 1957.*

affected by it. However it does seem to be a useful reflector on long circuits.

High-latitude $E_s$ also has little seasonal variation, but it occurs most often at night. Critical frequencies greater than 5 MHz are observed for more than 50% of the time prior to midnight and somewhat less often from midnight to dawn.

The different diurnal and seasonal variations at different latitudes tell us that $E_s$ is caused by different things at different latitudes. Simple deductions like these can save years of misguided effort since in this case, for example, we know not to try to explain all $E_s$ by the one grand theory.

## 3.6 The Equatorial Ionosphere

At first thought, we might expect the equatorial F region to be more or less the same as the mid-latitude F region, with perhaps higher critical frequencies because the sun is more directly overhead. We might also expect not much difference between winter and summer. However there are some basic differences between the mid-latitude and low-latitude ionospheres, the main one being the **fountain effect** which redistributes electrons at low latitudes, moving electrons from over the equator down to latitudes of 10° to 20°.

The fountain effect explains very well many of the observed features of the equatorial ionosphere [11, 12]. We start off with a background ionosphere which is more or less what we would have expected to see. To this we add an eastward electric field and the north-south magnetic field of the Earth, both of which are horizontal, or nearly so, in the equatorial regions. Then we start creating large numbers of electrons at the equator by photoionization due to the sun.

The combined effect of the electric and magnetic fields on the electrons is to cause them to rise (or drift) upwards, as illustrated in Figure 3.15 [13]. As they rise, they encounter the horizontal lines of force of the Earth's magnetic field. The electrons move (or diffuse) down these field lines and reenter the main body of the ionosphere where the field lines cut through the F region, giving rise to large clumps of electrons at latitudes 10° to 20° from the equator. These clumps are called the peaks or **crests** of the equatorial or Appleton (after Sir Edward Appleton) anomaly. Of course, since we now know what causes the crests, the **equatorial anomaly** is not really an anomaly any more.

The crests of the equatorial anomaly are most developed in the late afternoon and early evening, during the equinoxes, and at solar maximum. The critical frequencies associated with them often exceed 15 MHz, compared with values of something like 10 MHz at the equator. Values of $f_oF_2$ greater than 20 MHz have been observed at some stations under the anomaly crests at sunspot maximum. The height at which the maximum electron density occurs, $h_mF_2$, is less at the crests than it is at the equator. Figure 3.16 illustrates how $f_oF_2$ and $h_mF_2$ varied with latitude during the late afternoon on an Australia to Japan circuit on selected days in August 1970. The significant changes with latitude of $f_oF_2$ and $h_mF_2$ make it a little harder to determine the effects of the

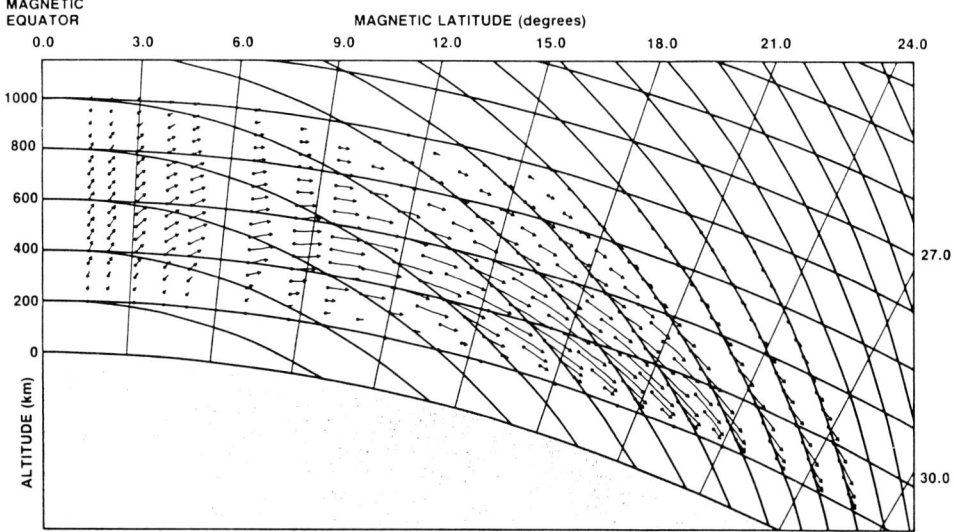

**Figure 3.15** The fountain effect in the low-latitude ionosphere. Electrons produced near the equator by photoionization drift upwards under the combined influence of horizontal electric and magnetic fields, then diffuse down along lines of force of the Earth's magnetic field towards lower altitudes and higher latitudes. Large numbers of electrons produced near the equator thus end up in the crests of the equatorial anomaly.

# The Ionosphere

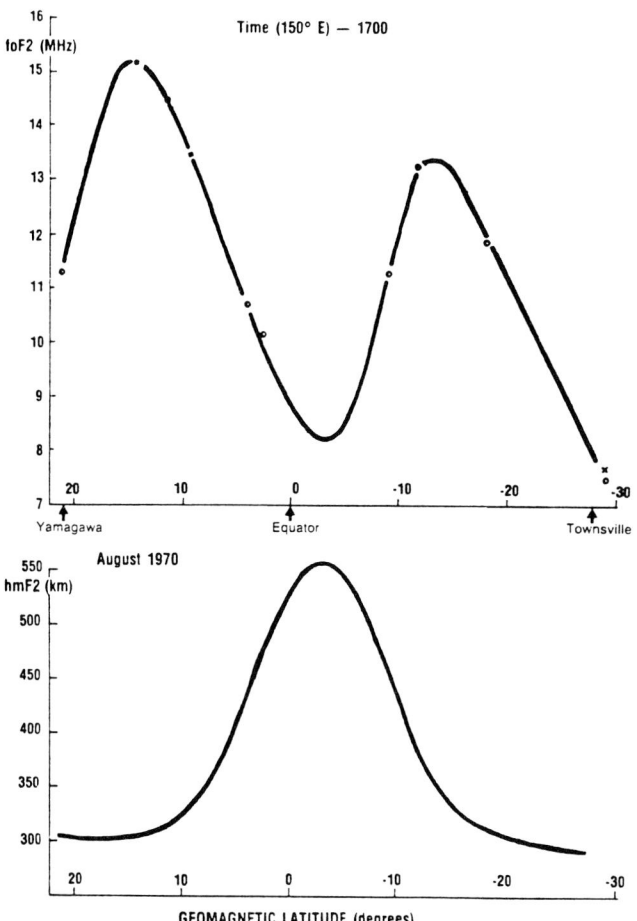

**Figure 3.16** The late afternoon variation for August 1970 of $f_oF_2$ and $h_mF_2$ (the height at which the maximum electron density occurs in the $F_2$ layer) along a circuit from Yamagawa in southern Japan to Townsville in northern Australia. The regions of high critical frequency occurring on either side of the geomagnetic equator are called the crests of the equatorial anomaly. $h_mF_2$ reaches a maximum value (of about 550 km) over the geomagnetic equator, where $f_oF_2$ has its lowest value.

equatorial ionosphere on HF propagation at any time, and also lead to some interesting propagation modes, as we shall see in Chapter 9.

The variations of the ionosphere at low latitudes are influenced by the Earth's magnetic field to such a large extent that, instead of looking at how the ionosphere varies with the normal geographic latitude, it is more fruitful to consider how it varies with geomagnetic latitude or with the dip angle of the Earth's magnetic field described in Section 3.8.

## 3.7 The Polar Ionosphere

The ionosphere over the north and south poles, alternatively called the polar or high-latitude ionosphere, is exceedingly complicated and discussion of it is best left until later. Using the ionosphere is not easy at high latitudes, as we shall see in Chapters 7 and 8.

## 3.8 The Earth's Magnetic Field

The Earth's magnetic field, or **geomagnetic field,** is not very strong compared to magnets in everyday use, but it has important effects on both the ionosphere and HF propagation. To a good approximation, the geomagnetic field is the same as that of a large bar magnet tilted at an angle to the geographic north-south axis of rotation of the Earth. This is illustrated in Figure 3.17. The southern end of the magnet, or pole of the magnet, is at approximately 79° S, 110° E, while the northern end is at 79° N, 70° W. The plane at right angles to the axis of the magnet cuts the surface of the Earth in a ring known as the **geomagnetic equator,** which is analagous to the more familiar geographic equator. The two equators do not coincide because the axis of the magnet is tilted (at an angle of 90−79 = 11°) with respect to the geographic axis of the Earth.

**Geomagnetic latitude and longitude** are measured in the same way as geographic latitude and longitude, but using the geomagnetic poles and equator. Figure 3.18 illustrates how these geomagnetic coordinates vary over the surface of the Earth when it is drawn in the common Mercator geographic projection. By definition, the 0° geomagnetic meridian is the meridian passing through the south geographic pole.

The strength of the geomagnetic field is conveniently measured in terms of the **electron gyro-frequency.**

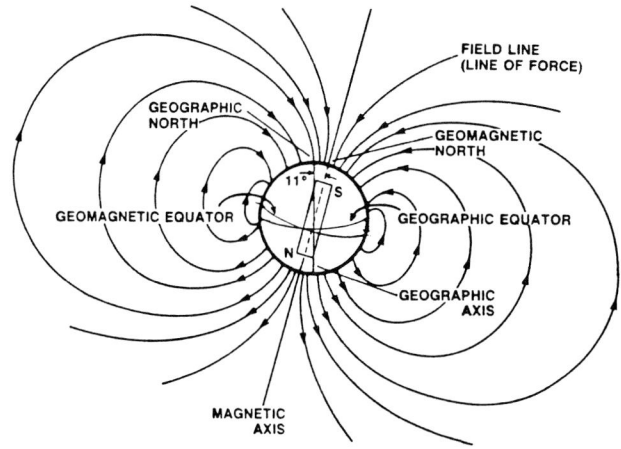

**Figure 3.17** Simple model of the Earth's magnetic field, in which it is assumed that the field is caused by a large bar magnet tilted 11° from the geographic axis of rotation. The plane to which the bar magnet is perpendicular cuts the surface of the Earth in a great circle called the geomagnetic equator. Note that the southern end of the equivalent bar magnet is actually a *north* magnetic pole, which attracts the *south* pole of a compass needle.

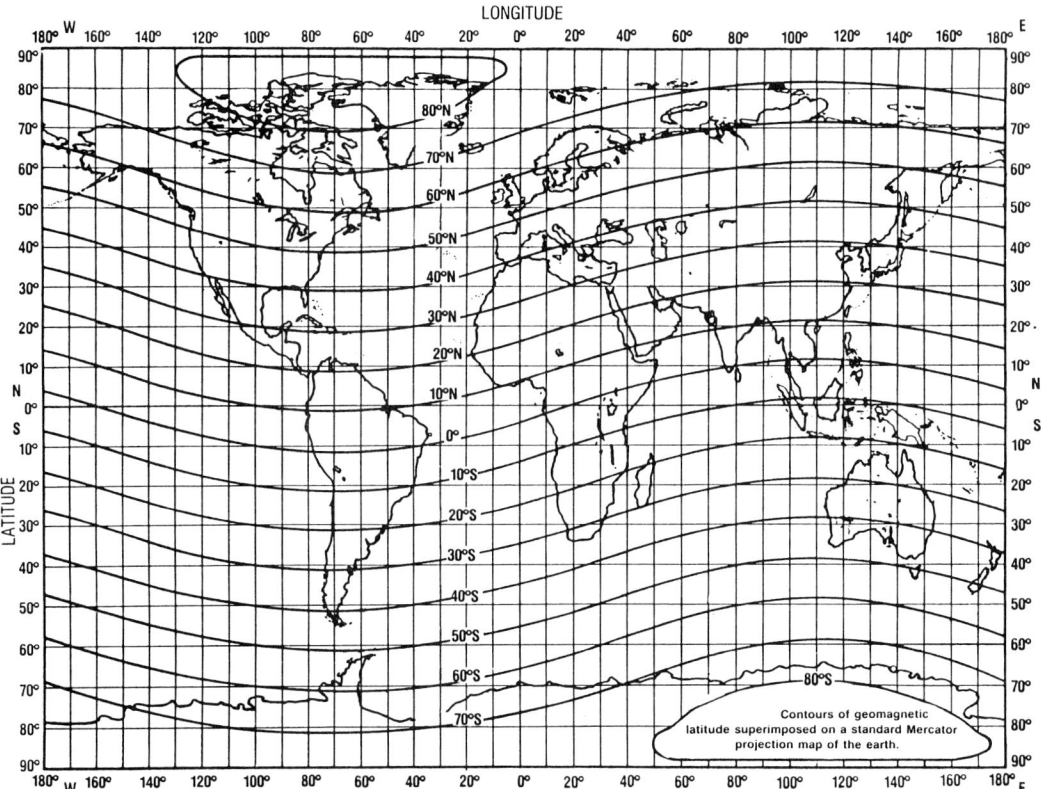

**Figure 3.18** Contours of geomagnetic latitude superimposed on a standard Mercator projection map of the Earth.

Charged particles such as electrons cannot move across a magnetic field line (in the absence of many collisions with other particles) but are forced to spiral or rotate around them. The rate at which they rotate is called the gyro-frequency and depends on how heavy they are, their electric charge, and the strength of the magnetic field. The lighter the particle and the stronger the field, the faster the particles rotate around the lines of force. For electrons in the geomagnetic field, the gyro-frequency is around 1 MHz and varies with latitude and longitude over the surface of the Earth. See Figure 3.19. It is least near the equator (about 1 MHz) and greatest near the poles (about 1.6 MHz). An interesting feature of the geomagnetic field is the low values of the gyro-frequency in the South Atlantic. This is known as the **South Atlantic anomaly** and leads to corresponding anomalies in the behavior of the ionosphere and of HF propagation. The bulk of the "glitches" (called single-event upsets) to the electronic circuits in satellites occur over the anomaly region, where the lower magnetic field offers a satellite less protection [14].

Another interesting feature of the geomagnetic field is the **dip angle** of the field. The field is horizontal only along a great circle around the Earth called the **dip equator.** At all other positions on the surface of the Earth, a suspended bar magnet free to move in any direction will point below the horizontal, as illustrated in Figure 3.20. The needles of magnetic compasses are specially mounted so that they stay reasonably horizontal, since they are mostly used to see how the field varies in a horizontal direction. Figure 3.21 shows how the dip angle varies over the surface of the Earth. The dip equator, which is where the dip angle is zero, does not coincide with the geomagnetic equator because the real geomagnetic field is not exactly the same as that of a simple bar magnet as assumed earlier. The dip angle reaches 90° at the poles, where a compass needle would stand up vertically if allowed to do so.

The behavior of the ionosphere at low latitudes depends to a large extent on the dip angle. For example the crests of the equatorial anomaly discussed in Section 3.6 lie at positions with dip angles of approximately 15°, independently of what the geographic latitude happens to be. At mid-latitudes there are significant differences between the ionosphere at geographically equivalent points in the northern and southern hemispheres simply because the geomagnetic field is different at the two points. Equatorial $E_s$ also illustrates the dominance of the magnetic field at low latitudes—this type of $E_s$ is confined to a narrow belt only a few degrees wide and centered on the *dip equator*.

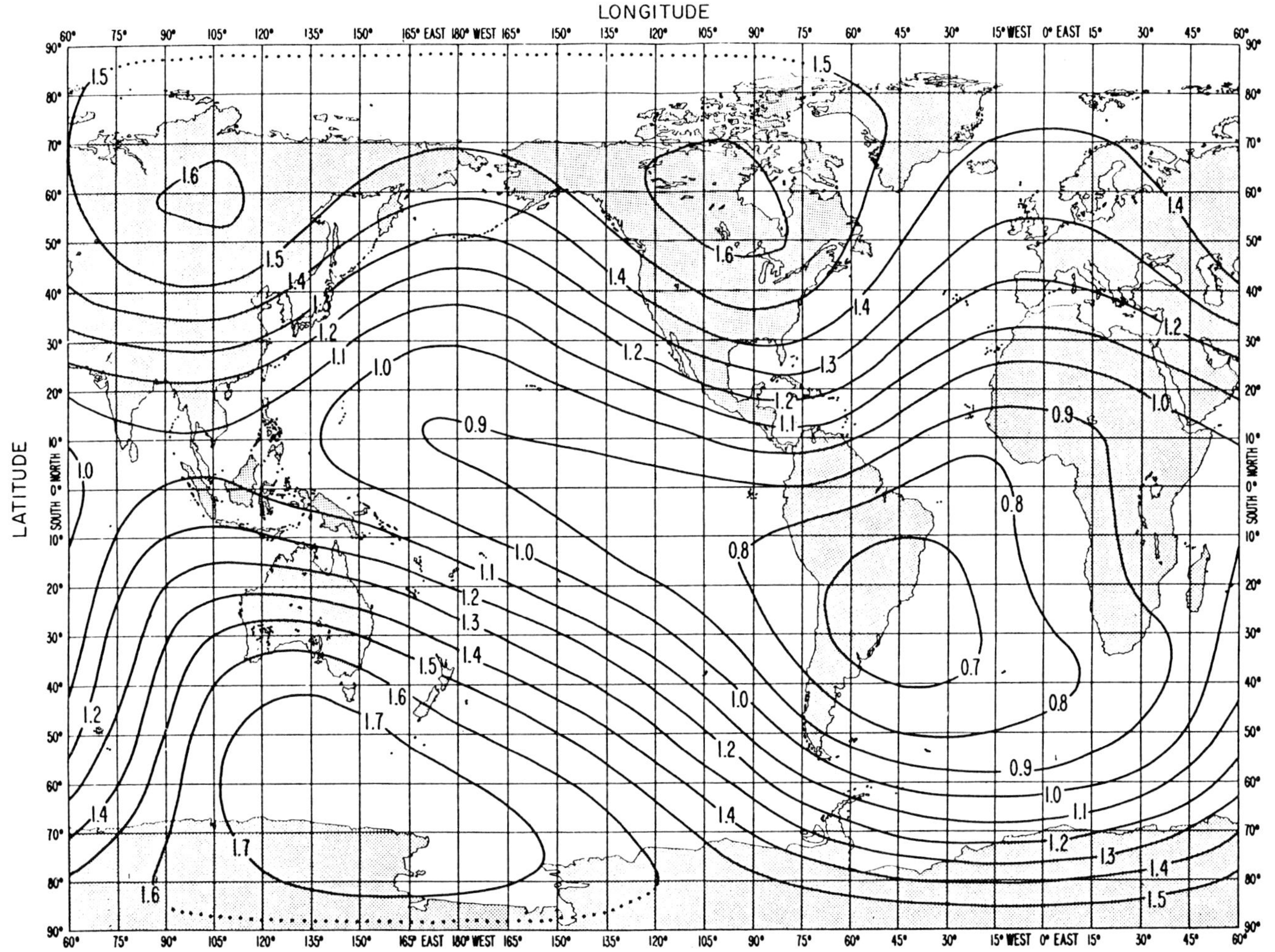

# The Ionosphere

**Figure 3.19** Contours of the electron gyro-frequency (at 100 km altitude) superimposed on a standard Mercator projection map of the Earth.

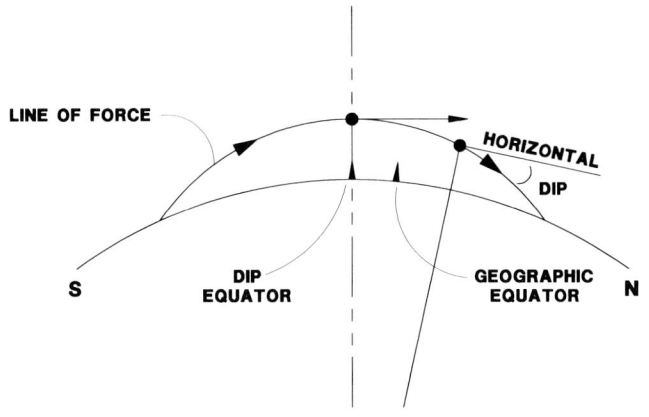

**Figure 3.20** Sketch of one line of force of the Earth's magnetic field, illustrating how the dip angle varies along the line of force and thus over the surface of the Earth. At the dip equator, the dip angle is zero and a suspended bar magnet would remain horizontal.

## 3.9 Irregularities in the F Region

As if the F region were not complicated enough, Nature has conspired to make it even more complicated at night by sometimes causing the ionization to break up into small bunches of electrons, rather than stay in a uniform, homogeneous sea of electrons. These bunches of electrons are known as **irregularities,** because when they exist the distribution of ionization in the F region is no longer a regular phenomenon. The processes which cause the nighttime F region to break up into irregularities are also the enemy in the fight to develop a reliable system of controlled thermonuclear fusion. One reason why physicists like to study the ionosphere is that it is an unconfined plasma, provided free by Nature. In the true style of the experimental scientist, they even try to perturb it to see what happens [15, 16].

Since electrons are free to move at will up and down along lines of force of the Earth's magnetic field, but not

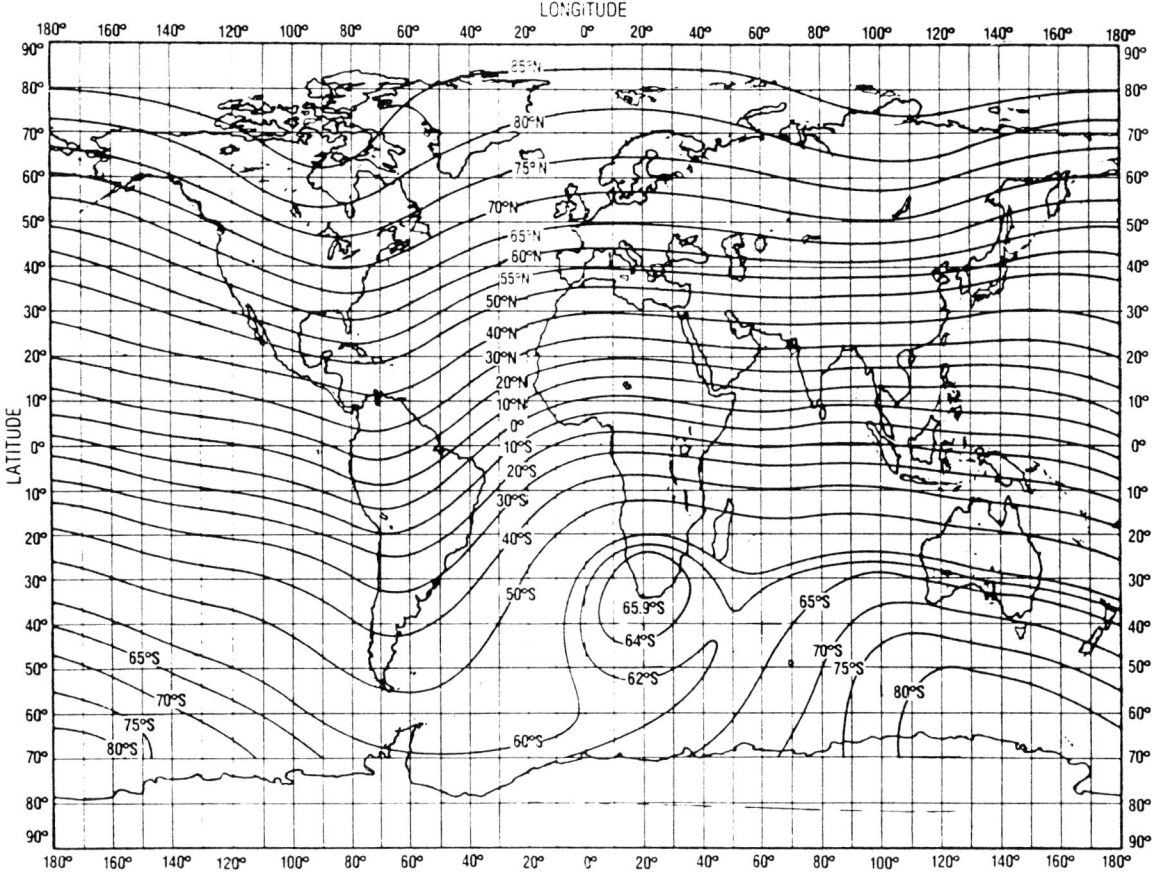

**Figure 3.21** Contours of the observed dip angle of the Earth's magnetic field, superimposed on a normal Mercator projection of the Earth's surface. The contour labels are given in the middle of the diagram. Note that the dip equator does not coincide with the geomagnetic equator. This is because the magnetic field of the Earth is only roughly similar to that of the bar magnet used to define the geomagnetic equator.

across them, the irregularities become stretched out along the lines of force soon after they are created. We therefore talk about **field-aligned irregularities**, or FAI, in the F region. When the FAI are present, a single pulse reflected from the ionosphere will be spread or stretched in time, echoes arriving with time delays significantly longer than those normally observed. We call such echoes from the F region **spread F** echoes and speak rather loosely of the irregularities as spread F, although this term strictly applies only to the echoes.

Spread F irregularities can lead to echoes which are stretched in range (which is equivalent to stretching in time), in which case they are described as **range-spread.** There is also the case in which the irregularities cause spreading at the critical frequency, which is no longer a simple frequency but instead covers a band of frequencies. The echoes in this case are described as **frequency-spread.** In general, spread F at low and mid latitudes occurs only at night, whereas at high latitudes it can also occur during the day. Range spreading and frequency spreading of echoes seem to have different causes because they have different variations with location on the Earth, level of solar activity, season, and time of day. Spread F will usually have detrimental effects on HF communications (Chapter 7), and also on ground-satellite links, but under some circumstances spread F is *necessary* for communications on particular circuits (Chapter 9).

## 3.10 Travelling Ionospheric Disturbances

Travelling ionospheric disturbances (TIDs) are wave-like motions of the ionosphere, attributed to the passage of waves through the neutral atmosphere and coupling between the ionosphere and neutral atmosphere [17]. The existence of waves in the ionosphere had been postulated in 1926, when Breit and Tuve suggested that the observed fading of a skywave was due to multiple reflections from a wavy surface "more or less as the flickering of light on a wavy surface of water." TIDs are especially important for HF communications at high data rates and for determining the location of HF transmitters at short ranges. In both cases they make life more difficult.

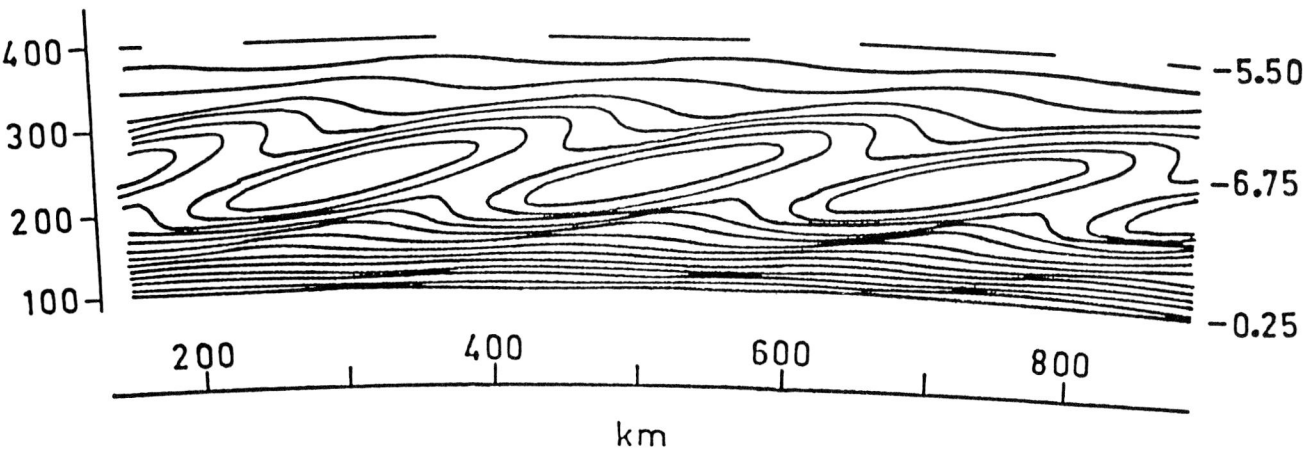

**Figure 3.22** Modelled plasma frequency profiles during the passage of a travelling ionospheric disturbance (TID). The disturbance had a wavelength of 200 km and a 10% variation in electron density. The heights are in km, and the plasma frequency contours in MHz. *P. L. George, private communication.*

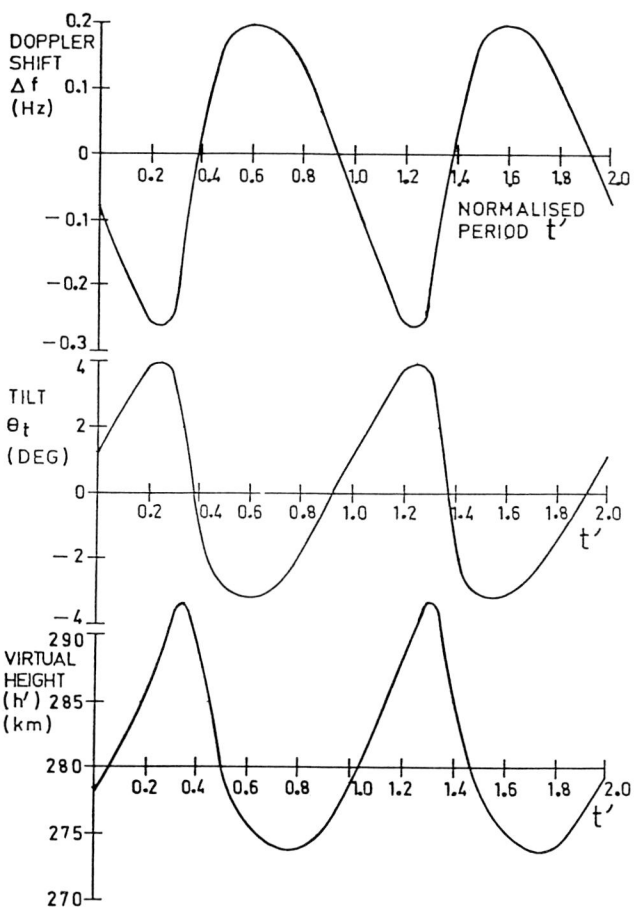

**Figure 3.23** Signatures of a TID, as seen in Doppler shift of the returning signals, variation of the apparent tilt of the reflecting surface, and variation of the virtual height. *P. L. George, private communication.*

### 3.10.1 Observed Characteristics

Moving waves of ionization can bring about changes in every parameter by which an HF wave is characterized—phase, amplitude, angle of arrival, frequency, and polarization. Each of these observables will change with space and time, as the wave travels from its source. Figure 3.22 illustrates a travelling ionospheric wave which is moving from left to right with inclined phase fronts. The slope of the individual iso-electronic (constant electron density) contours increases as the height increases. Figure 3.23 depicts the Doppler, angle-of-arrival and group-path signatures observed on the ground as the disturbance passes overhead. Note the close relation between the angle of arrival (measured in terms of deviation from the zenith) and Doppler signatures. In principle, observations of any one of these signatures at a number of spaced stations on the ground allow the measurement of the speed and direction of travel of the disturbance. The signatures are usually quasi-sinusoidal.

TIDs are present to some degree at all times of day and night, but are particularly noticeable during daylight hours. They have wavelengths of the order 50 to 500 km, travel apparently in a horizontal direction at speeds of 5 to 10 km per minute, and usually have a well-defined direction of travel which varies diurnally and seasonally in a reasonably predictable way. The electron density perturbations are rarely greater than 15%. The tilts of the iso-electronic contours, as observed over Australia, are 3° to 4° on average and are sometimes as large as 10°. The periods observed in the F region lie in the range 8 to 60 minutes, with the great majority of events having periods of the order of 12 to 25 minutes.

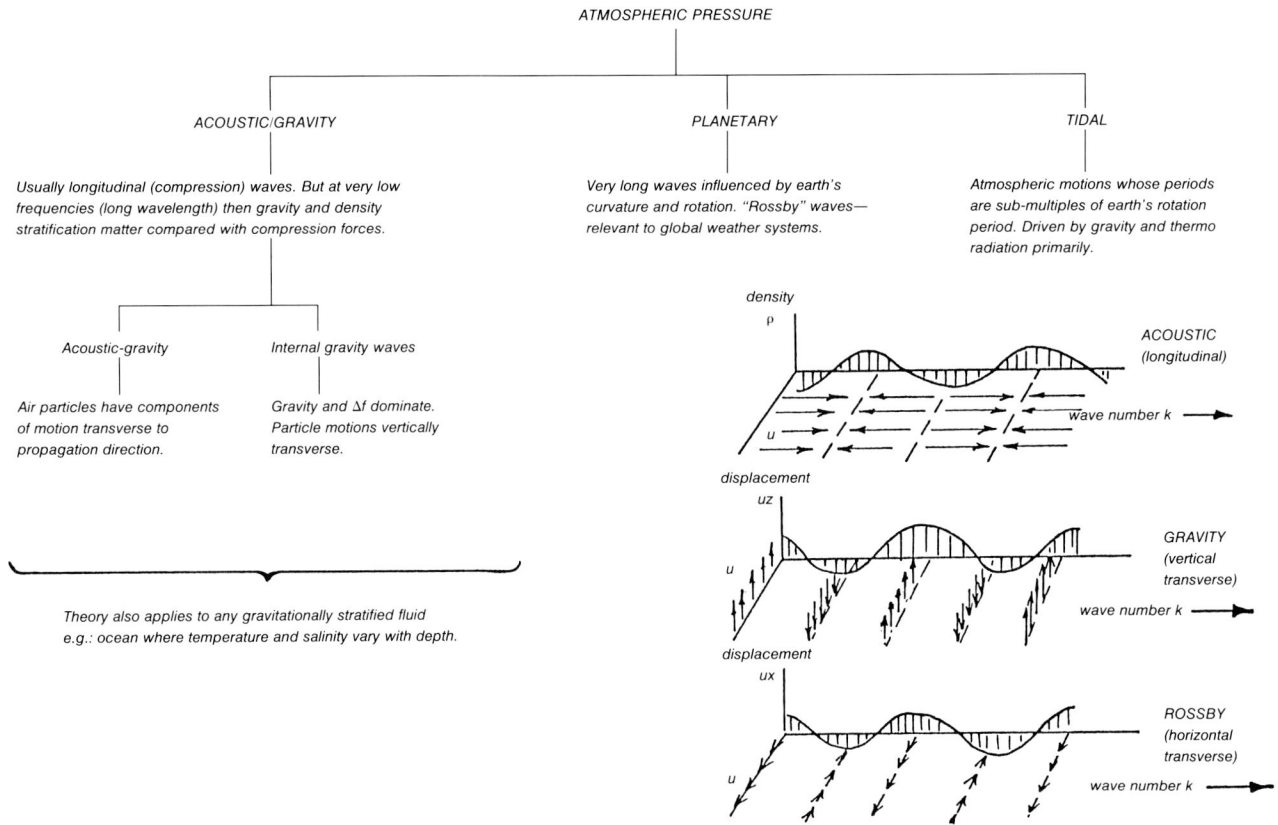

**Figure 3.24** Relationship between periodic atmospheric motions of different types. *P. L. George, private communication.*

### 3.10.2 Generation and Propagation of Atmospheric Waves

Probably the most familiar atmospheric wave process is the propagation of sound. Sound propagates as acoustic waves, where motion arises from an interchange between the kinetic energy of the atmospheric "fluid" and the potential energy stored in compression. In the terrestrial environment, gravity plays an important part, since in the presence of the gravitational force the stratified atmosphere gives rise to a buoyancy force which tends to restore the equilibrium of a displaced air "parcel." The atmosphere has a characteristic frequency of oscillation under such conditions, which is known as the **Brunt-Vaisala frequency.**

If the frequency of the wave motion is much higher than the buoyancy frequency, then gravitational effects can be ignored. However as the frequency is lowered towards the buoyancy frequency, the gravitational potential energy becomes as important as the compressional energy and kinetic energy. These waves are in the class known as **acoustic-gravity waves** (AGWs).

The relationship between these waves and other commonly occurring periodic atmospheric motions is summarized in Figure 3.24. Thus we see that atmospheric motions with periods which are submultiples of either the lunar or solar day are called atmospheric tidal oscillations. Recall that sea tides can be measured by noting the change in water height at a fixed location. Tidal motions in the neutral atmosphere, on the other hand, can be measured with a barometer (which effectively measures the weight of a column of air with a standard base size at that location).

For waves with longer periods of oscillation, the Coriolis force associated with the Earth's rotation becomes important. At sufficiently long periods, this force acts to convert gravity waves into rotational waves known as **Rossby or planetary waves.** These waves modify the background atmospheric circulation and, in doing so, influence global weather.

The particular branch of acoustic/gravity waves that is of concern to us is often referred to as **internal gravity waves** (IGWs). These waves carry energy upwards from the lower atmosphere into regions of lower gas density. Since fewer molecules exist at increasing heights, the wave carries its energy onwards by oscillating with increasing amplitude. Thus the wave grows larger as it rises above the source, just as an ocean wave increases in amplitude as the available water depth decreases as it approaches a sloping beach. In the case of IGWs, the amplification is enormous because of the very large change in atmospheric density. This intrinsic exponential growth of amplitude with height suggests strongly that IGWs may have their origin in very much less obvious waves at lower levels in the atmosphere.

Waves propagating in the Earth's atmosphere are both **anisotropic** and **dispersive.** The anisotropic nature, which leads to waves propagating differently in different directions, is a natural consequence of the directional nature of the force of gravity. The dispersive nature, which leads to different wave velocities at different wavelengths, has the consequence that long period waves, with their energy travelling essentially horizontally, can travel great distances and be observed many thousands of km from the source. This is true for large TIDs originating in the auroral zones following large events on the sun (such as described in Chapter 8). On the other hand, the observation of many short-period components may imply proximity of the observing station to the source. In this case, likely sources are thunderstorms and other weather related phenomena. The ionospheric response to the neutral atmospheric waves is driven by the direction of coupling into the Earth's magnetic field, since the free electrons cannot be driven across the field lines. Thus the response depends strongly on the magnetic dip angle.

## 3.11 Thermospheric Winds

The thermosphere is the region of the Earth's atmosphere lying above 100 km (at the mesopause) in which the ionosphere is embedded. Winds exist in the thermosphere, just as they do at sea level, and as they blow they attempt to drag the ionosphere with them. Thus the ionosphere at a specified location will depend on the neutral winds which happen to be blowing in the thermosphere at the time. At low altitudes, where the air density is much higher than the electron density, the ions and electrons move together in the direction of the wind. At high altitudes, the collision frequency is much lower than lower down, and the ions and electrons move in opposite directions to each other, and at right angles to the wind, producing an electric current [18].

Most of our knowledge of thermospheric winds is based on calculations made using semi-empirical dynamic models of the neutral atmosphere. The pressure gradients inherent in these density models provide the pressure forces that drive the atmospheric circulation [19, 20]. Direct measurements of the winds are difficult to make, and the variability of the winds is very large. The winds are different during quiet periods and geomagnetic storms (see Chapter 8). There is some hope that at least some of the global winds may be derived from ionosondes, because of the approximately linear relationship between the peak height $h_mF_2$ and the strength of the meridional neutral wind [21].

Because the winds are so difficult to measure (and currently impossible on any grand scale), their effect is usually treated as unknown, and serving to establish a limit to the predictability of the ionosphere. The day-to-day changes in the wind at a particular place and time are one cause of the day-to-day changes in the observed ionosphere and lead to the variability of parameters such as $f_oF_2$. It is the $F_2$ region which is most affected by winds.

## 3.12 References

1. Chamberlain, J. W. *Theory of Planetary Atmospheres; An Introduction.* Academic Press, New York, 1978.

2. Davies, K. *Ionospheric Radio.* Peter Peregrinus Ltd., London, 1990.

3. Hunsucker, R. D., *Atmospheric Gravity Waves and Travelling Ionospheric Disturbances: Thirty Years of Research.* Ionospheric Effects Symposium IES-90, Washington, D.C., May 1990.

4. Davies, K. *Ionospheric Radio. loc. cit.*

5. Davies, K. *Ionospheric Radio. loc. cit.*

6. Tascione, T. F. *Introduction to the Space Environment.* Orbit Book Co., Malabar, Florida, 1988.

7. Hargreaves, J. K. *The Upper Atmosphere and Solar-Terrestrial Relations.* Van Nostrand Reinhold, New York, 1979.

8. Wilkinson, P. J. *Monthly Ionospheric Indices.* Proceedings of the Workshop on Solar Terrestrial Physics, pp. 496–499, Meudon, June 1984.

9. Wilkinson, P. J. *Daily Indices.* Proceedings of the Workshop on Solar Terrestrial Physics, pp. 505–508, Meudon, June 1984.

10. Whitehead, J. D. *Recent Work on Mid-latitude and Equatorial Sporadic-E.* J. Atmos. Terr. Physics, 51(5), pp. 401–424, 1989.

11. Martyn, D. F., *Theory of Height and Density Changes at the Maximum of a Chapman-like Region, Taking Account of Ion Production, Decay, Diffusion and Total Drift.* Proceedings of the Cambridge Physical Society, London, pp. 254–259, 1955.

12. Klobuchar, J. A., D. N. Anderson, P. H. Doherty, *The Latitudinal Extent of the Equatorial Anomaly.* Ionospheric Effects Symposium IES-90, Washington D.C., May 1990.

13. Hanson, W. B. and R. J. Moffett. *Ionization Transport Effects in the Equatorial F Region.* J. Geophys. Res., 71, pp. 5559–5568, 1966.

14. Allen, J. H. *Solar Activity and its Consequences at Earth and in Space.* Paper presented at the DOD-DOE-NASA Seventh Symposium on Single Event Upsets. 1989.

15. Hunsucker, R. D. *Radio Techniques for Probing the Terrestrial Ionosphere.* Springer Verlag, 1991.

16. Bernhardt, P. A., W. A. Scales, M. J. Keskinen, H. L. Rowland, and L. M. Duncan. *Ionospheric Modification by Chemical Releases and High Power Radio Waves.* Ionospheric Effects Symposium IES-90, Washington, D.C., May 1990.

17. Hunsucker, R. D. *Atmospheric Gravity Waves and Travelling Ionospheric Disturbances: Thirty Years of Research.* loc cit.

18. Hargreaves, J. K. *The Upper Atmosphere and Solar-Terrestrial Relations.* loc. cit.

19. Jursa, A. (Ed). *Handbook of Geophysics and the Space Environment,* Air Force Geophysics Laboratory, Air Force Systems Command. NTIA Accession Number ADA 167000, 1985. Section 17.5.

20. Hargreaves, J. K. *The Upper Atmosphere and Solar-Terrestrial Relations.* loc cit.

21. Miller, K. L., P. G. Richards and D. G. Torr, *Using Ionosondes for Global Mapping of Thermospheric Meridional Winds.* Ionospheric Effects Symposium IES-90, Washington, D.C., May 1990.

## 3.13 Problems

1. How is the ionosphere formed?

2. What constituent of the ionosphere is responsible for the reflection of radio waves? Why aren't the other constituents of the ionosphere involved?

3. Electrons are lost from the ionosphere by several processes, depending on the altitude regime. What are the main loss mechanisms for each of the ionospheric layers?

4. Why is there more than one discrete or separate layer of electrons in the ionosphere? What wavelengths of radiation, and what types of neutral particle are involved for each layer?

5. Which of the four layers of the ionosphere disappear at night, and why?

6. What is the *critical frequency* of an ionospheric layer? Why is knowledge of the critical frequencies $f_oE$, $f_oF_1$, and $f_oF_2$ so useful?

7. What is the zenith angle of the sun as seen at noon by an observer situated at the Tropic of Cancer on 21 June?

8. If the sunspot number is 50 and it is noon on 21 March, what is the value of $f_oE$ at a location on the equator?

9. What are the five main variations of the ionosphere? Use figures given in the text to illustrate each of these variations.

10. Use Figure 3.10 to obtain the noon values of $f_oE$, $f_oF_1$, and $f_oF_2$ at Canberra in winter during a period of high solar activity.

11. Why do we choose to represent the values of the critical frequencies, such as $f_oF_2$, observed during a month by the median and decile values at each hour?

12. If we have 31 observations of $f_oF_2$ which are all different, how many would lie (a) above the median and (b) below the upper decile?

13. Use the calibration graph of Figure 3.13 to determine the value of $f_oF_2$ at noon at Washington during a month of December for which the sunspot number is expected to be 80.

14. Sunspots are actually relatively cool areas on the surface of the sun. Why then are they such useful indicators of what the ionosphere is doing as far as critical frequencies are concerned?

15. What are the main characteristics of the sporadic E layer?

16. Name and describe the most distinctive feature of the daytime equatorial ionosphere.

17. What is the electron gyrofrequency? What are typical values?

18. Give two lines of evidence indicating the importance of the Earth's magnetic field in the equatorial ionosphere.

19. At what times of day and at what latitudes would field-aligned irregularities be expected to occur in the F region?

# Chapter 4

# HF Radio Propagation

## 4.1 Introduction

Now that we have some idea of what it is that supports the propagation of HF radio waves, we can move on to the propagation itself. In this chapter we shall discuss a few things which determine whether or not a radio wave will actually propagate on a given circuit and how well it will do so. We shall start with a simple geometrical model of propagation. Figure 4.1 illustrates some of the ideas which we will need to discuss. Radio waves emitted by the transmitter T at an elevation angle E travel a distance D/2 before striking the ionosphere at the point P and being reflected back towards the ground, hitting it at the receiver R. In reality, the ray is not reflected at P, but is continuously **refracted** or bent towards the ground as it passes through the ionosphere. However for many purposes we can ignore this complexity and consider the ray to be *reflected* at P. The ionosphere at the point of reflection, P, is at a height h above the midpoint, M, of the circuit. The path of the waves, T-P-R, is called a **raypath.** The angle I between the raypath TP and the vertical PM is called the **angle of incidence** of the ray at the ionosphere. The distance along the ground between T and R is called the **ground range,** which we have denoted by D.

Figure 4.1 also illustrates the difference between the **ground wave** and the **skywave.** We are concerned in this book mainly with the sky wave, since it is the sky wave which goes up to the ionosphere and back, facilitating HF propagation over long distances. However for short distances, the ground wave is also useful and rather easier to use. Figure 4.2 illustrates the range covered by the ground wave for a typical circuit. It can be seen that the ground range decreases as the frequency increases. At 3 MHz the ground range is about 110 km. For lower frequencies, such as those used in the MF broadcast band, the ground range is somewhat greater. At 10 MHz, in the middle of the HF band, the ground wave will be useful out to about 60 km. The ground range increases as the electrical conductivity of the ground increases and is greater over the sea than over land. A detailed discussion of groundwave field strengths can be found elsewhere [1].

In practice, there are always at least two raypaths for a sky wave travelling between a given transmitter and receiver. Figure 4.3 illustrates what happens to rays of the same wave frequency leaving a transmitter at different elevation angles. For low elevation angles (ray 1), the signals arrive at the ground a long way from the transmitter. As the elevation angle increases (ray 2), the signals hit the ground at a point closer to the transmitter, that is at a shorter ground range, until the ground range is reduced to some particular value Z km (ray 3). Any further increases in the elevation angle cause the ground range to *increase* again (rays 4 and 5).

We can deduce several things from Figure 4.3. Firstly, there will in general be two raypaths between a given pair

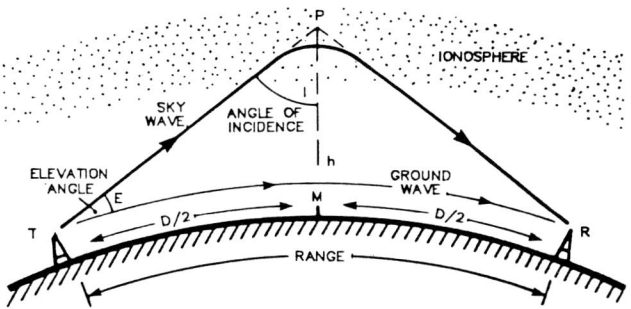

**Figure 4.1** A greatly simplified, but still very useful, view of the paths taken by a radio wave travelling between a transmitter (T) and receiver (R). The ground wave hugs the ground and will dissipate before reaching the receiver if the range of the circuit is too long. The sky wave, or ionospheric wave, is reflected from the ionosphere at a point P down towards the receiver. The angle between the ground and the path (known as a raypath) followed by the wave as it leaves the transmitter is called the elevation angle. The angle between the vertical at the point of reflection (P) and the raypath is called the angle of incidence. Note that the wave never actually reaches the point P; the wave is returned to the ground by a process of continual refraction (bending) within the ionosphere, rather than by the simple process of a single reflection. However, we get sufficiently accurate results if we think in terms of simple reflection.

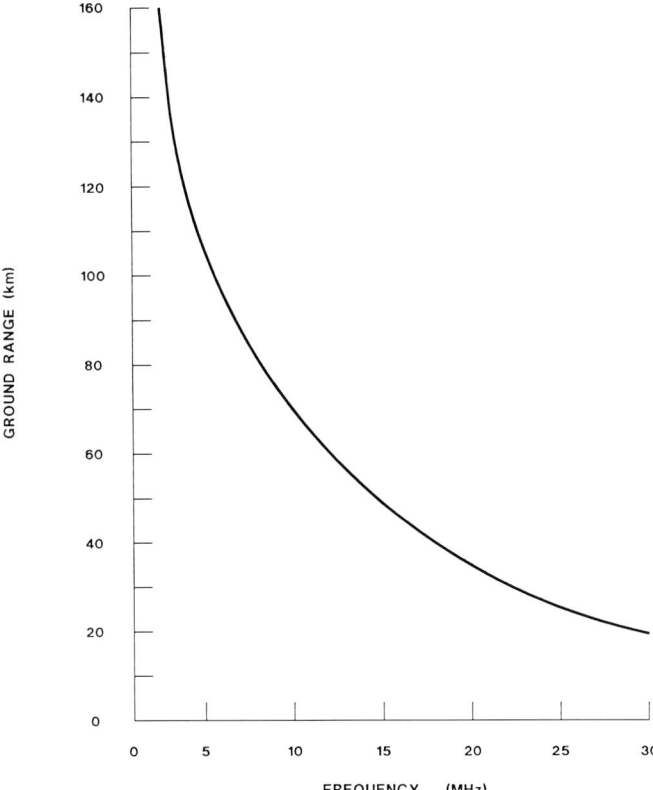

**Figure 4.2** The approximate range over which the ground wave may be used, as a function of frequency, for dry land. The ranges are much greater over the sea.

Pedersen ray) will be heavily attenuated and we will be left with only the low ray. The raypath 3 is unique because the corresponding high and low raypaths coincide exactly.

## 4.2 Polarization

Along with frequency, phase and direction of travel, polarization is one of the descriptors of a radio wave. Polarization of radio waves is best described by analogy, in view of the fact that we cannot "see" the radio waves themselves. Consider what happens to a long rope fixed at one end when it is flicked at the other end. If we flick the loose end up and down in a vertical plane, the rope will move up and down all along its length. All movement of the rope will be in a vertical plane—there will be no sideways motion. We can describe this movement of the rope as being **vertically polarized.** If we flick the rope from side to side, keeping our hand in a fixed horizontal plane, the motion of the rope will be confined to a horizontal plane and is described as **horizontally polarized.** The third thing we can do is to move our hand in a circular motion. All parts of the rope will then move in a circle and the motion of any point on the rope is therefore **circularly polarized.** When the motion is constrained to a single plane, such as with vertically and horizontally polarized motion, the motion is said to be **plane polarized.**

of transmitters and receivers. (Two more raypaths will appear in the next section.) For example, raypaths 1 and 5 connect the same two points on the Earth, as do raypaths 2 and 4. For any particular circuit, the two rays are called the **high ray** (rays 4 or 5) and the **low ray** (rays 1 or 2), corresponding to high and low angles of elevation. Under many circumstances, the high ray (also called the

In HF applications, antennas are also described as being polarized. The most common antennas used in HF communications are the vertical monopole (the whip antenna is one of these) and the horizontal dipole. Consider first the vertical monopole, as in Figure 4.4. In general, radiation leaves a length of wire acting as an antenna in a direction at right angles to the antenna, with the electric field of the wave vibrating (just like the rope) in a

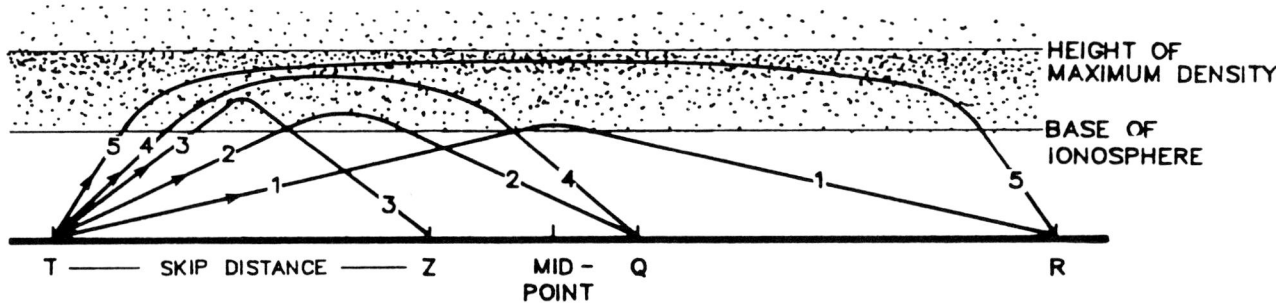

**Figure 4.3** Sample raypaths for a given ionosphere and fixed frequency as the elevation angle at the transmitter T is increased. It is found that increasing the elevation angle results in the reflected ray reaching the ground at a point which moves closer and closer to the transmitter (from R to Q to Z), until a point Z is reached after which increasing the elevation angle leads to increasing ranges. The rays 4 and 5 are called the "high" rays and have the same ground ranges as the "low" rays 1 and 2, respectively. The high and low raypaths coincide for the point Z and for the raypath 3. It is not possible for a sky wave to reach any point closer than Z to the transmitter, since changing the elevation angle from that corresponding to raypath 3 leads to an increased ground range, regardless of whether the elevation angle is increased or decreased. The distance TZ is known as the skip distance for the circuit (at the given hour, month, level of solar activity, and frequency).

# HF Radio Propagation

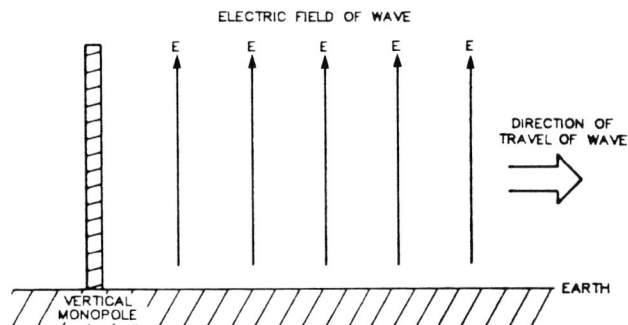

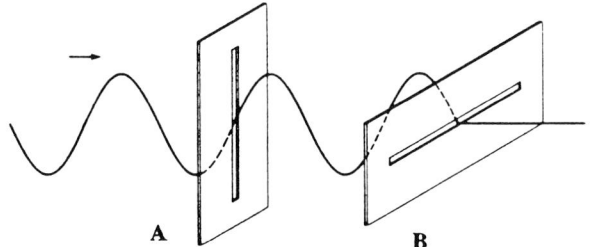

**Figure 4.4** Sketch of the electric field vector E of a radio wave emitted by a vertical monopole. The energy of the wave travels at right angles to the direction of E. The strength of the electric field oscillates in amplitude at a frequency equal to the wave frequency. Since a radio wave is also an electromagnetic wave, it will also have an oscillating magnetic field which is perpendicular to both the direction of the electric field and the direction of wave propagation.

**Figure 4.5** Passage of a plane-polarized wave through two slits held at right angles to each other. The plane-polarized wave passes untroubled through the first slit, which is vertical and matches the plane of polarization of the incoming wave. However, the wave cannot pass through the horizontal slit because this slit will accept only waves which oscillate in the horizontal direction, whereas the wave oscillates only in the vertical direction.

direction at right angles to the direction of travel. In the case of the vertical monopole, the electric field vibrates in a vertical plane and the energy radiated by the antenna is vertically polarized. The antenna is also described as being vertically polarized, because it transmits vertically polarized radiation. In a similar fashion, a horizontal antenna is said to be horizontally polarized.

What happens if we transmit on a vertically polarized antenna and receive on a horizontally polarized one? Think what would happen to our rope if we were flicking it in a horizontal plane and put two vertical posts astride it. The posts would stop the rope from moving side to side, which is the only motion which we have given to it, and the part of the rope between the posts and the anchor point will end up with no motion at all. In other words, there will be no energy transmitted from our hand to the anchored end of the rope. Similarly, if we cross our antennas, there can be no transfer of energy between them, unless the plane of polarization of the wave changes during the passage of the wave through the ionosphere. We end up drawing the same conclusions if we think in terms of what the electric field is doing. The important point in this case is that an electric field will induce a voltage in a conductor, provided the conductor is not at right angles to the direction of the field. This is the reverse situation to the antenna transmitting energy with the electric field parallel to the direction of the antenna. Polarized sunglasses give the same effect. Try looking through two pairs of Polaroid sunglasses when they are held at right angles to each other. If the glasses are well made, everything will appear very dark because very little light is being transmitted through the two lenses. The first lens transmits light polarized in only one direction and the second lens refuses to pass light polarized in that direction. This is illustrated in Figure 4.5.

Only in the case of perfectly polarized antennas will no radiation be received. In general, as with the usually imperfect sunglasses, there will be some radiation which will get through. In the rope experiment, increasing the slit width to several times the diameter of the rope will allow a range of polarizations to pass through the slit, rather than just the one parallel to the slit.

When a plane polarized radio wave hits the ionosphere, it splits into two **characteristic waves** which propagate independently through the ionosphere. These waves are known as the **ordinary** and **extraordinary** waves and are **elliptically polarized.** To get some feeling for what an elliptically polarized wave is, we can imagine ourselves travelling along with the wave, looking straight ahead. The electric field can be thought of as a vector at right angles to the direction of travel which rotates and changes its length as the wave progresses. It can rotate in either a clockwise or anti-clockwise direction corresponding to either the ordinary or extraordinary wave, depending on the orientation of the raypath relative to the direction of the Earth's magnetic field. If the vector keeps the same length as the wave progresses, we have the special case of circular polarization.

The question as to which wave is polarized in which sense is asked so often that it is worth making special note of it. The question tends to lead to confusion, since electrical engineers and physicists have different ways of looking at things. For example, when we talk about the sense of polarization, are we looking at the oncoming wave, or are we looking along the direction of travel? The key point is that *for a fixed observer looking in the direction of the imposed magnetic field (i.e., the Earth's), the electric vector of the extraordinary wave appears to rotate in a clockwise sense, regardless of whether the wave is coming towards the observer or going away* [2]. Recall that the Earth's magnetic field lines go from the south pole to the north.

## 4.3 The Maximum Usable Frequency (MUF)

One of the most important quantities in HF communications is the maximum usable frequency or MUF, which is the maximum frequency that will be reflected by the ionosphere for a given circuit. The MUF depends on just two things—the critical frequency, $f_c$, of the ionosphere at the reflection point, P, and the geometry of the circuit.

To a good approximation, the MUF is given by the formula

$$\text{MUF} = f_c / \cos I, \qquad (4.1)$$

where I is the angle of incidence. Since cos I can vary from 0 to 1.0, equation 4.1 indicates that the MUF is equal to the critical frequency of the ionosphere for vertical incidence (I = 0, cos I = 1.0) and much greater than $f_c$ for very large angles of incidence, which correspond to small elevation angles and large ground ranges. In practice, the world is round and the curvature of the surface prevents the angle of incidence from getting too close to 90°, so the MUF reaches a finite upper limit for a given ionosphere.

If the critical frequency is fixed, a higher MUF may be obtained by reflecting the signals from a lower layer. This is because a lower layer means a smaller value of cos I and thus a larger value of 1/cos I. It is a little tortuous to follow this path of turning things upside down, so we will write equation 4.1 in the somewhat easier form

$$\text{MUF} = f_c \cdot \sec I. \qquad (4.2)$$

Note that this is the limiting case of the "secant" law which is introduced in Chapter 12. The "equivalent vertical frequency" (see Chapter 12) has a maximum value of $f_c$, the critical frequency for the layer.

The factor sec I is usually called the **obliquity factor** for the circuit, because it relates the ionosphere at the reflection or midpoint of the circuit to what happens on the oblique circuit. Figure 4.6 illustrates how the obliquity factor varies with circuit length for reflection from the E layer (for h = 100 km) and F layer (for h = 300 km). As we have already seen, the obliquity factor is equal to 1.00 for very short circuits (D = 0). Note that Figure 4.6 actually shows the "corrected" obliquity factor, k sec I, where k is a correction factor which takes into account the fact that the Earth and ionosphere are both curved. The value of k is about 1.1 under most conditions.

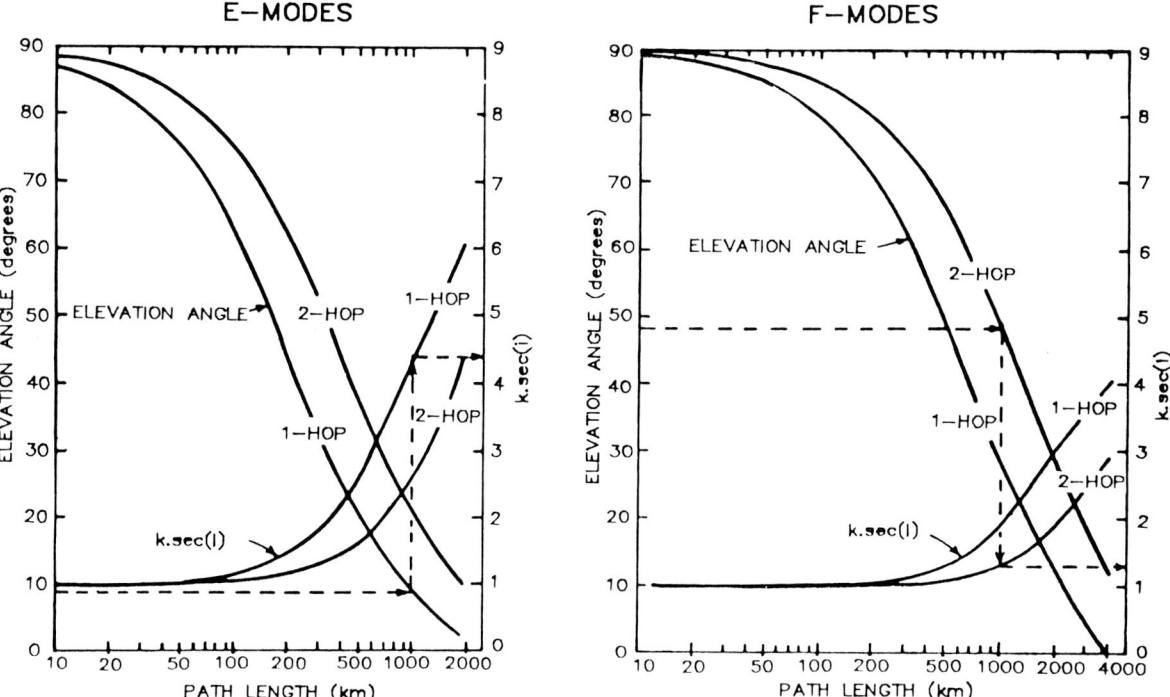

**Figure 4.6** The "corrected" obliquity factor k sec I and elevation angle as a function of circuit length for signals reflected from (a) the E layer, assumed to be at an altitude of 100 km and (b) the F layer, assuming that reflection takes place at 300 km. The two sets of curves correspond to one-hop and two-hop propagation modes. In the latter, the signals hit the ionosphere twice and hit the ground in between the two ionospheric reflections. These diagrams can be used to estimate the elevation angles for a circuit of given length and thence to help in the choice of appropriate antennas, as well as estimating the obliquity factor for the circuit. The obliquity factors are greater for E-layer reflection than for F-layer reflection. *Courtesy IPS Radio and Space Services.*

# HF Radio Propagation

The obliquity factor decreases as the altitude of the reflection level increases. Thus it is greater for E-layer reflections than for F-layer reflections. For the F layer, which is so thick that we must consider the effects of the different altitudes within it, signals reflected from low in the layer have higher obliquity factors than signals reflected from higher up. The corrected obliquity factor for an $F_2$ propagation mode for a standard distance of 3,000 km, $M(3000)F_2$, is derived routinely from ionograms, and mapped in much the same way as $f_oF_2$.

Equation 4.2 may be used to calculate the MUF for reflection at a given altitude in the F layer, provided the critical frequency $f_c$ ( = $f_oF_2$ in this case) is replaced by the plasma frequency at the reflection height, $f_N(h)$,

$$\text{MUF}(h) = f_N(h) \cdot \sec I(h), \quad (4.3)$$

where the "h" tells us that we are considering what happens if reflection occurs at the height h. In general, the higher the operating frequency, the higher the signals must penetrate into the ionosphere to find a plasma frequency high enough to reflect them. However penetrating further into the ionosphere is not a good thing from the point of view of the obliquity factor, which *decreases* as the height of the reflection level increases, and acts in opposition to the effect of increasing plasma frequency.

Figure 4.6 also shows the elevation angles corresponding to circuits of different lengths. For example, if the circuit length is 1,000 km and reflection is to be by the E layer, the elevation angle of the signals would have to be 9°. We can also use this figure to work out where our signals would land if our transmitting antenna puts most of its energy out at a particular elevation angle. For example, if a ray leaves the transmitting antenna at an angle of 28° and is reflected by the F layer at 300 km, it would travel 3,000 km.

Figure 4.6 can be used to great advantage in choosing the correct antenna for a circuit. HF communications will not be effective unless the antennas are chosen to match the path taken by the radio waves (see Section 4.7).

## 4.4 The Skip Zone

For propagation to be possible on a particular circuit, the operating frequency, $f_o$, must be less than or equal to the MUF for the circuit. At higher frequencies, the signals would simply penetrate the ionosphere. Now the MUF is defined as

$$\text{MUF} = Q \cdot f_c \quad (4.4)$$

where Q is the obliquity factor and $f_c$ is the critical frequency of the reflecting layer. Thus it is necessary that

$$f_o \leq Q \cdot f_c, \quad (4.5)$$

from which we can deduce that the obliquity factor, Q, must be greater than or equal to the quotient $f_o/f_c$.

Suppose the operating frequency $f_o$ is 20 MHz and the critical frequency $f_c$ is 10 MHz. It is then necessary that Q > 20/10, or Q > 2. Figure 4.6 tells us that for a one-hop F-layer propagation mode, Q > 2 requires that the path length be greater than or equal to 1,000 km (approximately). In other words, if we are using 20 MHz and the critical frequency of the ionosphere is 10 MHz, we can communicate with positions 1,000 km or more away from us, but we cannot communicate with any position closer than 1,000 km to us, using the sky wave. The 1,000 km zone around us is called the **skip zone**. Communication within the skip zone is still possible using the ground wave, but this peters out after about 50 to 100 km. When the operating frequency is less than the critical frequency, there is no skip zone. The mathematically minded reader will note that under these conditions, it is necessary that Q > N, where N < 1, but Q > 1 always because the secant of any angle is always greater than one. Another way of thinking of the situation is that, by definition, the ionosphere will reflect all waves at frequencies less than the critical frequency when these are incident vertically on the ionosphere, and oblique propagation *enhances* the ability of the ionosphere to reflect a signal with a given frequency.

The skip zone for a given transmitter will depend on the operating frequency (getting larger as the frequency increases) and on the critical frequency of the reflecting layer. The latter dependence means that the skip zone will have all the variations normally associated with the critical frequency—diurnal, seasonal, and solar cycle.

Figure 4.7 illustrates the skip zone for different operating frequencies for a transmitter situated at Alice Springs in Australia. Any location within a contour marked as "f" MHz (where f = 12, 13, 15, 20, 25 etc.) cannot be contacted via an F-region skywave using the frequency f. For example, if the operating frequency is 25 MHz, most of Australia lies in the skip zone.

It is important to note that if the receiver is on the edge of the skip zone at a particular operating frequency, that frequency must be the maximum usable frequency (MUF) for that particular circuit. Because energy arrives at a receiver on the edge of the skip zone by both the high and low rays, which coincide exactly at the MUF, the strength of the signal will be greater than at frequencies less than the MUF. One advantage of working at the MUF, and thus on the edge of the skip zone, is this enhancement of the signal strength. However, as we shall see in Chapter 7, working near the edge of the skip zone does have its problems.

The skip zone is usually a problem when it exists, but it can sometimes be put to good effect if secure

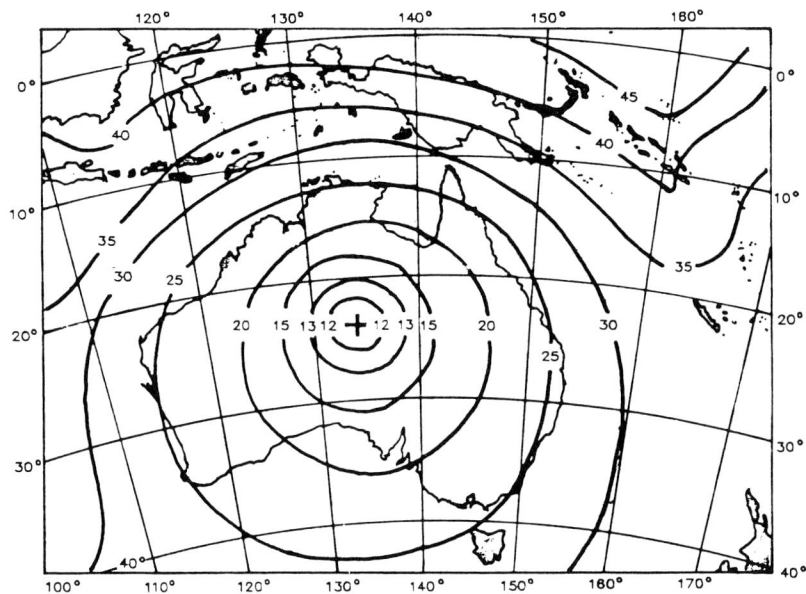

**Figure 4.7** The skip zone for different frequency transmissions from a site in Alice Springs, Central Australia. Within any contour labelled f, communication with a receiver inside the contour is not possible for frequencies greater than f MHz. For example, most of Australia lies within the 25 MHz skip zone. This means that a person in Alice Springs working on 25 MHz could not communicate with anyone in Australia apart from Tasmania, which lies outside the 25 MHz contour (near 42° south). Note that the skip zone varies as the ionosphere varies. This particular diagram corresponds to 05 UT, March, high solar activity. The value of $f_oF2$ at Alice Springs was 11 MHz. *Courtesy IPS Radio and Space Services.*

communications are required. If we do not want someone to hear our transmissions, we are sometimes able to ensure that he is within our skip zone.

## 4.5 Propagation Modes

A propagation mode is the path that a radio wave takes when travelling or propagating from the transmitter to the receiver. These paths are many and varied, and when a radio wave leaves a transmitting antenna, it will choose its own propagation mode, which may not be the one we would have preferred it to take. Some propagation modes are better to use than others and we should always be aware of the modes that our signals are probably taking, even if we cannot force the signals to travel via the best mode or mode of our choice. Propagation may be one-hop (one single reflection from the ionosphere), two-hop (two reflections from the ionosphere, with a reflection from the ground in between), and so on, and may be via any one or several of the layers of the ionosphere. If we ignore reflections from the D and $F_1$ layers, which are not usually important at HF, we need to consider only reflections from the E layer, a sporadic E layer, and the $F_2$ layer. Since we are ignoring the $F_1$ layer, we shall talk about the F layer, rather than the $F_2$ layer. Recall that at night only the $F_2$ layer exists and is simply called the F layer.

Figure 4.8 shows some of the propagation modes which are possible using just the three layers. Luckily, some modes lead to more attenuation of the signal than others and in practice we do not have to worry about them all. As a general rule, the higher the order of the mode, or the more hops it has, the lower its signal strength, because every reflection of the signal at either the ground or ionosphere results in loss of energy. For example, the 2-hop F-layer, or 2F, mode must be reflected from the ground once, and from the F layer once more than the 1F mode, suffering losses at each reflection. Consequently the losses on a 1F mode will always be less than those on a 2F or higher order (3F, 4F, . . . ) F-layer mode. Typical ground reflection losses for long hops (low elevation angles) are 3 dB for a poorly conducting ground and 0.5 dB for reflection from the sea.

Lower order propagation modes are also characterized by higher MUFs than those supported by higher order modes. For example, the MUF for a one-hop F mode (1F MUF) is greater than the MUF for a two-hop F mode (2F MUF) on the same circuit. This is because the obliquity factor for the 1F MUF corresponds to the full length of the circuit and is thus greater than for the 2F MUF, which corresponds to a hop length equal to half the full length of the circuit.

The E layer is a much poorer reflector than the F layer, and two reflections from it are normally enough to reduce the signal strength below the level of detectability, i.e. the signal sinks below the noise level. Sporadic-E layers, on the other hand, are often very good reflectors of radio

# HF Radio Propagation

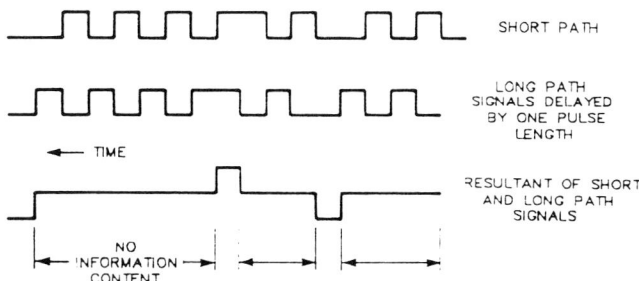

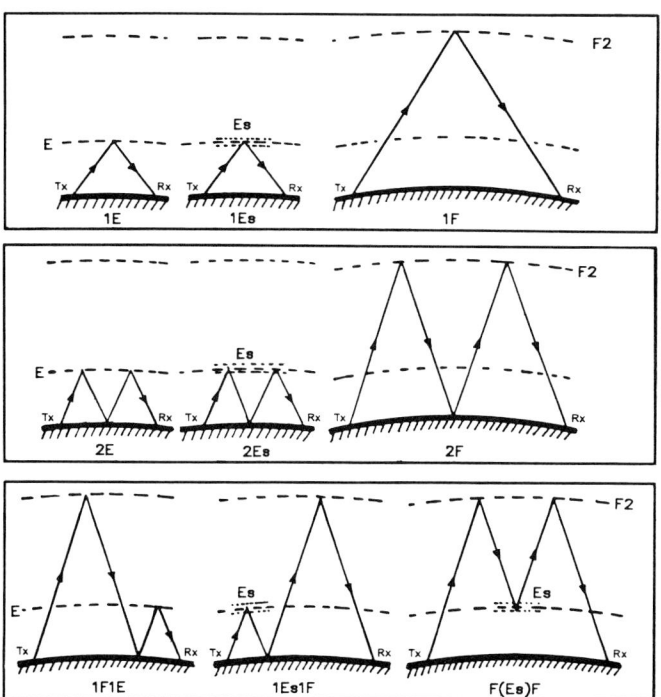

**Figure 4.8** Propagation modes which are possible for signals reflected from the $E_s$, E, or $F_2$ layers. The top panel illustrates one-hop modes (1E, $1E_s$, and 1F) which involve only one reflection from one ionospheric layer. The middle panel illustrates two-hop modes. The lower panel illustrates so-called "combination modes" which involve reflections from both an E layer (E or $E_s$) and the $F_2$ layer. Note that we have ignored modes involving reflections from the $F_1$ layer, which are not important under most conditions. *Courtesy IPS Radio and Space Services.*

wave energy and can be very useful, although they can also sometimes be a bit of a problem (see Chapter 7).

## 4.6 Multipath Interference

On any given circuit, it is usually preferable to have the radio waves propagate by only one propagation mode. If the signal arrives by two different modes, with approximately equal signal strengths, they can interfere with each other in what is known as multipath interference. This is because the signals usually have different travel times, the distances covered over the two propagation paths being slightly different. To see why having two signals is a bad thing, consider what happens to Morse code when the signals arrive by two different paths. This is illustrated in Figure 4.9, which shows what happens in the extreme (and unlikely) case in which the second signal takes exactly one pulse length longer than the first to arrive at the receiver. When we should have had no signal if we are going to make sense out of what we are receiving as signal 1, the signal coming via the second propagation mode has a pulse arriving. The result of the two signals adding together in our receiver is thus a continuous signal, with no breaks to indicate spaces between

**Figure 4.9** Pulse-coded signals arriving at a receiver by two different propagation modes whose path differences are such that the signals travelling over the longer path arrive exactly one pulse length later than those travelling via the shorter path. The resultant signal, illustrated in the lowest of the three sketches, is formed by adding the signals which arrive at the same time. It can be seen that the resultant signal often bears no relation to the transmitted signal, because pulses from the long path reach the receiver when a null is arriving from the short path. This is a particularly severe example of multipath interference. Pulse lengths are usually chosen to be much longer than the propagation time difference between likely propagation modes, and multipath interference is then not so severe. However, if high data rates are required, necessitating short pulse lengths, multipath inteference becomes very important and must be taken into account.

the pulses. All information content of the signal is therefore lost.

There are two ways to avoid multipath interference. The simplest is to choose a frequency that is too high to propagate by any mode other than the lowest order mode. Recall, for example, that the 1F MUF is always greater than the 2F MUF. The second way, which is not so simple, is to choose an antenna which favors one propagation mode. This is just one good reason for choosing an appropriate antenna for a given circuit.

Interference can also occur between the high ray and low ray on any circuit, if they have equal amplitudes. However there is no multipath interference at the MUF, since the signals propagating by the high and low rays follow exactly the same path and take exactly the same time to travel to the receiver. We shall return to this point in Chapter 7.

## 4.7 Choosing the Correct Antenna

Basically, there is just one reason for choosing the correct antenna—so that the signals propagate over the path that we want them to, both in terms of arriving at the receiver at all and then doing so by the best possible propagation mode. The choice of antenna type varies according to the intended application. One antenna may be ideal for one situation, but almost hopeless for another.

Before choosing an antenna, we must first work out the geometry of our circuit, especially the elevation angles, which are normally assumed to be the same at the transmitter and receiver. We can do this using Figure 4.6, reading off the elevation angles for each possible propagation mode. In practice, most propagation prediction programs (see Chapter 6) provide estimates of the elevation angles for a given circuit, using the same geometry as used to generate the curves in Figure 4.6.

Once we have deduced the elevation angles for the 1F and 2F modes, say, we can select an antenna which first of all has a radiation pattern with a maximum at the appropriate elevation angle. For example, for a 1,000 km circuit, the elevation angles for the two modes are 28° and 48°, so we would need to choose an antenna which radiates the bulk of its energy somewhere around those angles. An antenna which puts most of its energy out at an angle of 10°, for example, would be useless on this circuit; most of the energy would end up at a range of 2,000 km. In this case it would hit the ionosphere over the receiver, which is clearly not what is required.

Figure 4.10 illustrates the gain pattern for an antenna which would be quite suitable for a 1,000 km circuit, but which would be quite inefficient for a short circuit (including a vertical incidence one). Antennas must be chosen to match the intended circuit length. They must also be matched to the intended propagation mode, as illustrated in Figure 4.11. In this case most of the energy would travel via the 2F mode, so there is no point in attempting to work at a frequency which is supported by the 1F mode but is too high for the 2F mode.

Figures 4.10 and 4.11 also illustrate that, for the same operating frequency, signals on short circuits need to penetrate more deeply (higher) into the ionosphere before they can be reflected. At vertical incidence, a signal at a frequency equal to $f_oF_2$ would need to penetrate right through to $h_mF_2$, the height of maximum electron density. However on an oblique circuit, the same signal would be reflected at a lower level. The longer the circuit becomes, the lower the level of reflection, because the increasing obliquity factor means that a lower plasma frequency is adequate to reflect the signals.

If the obliquity factor is kept more or less fixed by considering a fixed circuit and the operating frequency is increased, the increasing frequencies would need to penetrate higher into the ionosphere for reflection to occur. This means that for a given circuit, the higher frequencies will leave the transmitting antenna at the higher elevation angles and follow higher raypaths to the receiver. This is illustrated in Figure 4.12. Note that signals at the MUF on an oblique circuit do not reach $h_mF_2$. Any attempt to use a higher frequency than the MUF, by trying to have the signals reflected from a higher altitude, is defeated by the fact that the obliquity factor decreases as the altitude increases. Such frequencies would therefore not be reflected, but would penetrate the ionosphere.

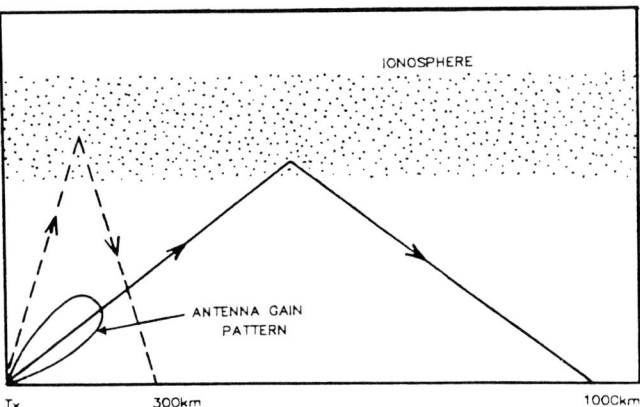

**Figure 4.10** The antennas should always be chosen to match the length of the circuit. In this diagram, the transmitting antenna is ideally suited to the 1,000 km circuit, but it is of no use for a 300 km circuit (or shorter) because virtually no energy goes off at the steep angles required for the short distances.

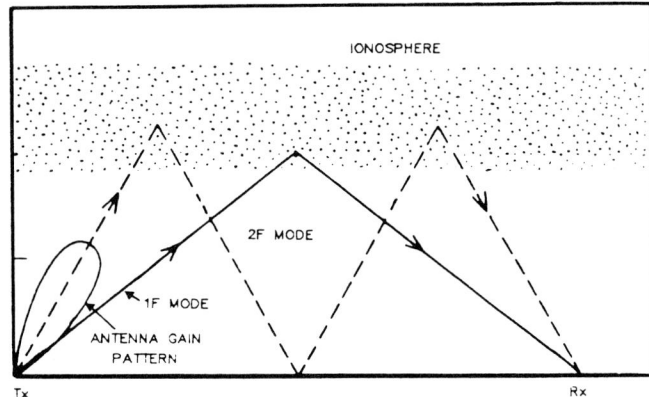

**Figure 4.11** The antennas should always be matched to the required propagation mode. In this example, most of the transmitted energy would travel via the 2F mode, so there is no point in choosing a frequency that is supported by the 1F mode, but not by the 2F mode. The failure of the antenna to excite the 1F mode as well could, however, be an advantage because of the decreased probability of multi-mode (multipath) interference.

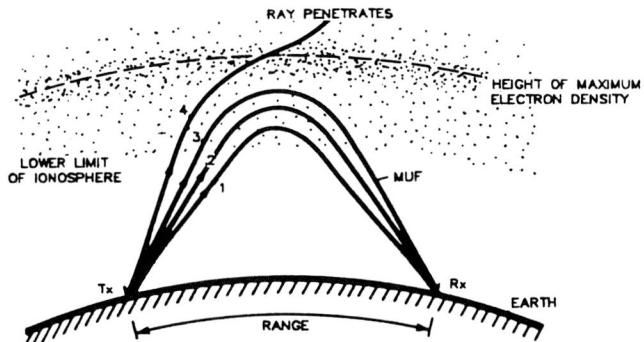

**Figure 4.12** Raypaths for increasing frequencies on a fixed circuit. The higher frequency rays (2 and 3) penetrate further into the ionosphere than the low frequency rays (ray 1) and leave the transmitting antenna at correspondingly higher elevation angles. Rays at frequencies higher than the MUF reach altitudes where the obliquity factor is too low for reflection to occur. They therefore penetrate the ionosphere.

# HF Radio Propagation

**Table 4.1** Examples of simple antennas which can be used on circuits of given lengths, and for given elevation angles. The propagation modes are assumed to be single-hop modes.

| Path length km | Required radiation (elevation) angles | Suitable simple antennas |
|---|---|---|
| 0-200 | 60°-90° | Horizontal dipole: broadside to required azimuth, 0.25 wavelength ($\lambda$) above ground. |
| 200-500 | 40°-70° | Horizontal dipole: broadside to required azimuth, 0.3$\lambda$ above ground. |
| 500-1000 | 25°-50° 10°-20° | 0.25$\lambda$ vertical monopole or horizontal dipole: broadside to required azimuth, 0.5$\lambda$ above ground. |
| 1000-2000 | 10°-30° and low angles | Vertical monopole: up to 0.3$\lambda$ long with ground screen. |
| 2000-3000 | 5°-15° and 20°-30° | Vertical monopole: up to 0.3$\lambda$ long. |
| >3000 | low angles | Vertical monopole: up to 0.6$\lambda$ long with ground screen. |

Table 4.1 is a good guide to the choice of an antenna appropriate to a particular circuit. For distances under about 1,000 km, a horizontal dipole antenna would be a good choice. A vertical monopole, or a whip antenna, would be a poor choice because most of the energy radiated from a vertical monopole leaves the antenna at low elevation angles. Vertical monopoles are most suitable for long circuits or ground wave propagation. For very short circuits, the best antenna is a half-wave horizontal dipole erected high above the ground. This antenna sends most of its energy straight up, or nearly so, where the ionosphere will reflect it back down to within an area centered on the transmitter. The reader with an interest in antennas is encouraged to follow up the references [3, 4, 5].

Choosing an antenna that will support only one of the possible propagation modes on a given circuit can be quite a difficult job and requires more complicated and expensive antennas than are normally available.

## 4.8 Absorption

Every radio wave reflected from the ionosphere is partially absorbed as it passes through the D region, both on the way up to the reflection point in the E, $F_1$, or $F_2$ layers and on the way back. It is therefore essential that we know something about absorption. Absorption is easily explained on what is called the microscopic level, which means that we look at what individual ions and electrons are doing. When an electron absorbs energy from a radio wave, it vibrates or oscillates to and fro at a frequency equal to that of the radio wave. The energy of the radio wave is thus transformed into kinetic energy or energy of motion. The oscillating electron re-radiates this energy in the form of radio waves (this is how a radio wave is transmitted from an antenna) so the energy of the radio wave is passed forward from one electron to the next. If, however, an oscillating electron collides with a heavy neutral atom, it will give up its energy to the atom. The atom does not then vibrate at the radio wave frequency, because it is not a charged particle, but uses its increased energy to travel a little faster. The energy which the electron took from the radio wave is thus lost as far as the radio wave is concerned, and we say that part of the wave's energy has been absorbed or that the signal has been attenuated.

Because it is the neutral atoms which take up radio wave energy from the electrons, we can deduce that absorption will be strongest where there are the most neutral atoms. This is at lower altitudes in the atmosphere, so the parts of the ionosphere giving rise to most absorption of radio signals are the D region and, to a lesser extent, the lower part of the E region.

To a first approximation, the absorption of a radio signal follows the same variations as the electron density in the D region. This is particularly the case for signals reflected obliquely from the F layer, which is our main area of interest here. For signals reflected vertically from the E region, extra absorption occurs near the level at which the signal is reflected, and this is related more to the way the electron density varies with height in the E region than to the underlying D region. The extra absorption is known as **deviative absorption** because it occurs when the ray is deviated by the ionosphere—in fact the ray is deviated completely around so that it retraces its upward path. In other words, the ray is reflected. The absorption suffered by the signal as it passes through the D region is called **non-deviative absorption,** because the ray passes essentially undeviated through the D region. It is the non-deviative absorption which follows the variations of the D region and which is important for HF communications. Non-deviative absorption builds up after sunrise, is greatest at noon, drops rapidly after sunset, and is almost zero at midnight. It is greater in summer than in winter, in general, and is greater at higher levels of solar activity. These variations are illustrated in Figures 4.13 and 4.14. For frequencies significantly greater than the gyrofrequency, the absorption varies inversely with the square of the frequency. For example, if the frequency is decreased by a factor of 2, the non-deviative absorption will increase by a factor of 4. This is one reason why higher frequencies are normally preferred to lower frequencies.

An interesting feature of absorption in winter at mid-latitudes is the occurrence on some days of very high values of absorption. These days are called **winter anomaly** days because of the anomalously or unusually high values of absorption, which can reach the levels normally encountered only in the summer. Fortunately, the effects of the winter anomaly can be overcome by the use of higher frequencies for communications since the critical frequency of the $F_2$ layer is higher in winter than summer. This is the mid-latitude seasonal anomaly described in Chapter 3.

The absorption of a wave depends on the sense of its polarization. For example, an extraordinary wave is more heavily absorbed than an ordinary wave. This can be explained quite simply on the microscopic level. In the absence of collisions, the electrons in the ionosphere cannot move across the field lines of the Earth's magnetic field, but instead circle or gyrate around them. The rotating electric field of a circularly polarized wave can act to either speed the electrons up or slow them down, depending on the direction of rotation of the electric field. The ordinary wave is the one whose electric field slows the electrons down, while the extraordinary wave gives them extra energy and speeds them up. The more energy the electrons get from the wave, the larger the radius of the circle which they travel in, and the larger the chance of their striking a neutral atom and losing all of their energy. Consequently the extraordinary wave is more heavily absorbed than the ordinary wave. When the wave frequency of the extraordinary wave is equal to the gyrofrequency (the frequency at which an electron gyrates around a field line—see Chapter 3), all the energy of the wave is transferred to the neutral atoms and the wave is completely absorbed. This does not happen with the ordinary wave.

## 4.9 The Lowest Usable Frequency

We saw in Section 4.3 that the highest, or maximum, usable frequency depends on the critical frequency of the ionosphere at the reflection point and on the geometry of the circuit, which determines the obliquity factor. There is also a lowest usable frequency for a given circuit at a given time, which depends on (1) the efficiency of the antennas, (2) the transmitter power, and (3) how much absorption the signal suffers.

At low frequencies, the optimum size of an antenna can get prohibitively large. For example, at 2 MHz, a half-wave dipole is 75 m long. It is therefore common practice to use smaller, less efficient antennas which lead to less energy actually being radiated for a given transmitter power than is possible for an ideal antenna. This inefficiency can be overcome to some extent by increasing the transmitter power to get the received signal up to the required signal strength. Since the efficiency of a given antenna decreases with frequency, a point will be reached as the frequency is decreased when the signal arriving at the receiver cannot be detected above the background noise level. This frequency is called the lowest usable frequency or LUF.

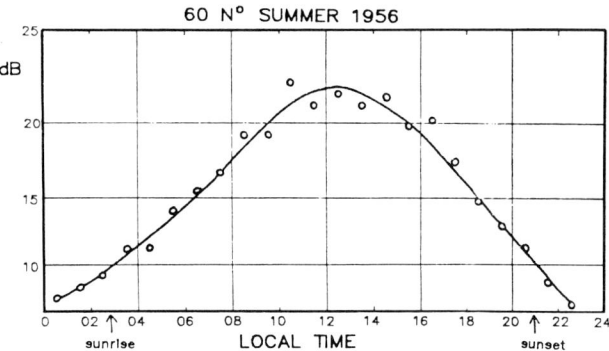

**Figure 4.13** The diurnal variation of absorption due to the D layer at a station 60° N in summer at high solar activity. The absorption peaks at noon and drops off to zero during the very short night.

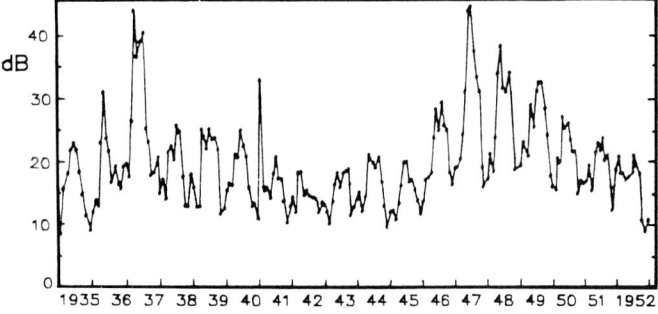

**Figure 4.14** The variation of absorption of a vertically incident radio wave at 4 MHz, for Slough (near London), from 1935 to 1952. The absorption shows an annual variation, generally being greater in midyear (summer) than at the end of the year (winter). It also shows a solar-cycle variation, being greatest in years of high solar activity, such as 1937 and 1947. The January 1941 results show a winter anomaly period during which the absorption exceeded even the normal summer levels.

The LUF is also heavily affected by the ionosphere, in particular by the amount of absorption that a signal suffers as it traverses the D region. We have seen that absorption depends inversely on the square of the frequency, so that as the frequency goes down, the absorption increases and the signal at the receiver gets correspondingly weaker. The effect on propagation is obviously very similar to that of an inefficient antenna, both acting to decrease the signal strength as the frequency decreases. The LUF on any circuit is set by the two factors acting together.

Because the LUF on a circuit depends significantly on the amount of absorption that a signal suffers, it will vary in much the same way as absorption itself varies. In other words, the LUF on a given circuit will be highest during the day, during the summer, and at solar maximum. At night, when the absorption is very small, the LUF decreases to very low values and is controlled only by the inefficiency of the antennas, provided the transmitter power is sufficient to give an adequate signal-to-noise level at the receiver.

## 4.10 Variability of the MUF

We saw in Chapter 3 that the ionosphere varies significantly from one day to the next and that this variability is described in practice by the statistical terms lower decile, median, and upper decile. Because of this variability, the MUF defined by equation 4.2 will also vary from day to day and must be described in statistical terms. In practice, the term MUF is used in two senses:

1. As the maximum usable frequency corresponding to a given critical frequency and circuit (equation 4.2)
2. As the monthly median value of the maximum usable frequency on a given circuit at a given hour for a given month

The term is used in the latter sense when talking about predicted values of the maximum usable frequency. When predictions of the MUF are made for a particular month on a given circuit, the frequencies specified are normally the expected median values of the individual MUFs for the month. Recall that the median value is the value which is greater than half of the individual values. If we were to use the predicted MUF at a particular hour for communications on the corresponding circuit, we would expect to get through on that frequency on 50% of the days of the month.

Although HF communication has its failings, we can certainly do better than a 50% success rate for communicating on a given circuit. If we use a frequency which is lower than the predicted median MUF (and if our predictions have no errors!), there will be more than half the days on which our working frequency is below the actual daily value of the MUF. Some communicators work on a frequency which is 15% less than (or 85% of) the predicted MUF, with a view to achieving substantially higher success rates than 50%. However it is more logical to use a frequency which is equal to the lower decile value of the individual MUFs. We would then expect to get through on that frequency on 90% of the days of the month (at a particular hour), without having to guess how often communications at a frequency equal to 85% of the MUF would be successful.

The frequency which is equal to the lower decile value of the 30 or 31 individual MUFs for the month is known as the **optimum working frequency** (OWF) or *frequency optimum travail* (FOT). The OWF or FOT is the internationally agreed standard for the "best" or "optimum" frequency to use at a given hour on a given circuit. Its use will result in successful communications, at least as far as the correct choice of frequency is concerned, on 90% or 27 days of the month.

## 4.11 References

1. Braun, G. *Planning and Engineering of Shortwave Links*. Siemens Aktiengessellschaft, Heyden & Son Ltd., London, 1982.

2. Davies, K. *Ionospheric Radio Waves*. Blaisdell Publishing Co, Waltham, Mass., 1969.

3. Kuch, J. A. *Field Antenna Handbook*. ECAC-CR-83-200 DoD Consulting Report, June 1984.

4. Braun, G. *Planning and Engineering of Shortwave Links*. loc. cit.

5. Radford, M. F. *High Frequency Antennas*. The Handbook of Antenna Design. Eds. A. W. Rudge, K. Milne, A. D. Oliver, and P. Knight. Volume 2. Peter Peregrinus Ltd., 1983.

## 4.12 Problems

1. In general there are three ways that a signal can travel between any two points—via the *direct* wave, the *ground* wave, and the *sky* wave. Under what conditions would you be able to use each of these waves? What is the main advantage of the sky wave, or ionospheric wave, over the ground wave? Which wave would offer the best HF communication path?

2. There are four raypaths which normally connect a transmitter and receiver, even for a single propagation mode. Explain how these arise. Is there one raypath which dominates over the other?

3. Describe the factors which influence the maximum usable frequency for a given circuit at a given time, and explain how the MUF may be determined.

4. If reflection takes place via the $F_2$ layer for which $f_oF_2$ is 10 MHz, what is the MUF for vertical incidence?

5. Use Figure 4.6 to determine the elevation angle and obliquity factor for an F-layer reflection on a one-hop 1,000 km circuit. If there was in fact an $E_s$ layer at the reflection point and the elevation angle for propagation remains the same, at what range would your signals return to the ground?

6. Assume that $f_oF_2 = 10$ MHz throughout the whole $F_2$ layer. What is the radius of the skip zone, in other words the skip distance, for operating frequencies of 5 MHz, 10 MHz, 20 MHz, and 100 MHz?

7. Various governments around the world have allocated the citizens' band a frequency of 27 MHz. Why was this particular frequency chosen? Give reasons why the choice of frequency was a good or bad one.

8. List in probable order of decreasing signal strength the following propagation modes: 1F, 2F, $1E_s$, 1E, 2E.

9. Describe multipath interference. Explain why it should be avoided and how this can be accomplished.

10. Use Figure 4.6 and Table 4.1 to select suitable antennas for the following mid-latitude circuits and conditions:
    a. Noon, 1,000 km circuit, operating frequency 10 MHz
    b. Same as (a), except at midnight

11. Indicate under which conditions non-deviative absorption of an HF signal is greater:
    a. Noon or midnight
    b. Summer or winter
    c. Low solar activity or high solar activity
    d. 4 MHz or 8 MHz
    e. 2 MHz or 4 MHz
    f. Winter anomaly days, or summer days
    g. 3 MHz ordinary or extraordinary mode

12. What two factors determine the lowest usable frequency (LUF) on a given circuit? How would you go about improving the low-frequency performance of your circuit? Are there any practical advantages in using as *low* a frequency as possible, rather than as *high* a frequency as possible?

13. If you are working at (a) the MUF or (b) the OWF or (c) the FOT, on how many days of the month would you expect to have successful communications?

# Chapter 5

# Ionograms and Their Interpretation

## 5.1 Introduction

An ionogram is a plot of the apparent length of the propagation path as a function of frequency, for radio wave energy emitted by a swept-frequency transmitter (Tx) and received by a receiver (Rx). We will be concerned in this book with three types of ionograms: vertical incidence (VI) ionograms, oblique incidence (OI) ionograms, and backscatter (BS) ionograms. As suggested by Figure 5.1, the three types can be considered to be special cases of the one general class. If D is the distance from the transmitter Tx to the location where the signals first hit the ground, we have :

1. D = 0; the Tx and Rx are co-located; VI ionograms.

2. D = fixed; the Tx and Rx are separated, and the signals hit the ground at the receiver; OI ionograms.

3. D = variable; the Tx and Rx are nominally co-located, but D is defined only by the requirement that the signals reach the receiver after being scattered back from the Earth; BS ionograms.

There is another class of ionograms, known as **topside ionograms,** obtained using an ionosonde in an orbiting satellite (with D = 0). These ionograms give information regarding the variation of electron density with altitude above the $F_2$ layer peak, and down to the peak itself. Values of $f_oF_2$ scaled from topside ionograms, such as that shown in Figure 5.2, have been used to generate 3-month averaged maps of the worldwide variation of $f_oF_2$. The averaging is required because it takes this long for the satellite to sample the diurnal variation of the ionosphere at any geographic location. Topside ionosondes have their frustrations, since the satellites cannot be directed to be at a particular place at a particular time. However this disadvantage is outweighed by the sounder's unique ability to observe the ionosphere over the remote and inaccessible areas of the world. The ionograms themselves are of little interest to us here, since we are concerned only with what happens *below* the $F_2$ peak.

## 5.2 Vertical Incidence (VI) Ionograms

Vertical incidence or VI ionograms have historically been used mainly to study and quantify the ionosphere. VI ionosondes have been sounding on the quarter hour all around the world for over 50 years and through four solar cycles. A conservative estimate of the number of ionograms recorded to date around the world would be 100 million. We shall consider just a few examples, in conjunction with a discussion of different ionosondes, after a little mathematical digression. Firstly, however, it is useful to discuss the "ball and hill" analogy for ionograms, which helps to provide some physical feeling for something we cannot see.

### 5.2.1 The Ball and Hill Analogy

This discussion refers to Figure 5.3. Referring to part B of the figure, a ball is rolled up the hill with a given initial velocity, v, and the time, t(v), it takes to return to its initial position measured. This is done for a range of velocities, from very small velocities up to the velocity which is just large enough to send the ball to the top of the hill, where it stays. The bottom plot in part B is a graph of

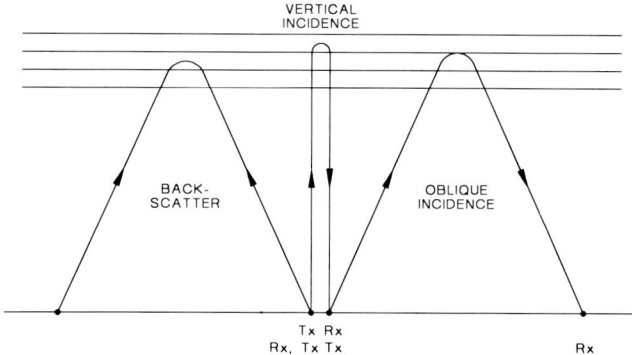

**Figure 5.1** Sketch of the raypaths for ionograms corresponding to vertical incidence, oblique incidence, and backscatter. The three cases can be considered as essentially the same situation, except for the definition of the ground range D.

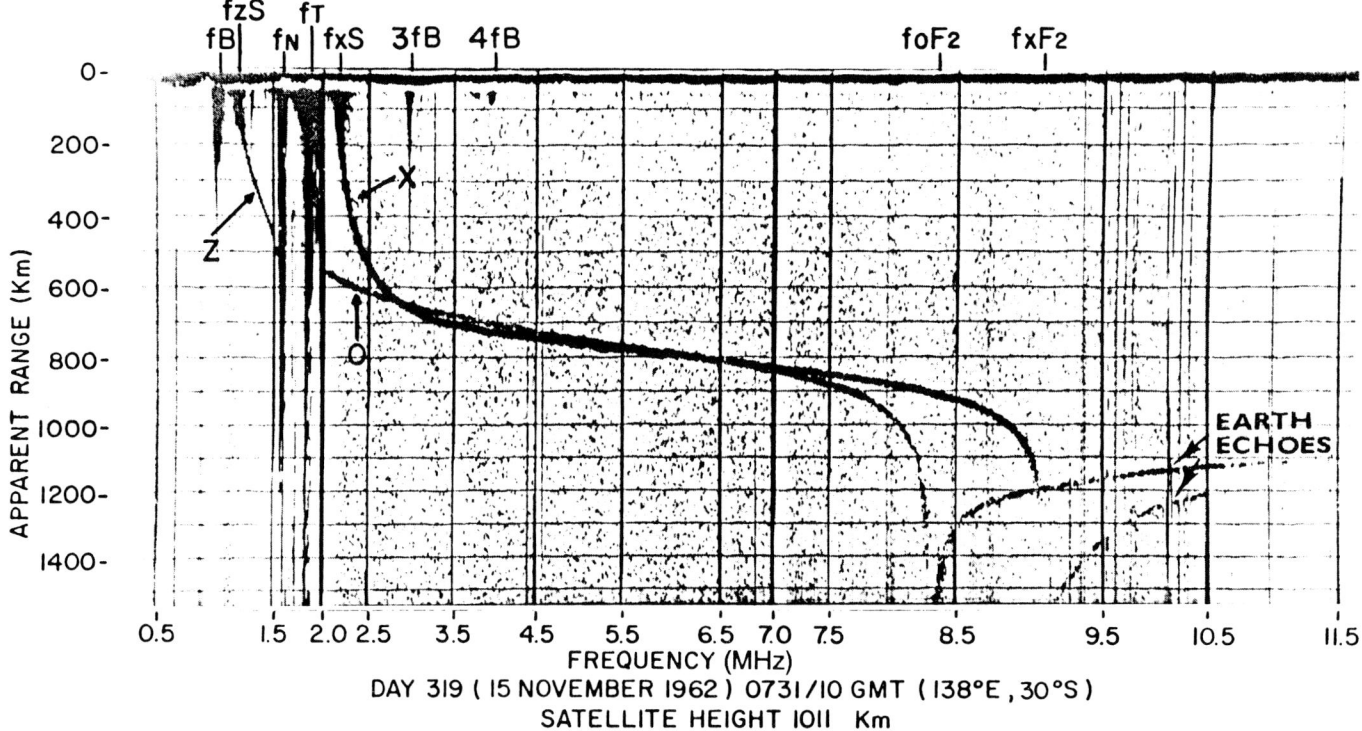

**Figure 5.2** A topside ionogram obtained with the Alouette I satellite. A topside ionogram can provide the plasma and gyro frequencies at the satellite, as well as the apparent range (time delay) as a function of frequency. Observed values of $f_oF_2$, including the only direct observations over the large expanses of ocean, have been used to generate worldwide maps of that parameter.

the time delay versus the initial velocity, which illustrates that the time delay increases quite dramatically with velocity, until a critical velocity is reached for which the ball just reaches the top of the hill. If the initial velocity exceeds this critical velocity, the ball will go right over the hill and never come back. We can say that the ball is "reflected" for velocities below the critical velocity and "penetrates" the hill for velocities greater than the critical velocity.

Part A of Figure 5.3 shows the ionogram equivalent of the ball and hill. As the frequency is increased, the time delay for a signal travelling vertically increases until the frequency is equal to the critical frequency of the E layer, at which stage the signal will just penetrate the E layer. For frequencies just above $f_oE$, the time delay decreases with frequency, since the signals at these frequencies find it increasingly easy to penetrate the E layer. However as the critical frequency for the $F_1$ layer is reached, the signals start to slow down again as they approach penetration. The same decrease followed by an increase in delay time happens for the $F_2$ layer. When the frequency exceeds the critical frequency for the $F_2$ layer, $f_oF_2$, the signals penetrate the whole ionosphere and go on into space. If the frequency is too low, there will be no returning signal, because of absorption. In terms of the ball and hill analogy, this would correspond to a soft, boggy approach to the bottom of the hill.

### 5.2.2 Calculation of VI Ionograms

Operationally, VI ionograms are the experimental observations from a swept frequency HF radar which looks vertically up. The time delay for energy to return to the radar is converted into an apparent or "virtual" height, from which the signals appear to have been reflected. Ideally, the ionograms would be used directly (since no assumptions or approximations then need to be introduced), as in characterizing the ionosphere in terms of the various critical frequencies, or in determining the effective height of reflection of an obliquely propagating signal (Chapter 12). Sometimes it is necessary to "invert" the observed variation of virtual height with frequency to determine the variation of plasma frequency or electron density with height. This is a very tedious and often uncertain business, as we shall see in Section 5.5. It is also sometimes necessary to calculate an ionogram from a given electron density profile, which is also tedious, but more certain. We present some details of such calculations here, mainly with a view to clarifying some of the ideas involved.

To calculate an ionogram, we use mathematics to simulate the propagation of a radio wave through the ionosphere. We consider a pulse of radio wave energy transmitted vertically upwards from a transmitter, being reflected by the ionosphere and returning to the receiver

Ionograms and Their Interpretation

some time T later. The group height, or **virtual height,** $h'$, at a specified wave frequency is the distance that a radio wave at that frequency would have travelled in half the elapsed time, $T/2$, if it had travelled at the speed of light in free space, $c$. Thus

$$h' = c T / 2.$$

This can be written as

$$h' = \int c \, dt / 2,$$

where the integral is taken over the total elapsed time. Now the velocity at which the energy of the signal travels, $v$, is related to the velocity of light by

$$v = c / \mu',$$

where $\mu'$ is the **group refractive index.**

Therefore,

$$h' = \int \mu' v / 2 \, dt,$$

where the integration is from 0 to T. Recall that $\mu'$ (and $v$) varies along the propagation path, which is why we have to work in small segments and integrate over the whole path.

Since $dh/dt = v$, we have

$$h' = \int \mu' dh,$$

where the integration limits are now $h = 0$ (at the ground) and $h = h_r$, the reflection height. Since $\mu' = 1$ in the free space below the base of the ionosphere, $h_0$, we can write this as

$$h' = h_0 + \int \mu' \, dh, \qquad (5.1)$$

where the lower integration limit is now $h_0$.

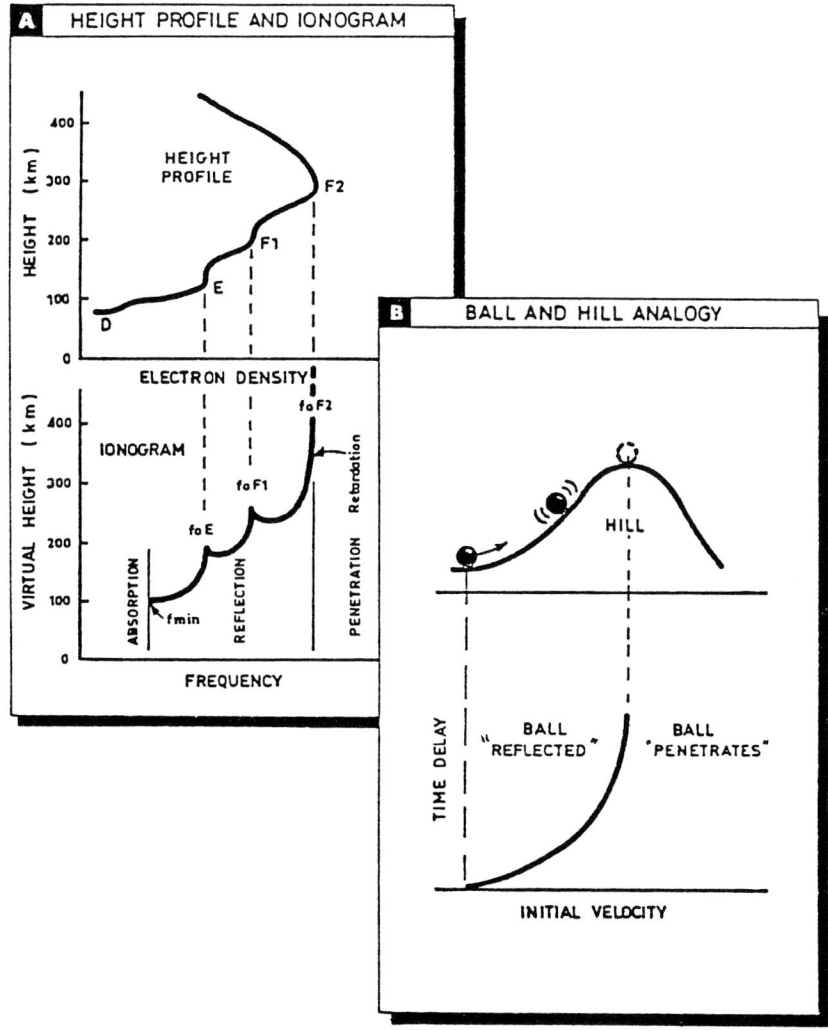

**Figure 5.3** The ball and hill analogy of how an ionogram is created. The time it takes for a ball to roll up the hill and back again is a function of its original velocity. This is compared with the time it takes for vertically incident radio waves of increasing frequency to return to a transmitter. *Courtesy IPS Radio and Space Services.*

For the special case of no magnetic field and no collisions, the **phase refractive index** for a wave frequency f and a plasma frequency $f_N$ is given by [1]

$$\mu^2 = 1 - (f_N/f)^2,$$

and the group refractive index $\mu'$ follows from the relation $\mu \mu' = 1$.

Since the refractive index is a function of the plasma frequency, we can write

$$h' = h_0 + \int \mu'(f_N) \cdot (dh/df_N) \cdot df_N, \quad (5.2)$$

where the integration is now from $f_N = 0$ to $f_N = f$, where f is the operating frequency, since the ray is reflected at the height where the plasma frequency is equal to the wave frequency.

This form of the integral shows us that we will get an infinite result for $h'$ if the gradient of plasma frequency with respect to height, $df_N/dh$, ever goes to zero. This is exactly what happens at the peak of a layer, leading to the cusps observed at each critical frequency on the ionogram. The cusps do not usually extend to great virtual heights in practice, because the returning signals are heavily attenuated near a critical frequency.

If we change the variable of integration from $f_N$ to X, where $X = (f_N/f)^2$, we get

$$h' = h_0 + \int \mu'(X)(dh/df_N)(f^2/2f_N)dX, \quad (5.3)$$

where the integration limits are now $X = 0$ and $X = 1$.

Since we are ignoring the magnetic field, $\mu\mu' = 1$, so that

$$h' = h_0 + \int \frac{(f^2)(2f_N)(dh/df_N)}{1-X} dX. \quad (5.4)$$

If $dh/df_N$ is a reasonably simple function of $f_N$, we will be able to perform the integration analytically. If it is not, we need to perform the integrations using some numerical scheme such as Gaussian quadrature [2].

We will have difficulty evaluating the integral accurately if it is not analytic, because of the singularity (that is, the integrand goes to infinity) at $X = 1$. However changing the variable to

$$t^2 = 1 - (f_N/f_r)^2,$$

where $f_r$ is the plasma frequency at which reflection occurs, keeps the integrand finite for the range of t, which is from 1 to 0. Determination of the expression for $h'$ in terms of t is left as an exercise for the reader. See also problem 16.3 of Chapter 16.

A complete numerical ionogram can be obtained by evaluating the above integrals at different wave frequencies. The effects of the Earth's magnetic field can be taken into account by using the appropriate formula for the refractive index, which is much more complex than the no-field one used above [3]. When the field is included, all hope is lost of being able to perform the integrations analytically.

### 5.2.3 Examples of VI Ionograms

Even to include just a small fraction of the more interesting VI ionograms recorded to date would be prohibitive here, since the ionosphere is so complex and ionograms have such a wide diversity of forms and features. We have therefore opted to include just a few, referring the reader to other sources for more [4, 5, 6, 7, 8]. Several ionograms are illustrated and discussed in Chapter 12, in connection with their use for the single station location of an HF transmitter.

Figure 5.4 gives an idealized daytime ionogram which illustrates the effects of the various layers and critical frequencies introduced in Chapter 3. Recall that there are in general two traces for each layer of the ionosphere, since the presence of the Earth's magnetic field gives rise to the ordinary and extraordinary rays. The vertical asymptotes for $f_oF_2$ and $f_xF_2$ are separated by approximately half the gyrofrequency, $f_B$, since the O and X frequencies reflected at the same height (and plasma frequency) are related by

$$f_o^2 = f_x(f_x - f_B).$$

By completing the square, we find that

$$f_o^2 = (f_x - f_B/2)^2 - (f_B/2)^2.$$

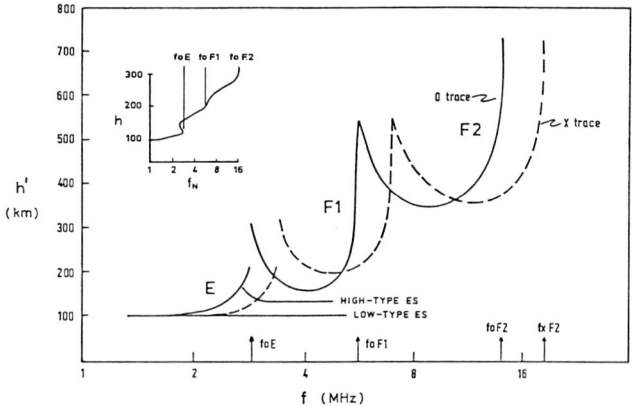

**Figure 5.4** Highly idealized ionogram corresponding to the plasma frequency profile in the inset. The critical frequencies $f_oE$, $f_oF_1$, and $f_oF_2$ correspond to the plasma frequencies of the local maxima in the plasma frequency profile. Note that the extraordinary trace would not normally be seen at frequencies below about 4 MHz because of absorption.

# Ionograms and Their Interpretation

Thus if $f_B$ is small compared with $f_x$, we have

$$f_o = f_x - f_B / 2.$$

The inset in Figure 5.4 is a typical ionospheric profile, $f_N(h)$, which would lead to the ionogram illustrated. Note that in practice, the extraordinary ray trace would not extend so low in frequency; absorption would probably limit it to above 4 MHz or so.

Figure 5.5 gives two typical (film) ionograms for the mid-latitude station White Sands, New Mexico, for midnight in summer and winter. These ionograms present no difficulties in interpretation. The high-latitude ionograms of Figure 5.6, which are from Syowa, Antarctica (magnetic dip = 68°), are also simple to interpret, which is often not true for high-latitude ionograms.

Figure 5.7 gives two daytime ionograms for Huancayo, Peru, which is very nearly on the magnetic equator (the dip is 1.2°). The most interesting features of these ionograms are the spread $E_s$ traces at 100 km, and the spreading of the second-hop F-layer traces. The $E_s$ is typical of equatorial stations and is known as equatorial $E_s$. Figure 5.8 shows more examples of the incredible variety of $E_s$ recorded at Huancayo.

The spread F which was discussed in Chapter 3, in association with the presence of irregularities in the ionospheric plasma, manifests itself in many weird and wonderful ways on ionograms. Figure 5.9 shows equatorial spread F observed at Huancayo at different times on the one night, while Figure 5.10 shows some high-latitude spread F recorded at Thule, Greenland. The very high magnetic dip angle (86°) for Thule means that the Thule ionosphere is very susceptible to the effects of particle fluxes from the sun.

In spite of the wealth of knowledge gained from them, the film ionograms reprinted in Figures 5.5 to 5.10 represent a bygone era of technology. Conventional ionosondes which yield film ionograms are slowly being replaced by digital ionosondes which present the ionograms in digital form for later schemes of digital processing and analysis. In contrast to restrictions placed on what information can be obtained from an ionogram by a human operator, the only restrictions placed on the use of digital ionograms are set by the initial design and capabilities of the ionosonde itself. Figure 5.11 gives a printout of a Digisonde 256 [9] ionogram obtained at Goose Bay, Labrador, during relatively quiet conditions. The O and

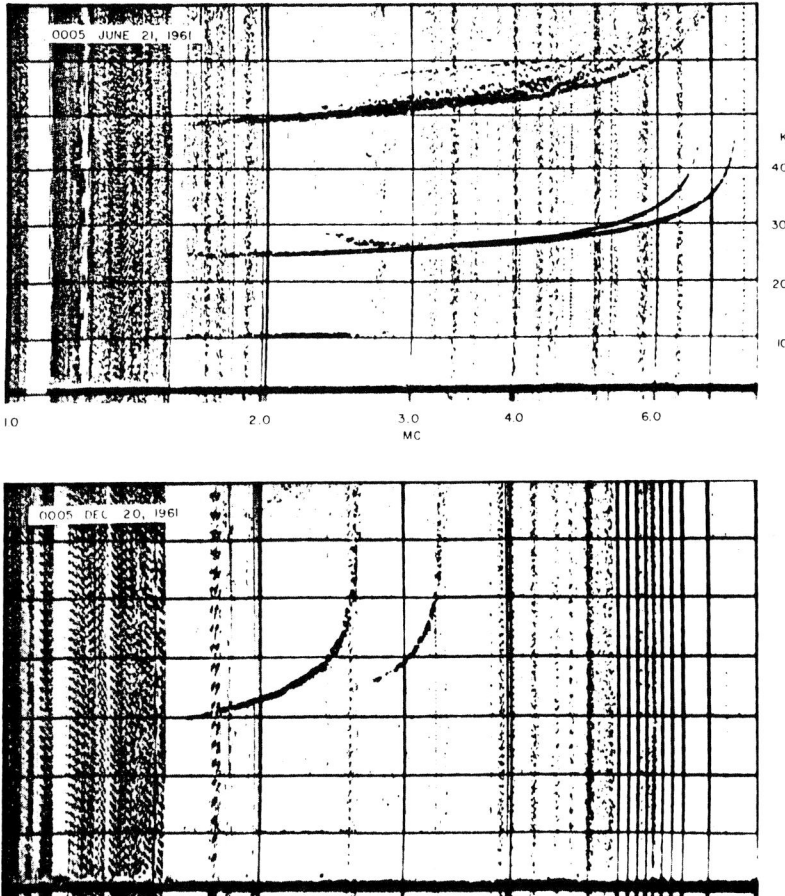

**Figure 5.5** Film ionograms for White Sands, New Mexico, which has a magnetic dip angle of 60.2°. Both were observed at local midnight, but the critical frequencies are much greater in the summer (June) than in winter (December). *After UAG Report 10.*

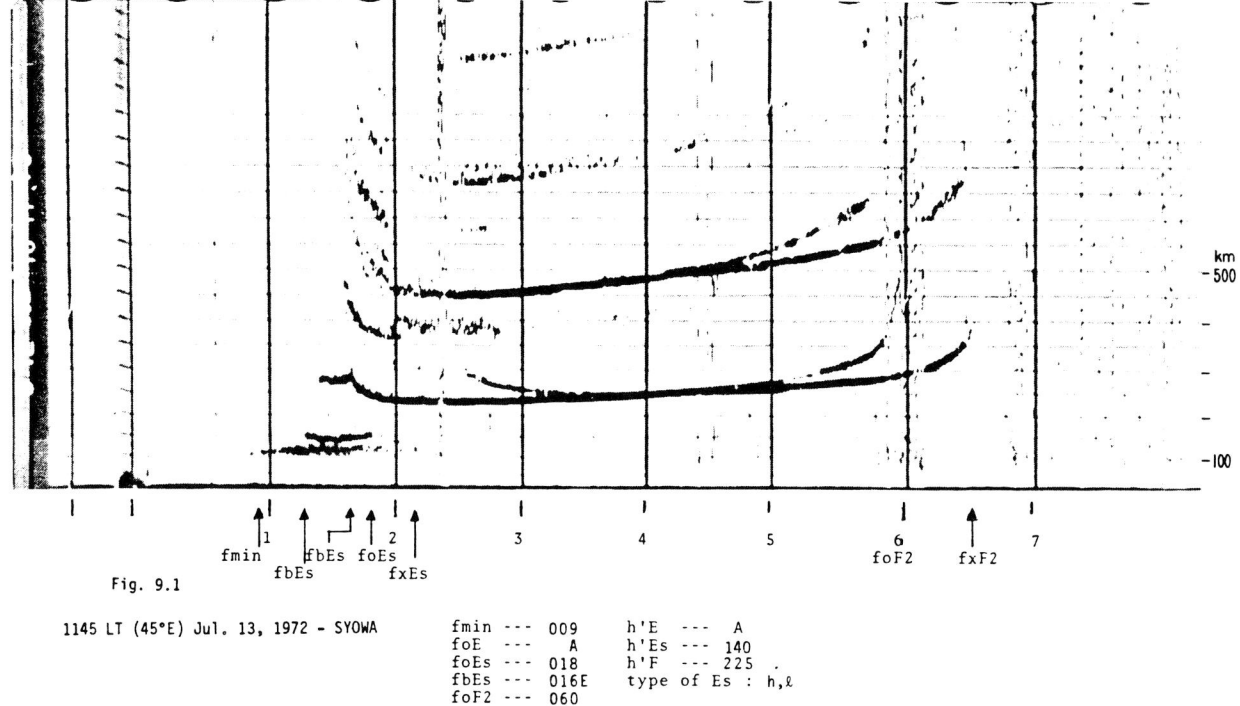

DAYTIME IONOGRAM IN WINTER - SYOWA STATION

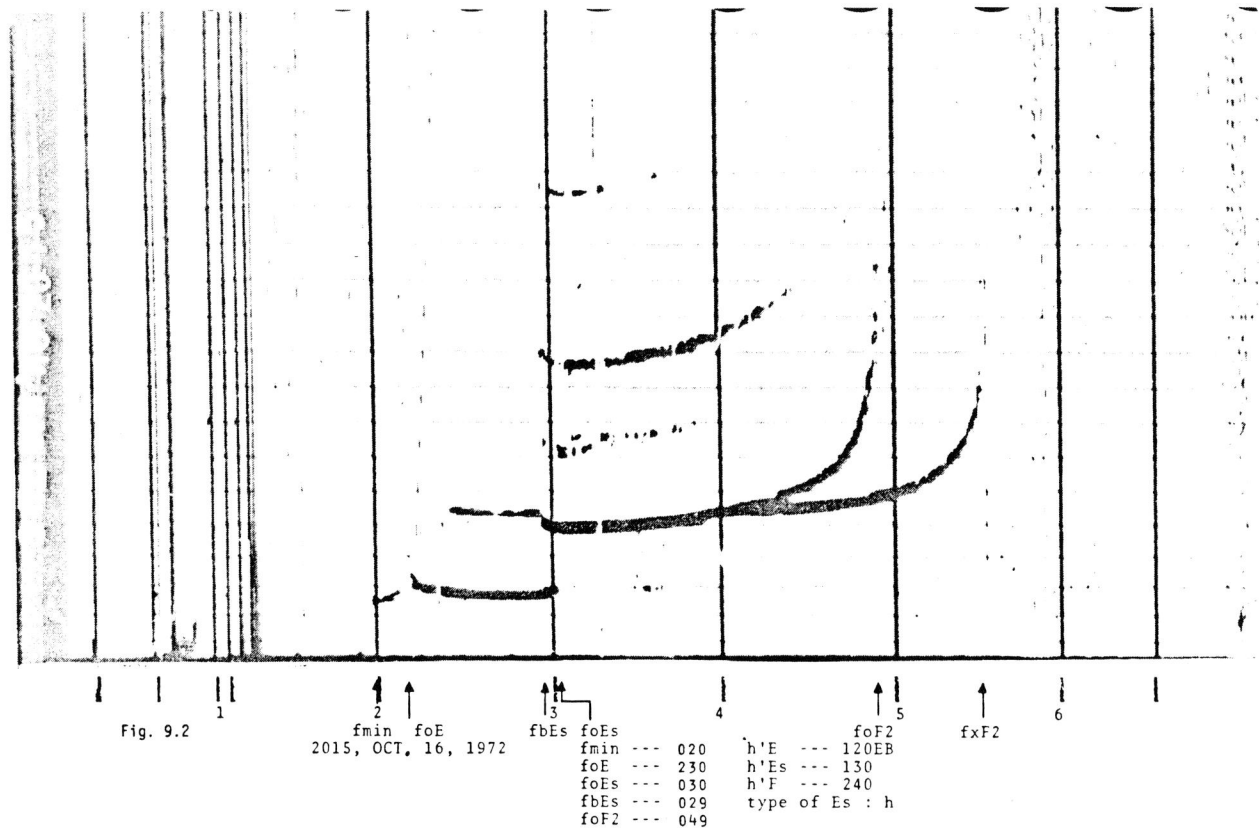

NIGHTTIME IONOGRAM IN WINTER - SYOWA STATION

**Figure 5.6** Day and night ionograms at Syowa station, Antarctica, dip = 68°. The dip at Syowa is not much higher than for the mid-latitude station White Sands (Figure 5.5). *After UAG Report 10*.

# Ionograms and Their Interpretation

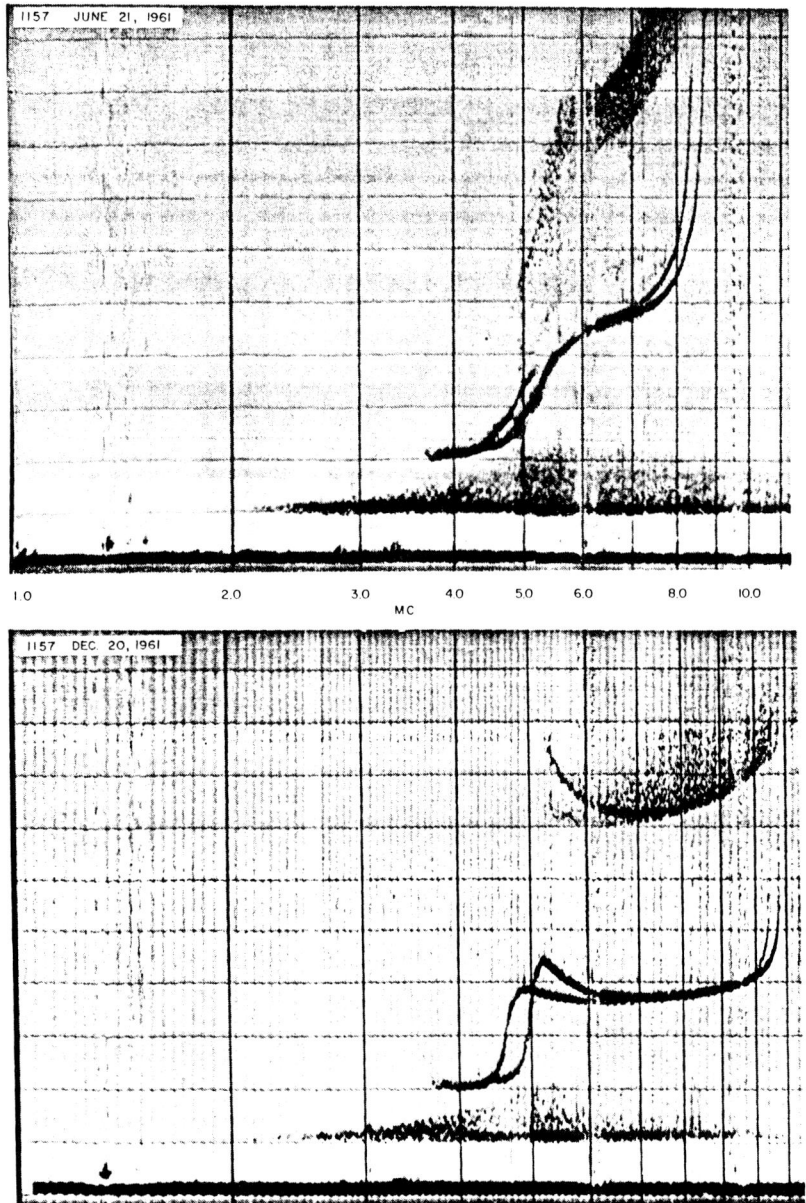

**Figure 5.7** Summer and winter noon ionograms for Huancayo, Peru, which is almost on the dip equator (dip = 1.2° N). Both ionograms show range-spread equatorial $E_s$ at a height of 100 km. *After UAG Report 10*.

X traces now come conveniently tagged, and the ionogram has been scaled automatically using ARTIST to yield the critical frequencies and heights normally read off by a human operator [10]. The leading edges (lowest group delay) for the two traces have also been determined automatically and used to derive the corresponding electron density profile.

## 5.3 Oblique Incidence (OI) Ionograms

Oblique incidence or OI ionograms are obtained using separated but synchronized transmitter and receiver on a given circuit and have historically been used to quantify the propagation conditions existing on that circuit at different frequencies. In particular, they provide direct observations of all the propagation modes being supported by the ionosphere at the time and of the MUF for each mode. OI ionograms collected on a routine basis led to the discovery or interpretation of some of the unusual propagation modes described in Chapter 9. As with the VI ionogram, the OI ionogram is a display of group path versus frequency. Mathematically, the group path $p'$ can be defined as the integral of the group refractive index over the propagation path :

$$p' = \int \mu' \, ds. \qquad (5.5)$$

The integral is analytic only in special cases and when the Earth's magnetic field is ignored.

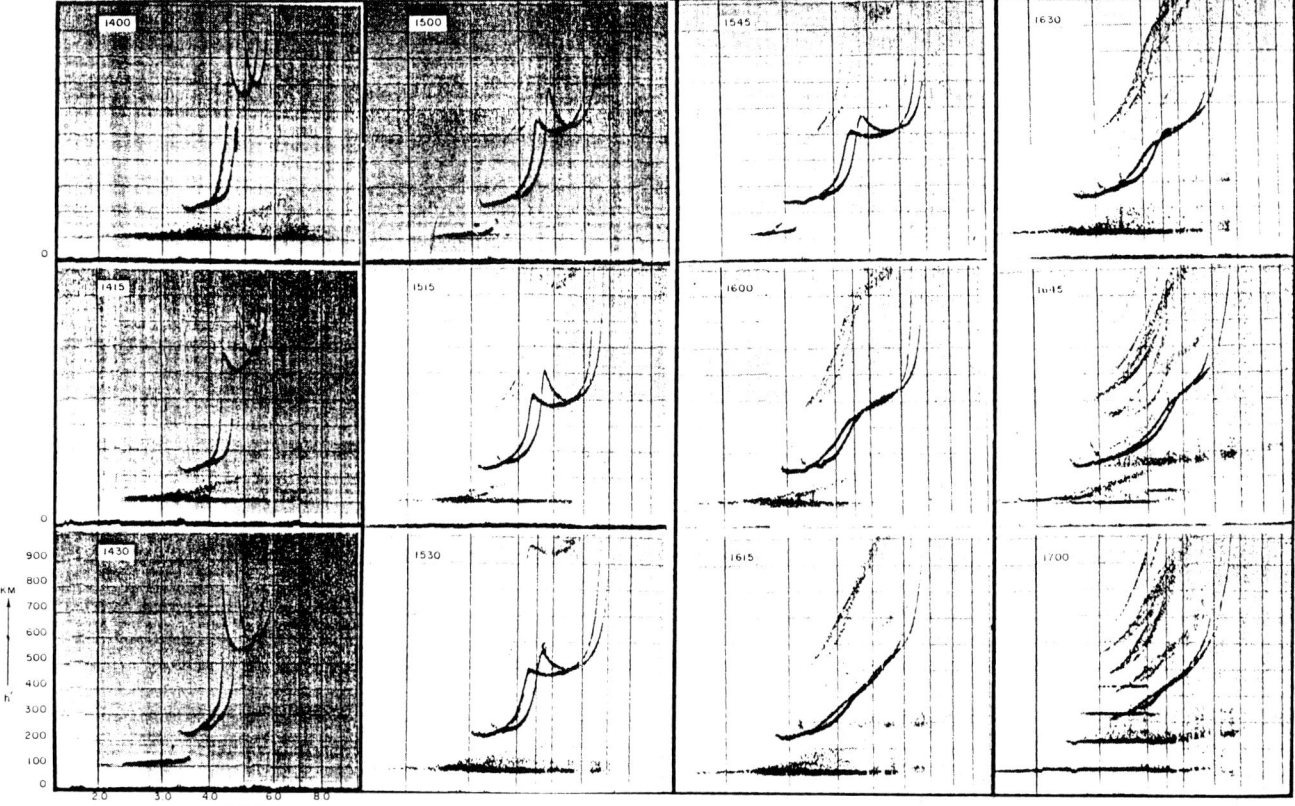

**Figure 5.8** Three hours of ionograms from the equatorial station Huancayo, Peru, showing the incredible variety of $E_s$ traces observed at that station. Equatorial $E_s$ ($E_{sq}$) occurs at 1400, 1415, 1515, 1530, 1600, 1615, and 1630. Slant $E_s$ occurs at 1400, 1415, 1530, 1600, 1630, and 1645. *After UAG Report 10*.

It is instructive to see how an OI ionogram may be deduced mathematically from a VI ionogram obtained at the midpoint of the circuit. The basis of the transformation will be described in some detail in Chapter 12, since it is also closely related to the problem of the single station location of HF transmitters. Assuming that the Earth and ionosphere are flat, and ignoring the Earth's magnetic field, the group path at a frequency $f_{ob}$ for an OI path is related to the virtual height $h'$ at a frequency $f_v$ on a VI ionogram taken at the path midpoint by

$$p' = 2 h' \sec X$$

where X is the angle of incidence at the base of the ionosphere, and the "equivalent vertical frequency" $f_v$ is related to the oblique frequency by

$$f_{ob} = f_v \sec X.$$

The angle X can be determined from the equation

$$\tan X = (D/2) / h'.$$

Any point $(f_v, h')$ on the VI ionogram can thus be transformed into a point $(f_{ob}, p')$ and the OI ionogram generated, as illustrated in Figure 5.12. Note, however, that the transformation is only approximate because of the assumptions made in its derivation. The inverse transformation can be used to derive an equivalent VI ionogram at the circuit midpoint, given the OI ionogram, but is again only approximate (see later).

Figure 5.12 shows that the group path is sometimes a double-valued function of frequency, with two values of $p'$, at the higher frequencies, corresponding to the low and high rays. The MUF for the circuit is the frequency at which the low and high rays travel the same physical path and consequently have the same group path. We saw in Chapter 4 that the MUF is equal to the product of the critical frequency for the layer, $f_c$, and the obliquity factor. The fact that there is a highest frequency supported on the circuit (the MUF) arises from the competing influences of a decreasing sec X and an increasing plasma frequency as the operating frequency increases and the signals penetrate further into the layer. The ray corresponding to the MUF does not penetrate up to the height of maximum plasma frequency in the layer except for the special case of vertical incidence.

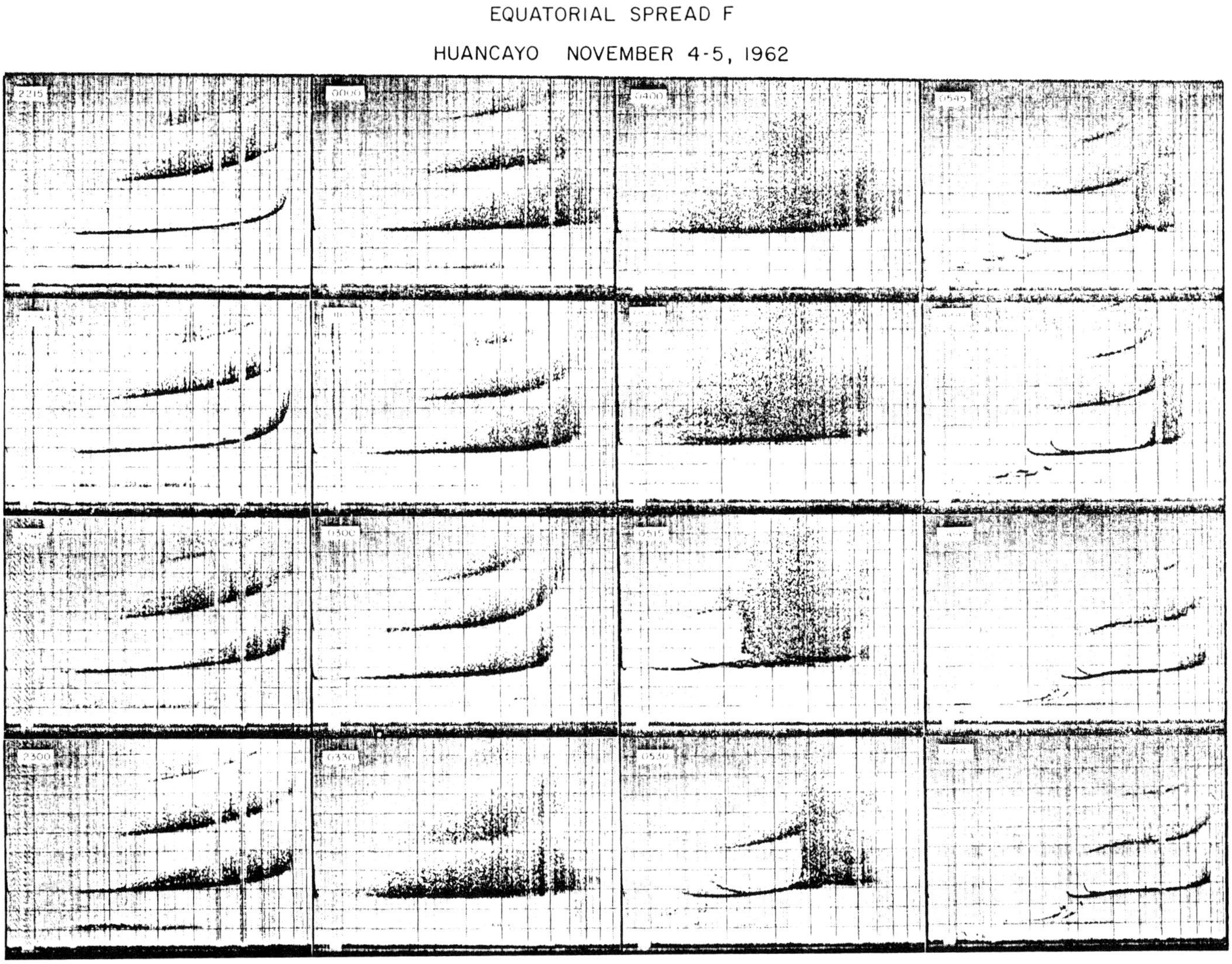

# Ionograms and Their Interpretation

**EQUATORIAL SPREAD F**

**HUANCAYO NOVEMBER 4-5, 1962**

**Figure 5.9** Huancayo, Peru, ionograms illustrating the development of equatorial spread F over the night of November 4–5, 1962. *After UAG Report 10.*

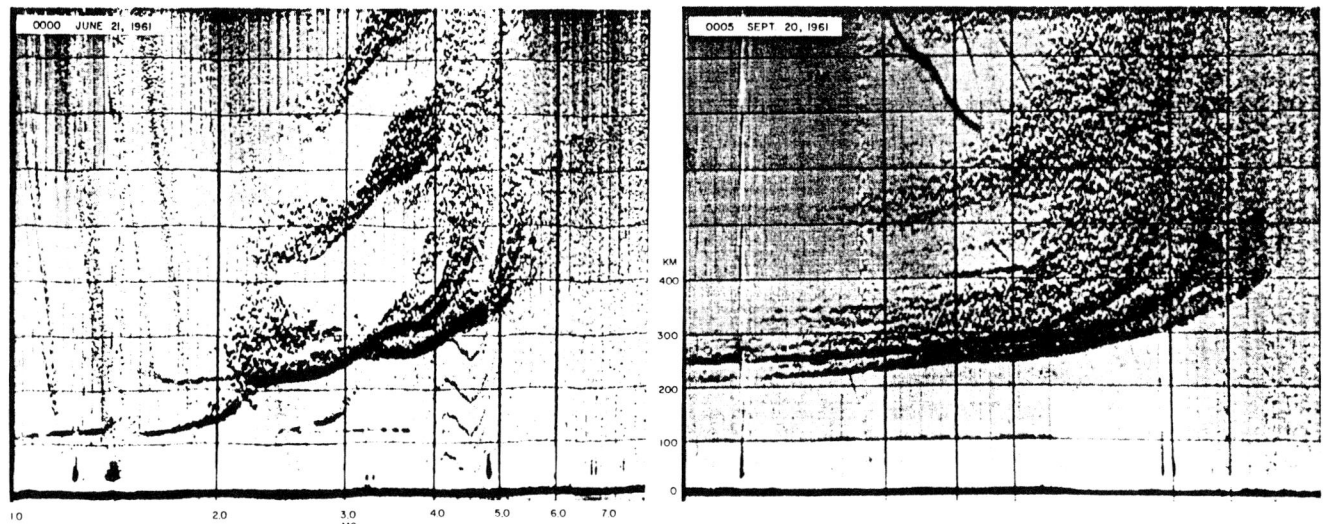

**Figure 5.10** Midnight ionograms for Thule, Greenland, dip = 85.8°. The extremely spread traces are typical of high latitude spread F. *After UAG Report 10.*

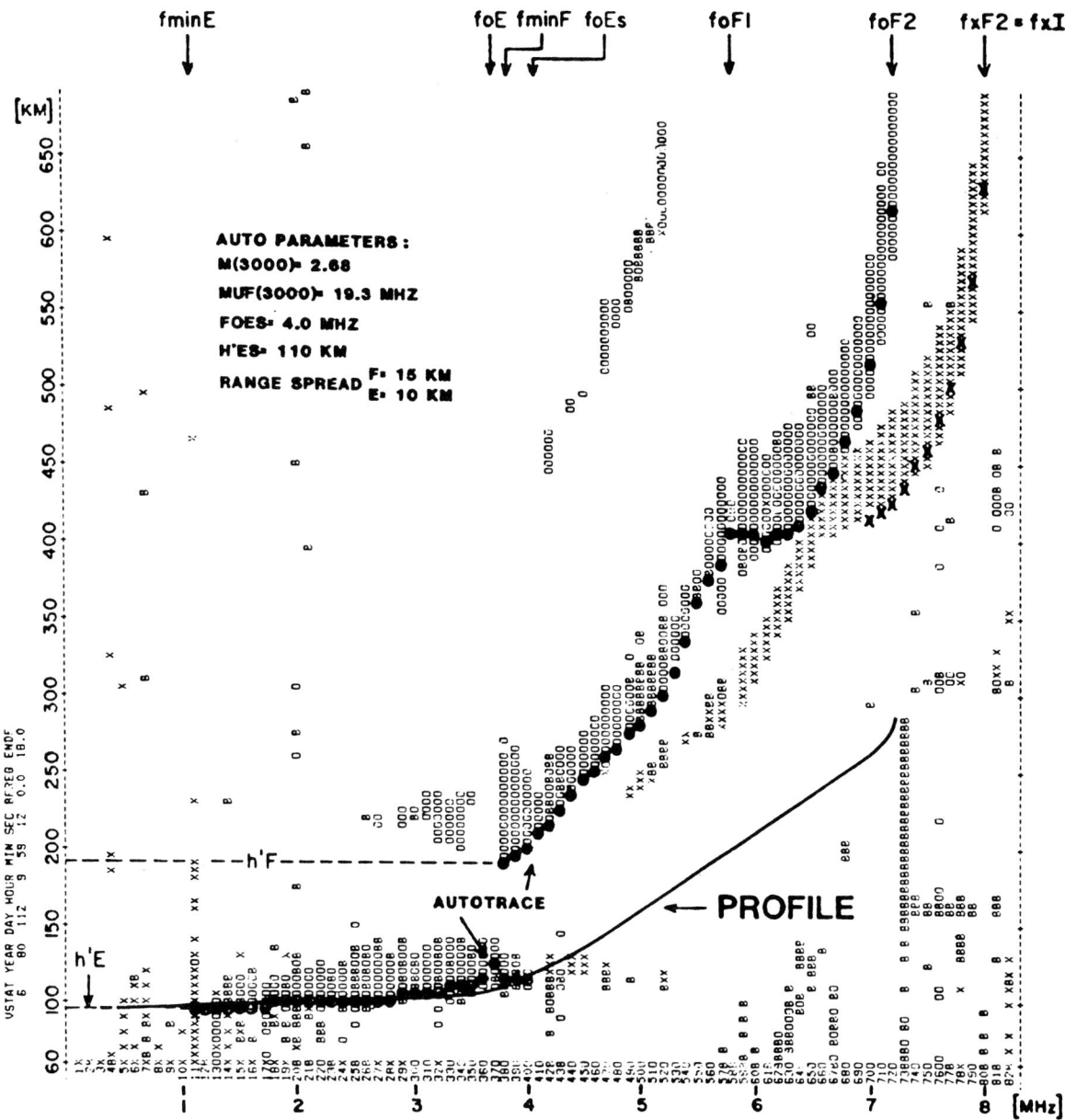

**Figure 5.11** Digisonde ionogram for magnetically quiet conditions at Goose Bay, Labrador (dip = 76.9°). The ionogram traces have been automatically identified and scaled by ARTIST, and a plasma frequency profile deduced. *Courtesy University of Lowell Center for Atmospheric Research.*

### 5.3.1 Simulated OI Ionograms

Simulated OI ionograms may be derived for any specified model of the ionosphere by performing the integration in equation 5.5. The ionograms may even be evaluated analytically if certain simplifying assumptions are made. Figure 5.13 shows the OI ionogram calculated for the Adelaide to Canberra circuit in Australia, using a multisegmented quasi-parabolic model of the N(h) profile (see Chapters 12 and 15) based on predicted values of the ionospheric parameters. The calculation proceeded by deriving the ground ranges and group paths for rays at all angles for which the model supported the simulated propagation. This was done at all frequencies for which propagation was possible. Interpolation in the array of ground range versus elevation angle at each frequency yielded the elevation angle, $\beta$, corresponding to the required ground range D. There are, of course, two angles which yield a specified ground range, corresponding to the low ray and the high ray. Subsequent interpolation in the array(s) of group path versus elevation angle then gave the group path corresponding to $\beta$, and thus to D at that frequency. The figure shows the low and high rays for six propagation modes, 1E, 1F$_1$, and 1F$_2$, as well

# Ionograms and Their Interpretation

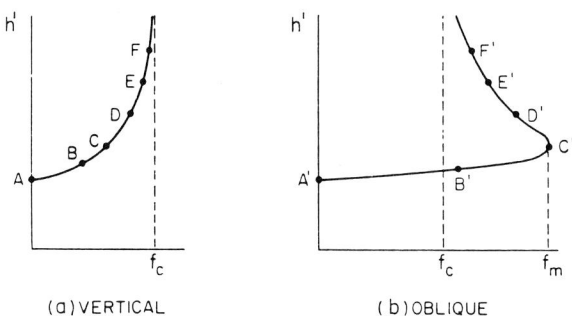

Corresponding vertical and oblique ionograms.

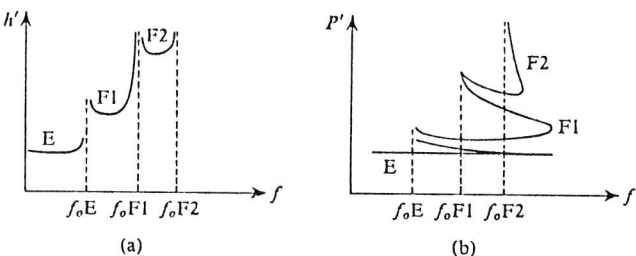

Corresponding vertical and oblique $P'f$ traces for a multilayered ionosphere. (a) Vertical, (b) oblique.

**Figure 5.12** Sketches illustrating the conversion of a vertical incidence ionogram into the equivalent oblique incidence ionogram. *After Davies (1965, 1969)*.

as the 2-hop modes. Note that the transition between two layers is marked by a very extended cusp extending backwards to the top left of the diagram. These are the analogues of the cusps in the VI ionogram. The calculated traces would be very model-dependent in the region of the cusps. As with the VI ionograms, the extreme parts of the cusp will not usually appear on an observed ionogram because the signals would be heavily attenuated.

Figure 5.13 illustrates the fact that the propagation conditions can be quite different at different frequencies. For example, at 9 MHz propagation is possible by the 1E, $1F_1$, $1F_2$, and $2F_2$ modes, while for frequencies greater than 10.6 MHz, only the $1F_2$ mode is supported. The frequency range from 10.6 to about 13.0 MHz would be especially good for HF communications, since the signals would travel by only the one path—the low-ray $1F_2$ path—and there could be no interference between signals arriving by different paths and with different time delays.

Figure 5.14 shows another simulated ionogram based on an ionospheric model, with 1, 2, and 3-hop traces. The circuit length was 3,000 km, so the hop lengths were 3,000, 1,500 and 1,000 km respectively. It can be seen that as the hop length gets shorter, the OI ionogram begins to look more like a VI ionogram. The value of this diagram is that it shows the elevation angles corresponding to each part of the traces. We see, for example, that the low and high rays for the 1-hop mode at 24 MHz correspond to 3° and 14°. We can also see that the angles corresponding to the MUFs for the different orders of propagation increase from about 8° to 20° to 30° for 1, 2, and 3-hop modes. Unfortunately, real OI ionograms do not come tagged with elevation angle.

### 5.3.2 Sample Oblique Ionograms

Most of the OI ionograms obtained on a routine basis over the last 30 or so years have been obtained with the Granger pulsed oblique sounder (Figure 5.15) and the Barry FM-CW chirpsounder (Figure 5.16). Figures 5.17 and 5.18 show corresponding Alice Springs VI (bottom panels) and Darwin to Alice Springs OI ionograms obtained using an FM-CW ionosonde in support of the Australian Jindalee Over the Horizon Backscatter radar system [11], just after sunset and at noon (October 1986). Jindalee ionograms are digital, with full amplitude information, and are usually displayed in color (which means that the present figures do not do them justice). The correspondence between VI and OI ionograms has been illustrated previously (Figure 5.12). Figure 5.19 shows an oblique ionogram obtained using two University of Lowell Digisondes. Modern digital ionosondes offer a quantum leap forward in our ability to investigate propagation conditions along a fixed circuit. Interpretation of the traces in the various figures is left to the reader.

Figures 5.17 to 5.19 illustrate the effects of the magnetic field, which was not taken into account for Figures 5.13 and 5.14. The separation between the MUFs for the O and X waves for Figure 5.19, for example, is about 0.7 MHz, with the X MUF being the higher of the two. The separation between the O and X MUFs depends on the length of the circuit and the orientation of the circuit with respect to the Earth's magnetic field [12]. When the circuit length drops to zero, the separation is just half the gyrofrequency, since the ionogram would then be a VI ionogram.

## 5.4 Backscatter Ionograms

We saw in Section 5.3 that OI ionograms are plots of group path versus frequency for a fixed circuit length. For BS ionograms, the circuit length is not specified. All that matters is that at least some signals from the transmitter return to the receiver, having been scattered at the surface of the Earth back along the original direction. Simulated BS ionograms can also be calculated by integrating the group refractive index along the total physical

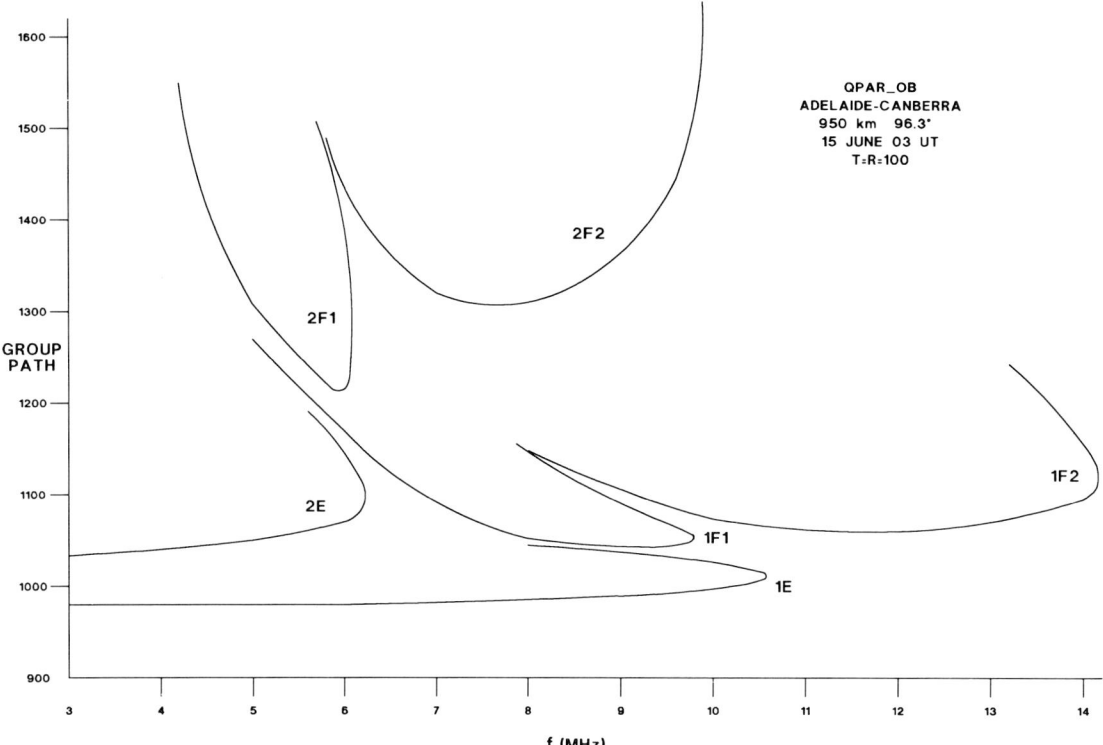

**Figure 5.13** Simulated oblique ionogram for propagation between Adelaide and Canberra, 03 UT, 15 June, T = 100. The circuit length was 950 km. The figure illustrates the 1E, $1F_1$, $1F_2$, 2E, $2F_1$, and $2F_2$ propagation modes. The high frequency noses of each trace are where the low and high rays travel the same path and are at the MUF for that mode. The 2-hop traces, which correspond to the relatively short hop length of about 475 km, do not look greatly different from a vertical incidence ionogram.

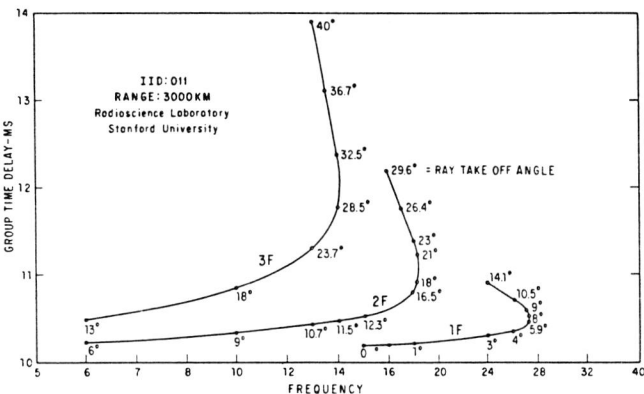

**Figure 5.14** Simulated oblique ionogram for a 3,000 km circuit, showing the elevation angles corresponding to each part of the traces. The angles clearly illustrate the difference between low and high rays, and the increase of the angles as the hop length decreases.

path, for each frequency and elevation angle. All frequencies and angles for which propagation is possible to any range from the transmitter are represented in a full BS ionogram. The most interesting and useful part of the BS ionogram is the **leading edge,** which corresponds to the minimum group path at a given frequency. It is determined in simulation by interpolating in the array of p' versus elevation angle to determine the minimum value of p'. BS ionogram calculations usually also include the determination of the ground range at this same elevation angle, in order to define the relation between group path and ground range, which is required for coordinate registration for over-the-horizon radars (see Chapter 15).

Figure 5.20 shows how the calculated group path and ground range varied with elevation angle for propagation due north at 20 MHz, with the transmitter at Alice Springs (25 March, 0037 UT, T = 40, R = 20). The group path has a minimum value of 2,025.4 km at 15.35°, while the ground range has a minimum value of 1,868.7 km at 15.7°. The minimum in ground range corresponds to the skip distance at 20 MHz, and to a group path of 2,027.8 km. The minimum in the group path corresponds to the leading edge of the BS ionogram at 20 MHz. The third curve in the figure shows the variation of the ionospheric illumination power (IIP) with elevation angle [13]. The IIP is the ionospheric component of the expression for the returned power of a pulse backscattered at oblique incidence in a spherically stratified ionosphere. The IIP has two sub-maxima, at the angles corresponding to the leading edge and to the skip distance. The main peak increases by two orders of magnitude in intensity, as the leading edge is approached. The sub-maxima are due

# Ionograms and Their Interpretation

**Figure 5.15** Oblique incidence ionogram obtained with the Granger sounder on the Townsville to Adelaide circuit. This sounder was a pulsed sounder. The ionogram shows traces for hops up to the fourth order.

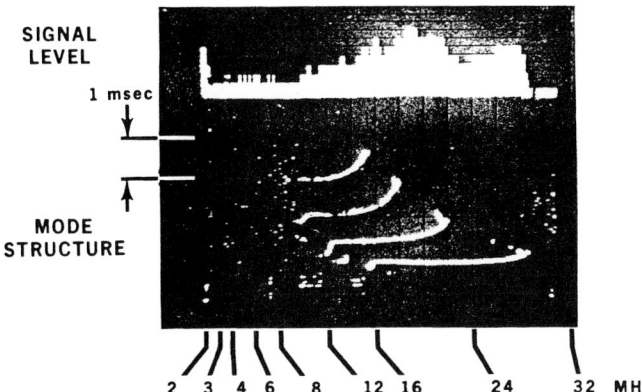

**Figure 5.16** Oblique incidence ionogram obtained with the Barry Research chirpsounder on a circuit from Hawaii to California. Four propagation modes may be resolved.

to focussing, the energy from a *cone* of rays ending up with essentially the same group path, or ground range, rather than just one ray. For example, all rays within ± 0.5° of 15.35° yield group paths within 4 km (<1%) of each other.

Figure 5.21 shows the simulated BS leading edges for a transmitter at Alice Springs transmitting on an azimuth of 345.9° (03 UT, April, sunspot maximum). There is a leading edge for each propagation mode, 1E, 1F$_1$, and 1F$_2$. In this particular case, the F$_2$ leading edge would be *the* leading edge for the actual ionogram for frequencies greater than 7 MHz. In general, the 1E and 1F$_1$ modes will take over this role at lower frequencies. Also shown on Figure 5.21 are two oblique ionograms calculated for the same azimuth, with receivers at ranges of 1,303 km (Darwin) and 3,000 km. It can be seen that the BS leading edges are just the loci of the noses of the individual OI traces. The leading edge is, strictly speaking, tangential to the nose just below the MUF and on the low ray, since the leading edge has a finite positive slope. The very peak of the nose of the oblique ionogram is vertical and has an infinite slope. The leading edges correspond to different elevation angles for each propagation mode. For example, at 16 MHz, the E, F$_1$, and F$_2$ leading edges correspond to elevation angles of 7°, 12°, and 31°. At the longer range (3,000 km), only F$_2$ propagation at higher frequencies is supported (apart from propagation via the

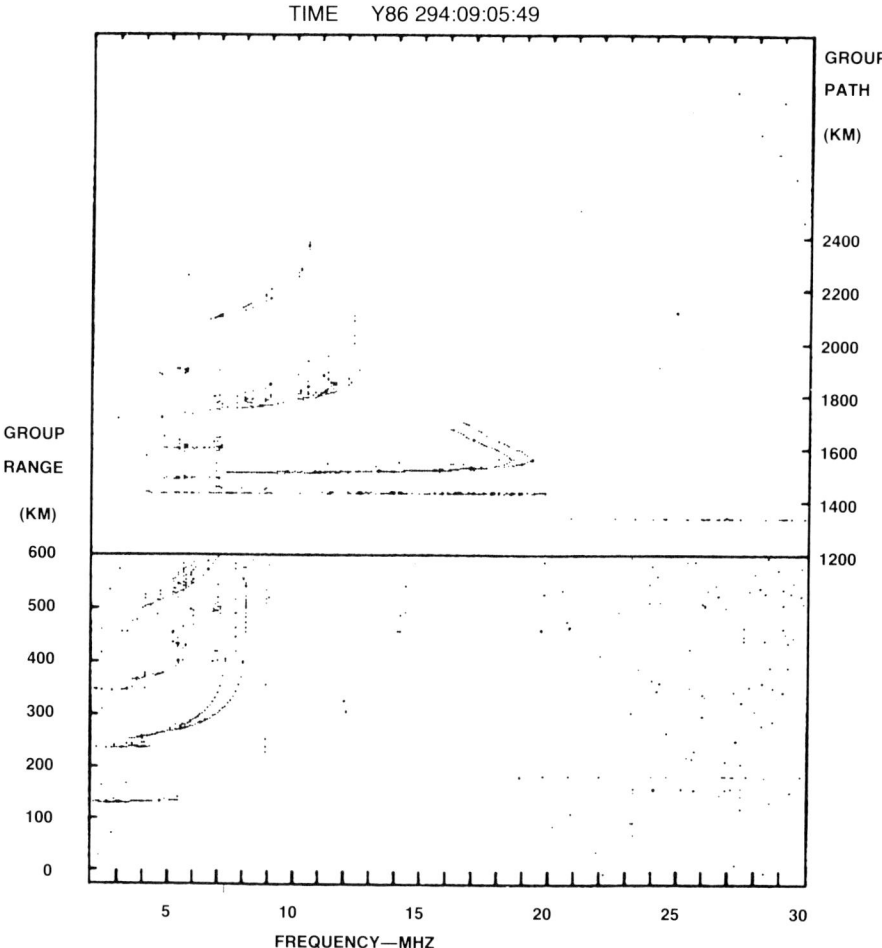

**Figure 5.17** Sunset FMCW oblique incidence ionogram for the Alice Springs to Darwin circuit and a simultaneous vertical incidence ionogram for the Alice Springs area at 09 UT in September 1986. One hop $E_s$ and $F_2$ modes can be clearly seen in both ionograms. *Courtesy Australian Defence Science and Technology Organization.*

$F_1$ layer at angles of 3° or lower), and the elevation angle at the MUF decreases—to 8° for the range of 3,000 km. The curve drawn to the right of the $F_2$ leading edge is the ground range at that frequency, corresponding to the angle which gives the minimum group path. Thus at 22 MHz, for example, the ground range to the reflection area corresponding to the leading edge is 1,303 km, the range to Darwin.

Broadly speaking, the BS ionogram will consist of echoes at all group paths greater than that of the leading edge. However there will be a cut-off at low angles, and high values of group path, because the transmitting and receiving antenna arrays have negligible gain at very low angles. The process of backscatter itself is also very inefficient at grazing incidence, with virtually all of the energy being reflected from the Earth in the forward direction. Figure 5.22 shows the longest group path for any of the possible propagation modes, as a function of frequency, for elevation angles of 3° or greater. If there were no angle-dependence of the observed signal strength, this curve would correspond to the trailing edge. It can be seen that the trailing edge is defined by each of the three layers at different frequencies. At low frequencies, the trailing edge is set by the E layer; the only signals which can penetrate the E layer to reach the $F_2$ layer are the steeper angles with shorter ranges. At high frequencies, only $F_2$ propagation is possible.

Figure 5.23 shows the calculated leading edges for different ionospheric indices and for midday (03 UT) and dawn (20 UT) in the same month. The shapes are basically the same for the different indices at a given time. However the frequency range supported by the ionosphere is dramatically lower at dawn and at solar minimum. The highest usable frequency is 11 MHz at dawn (T = 100) and >40 MHz in the middle of the day. To mount surveillance over an area at and beyond a given fixed range will require the use of much lower frequencies at solar minimum. This necessitates the use of much

# Ionograms and Their Interpretation

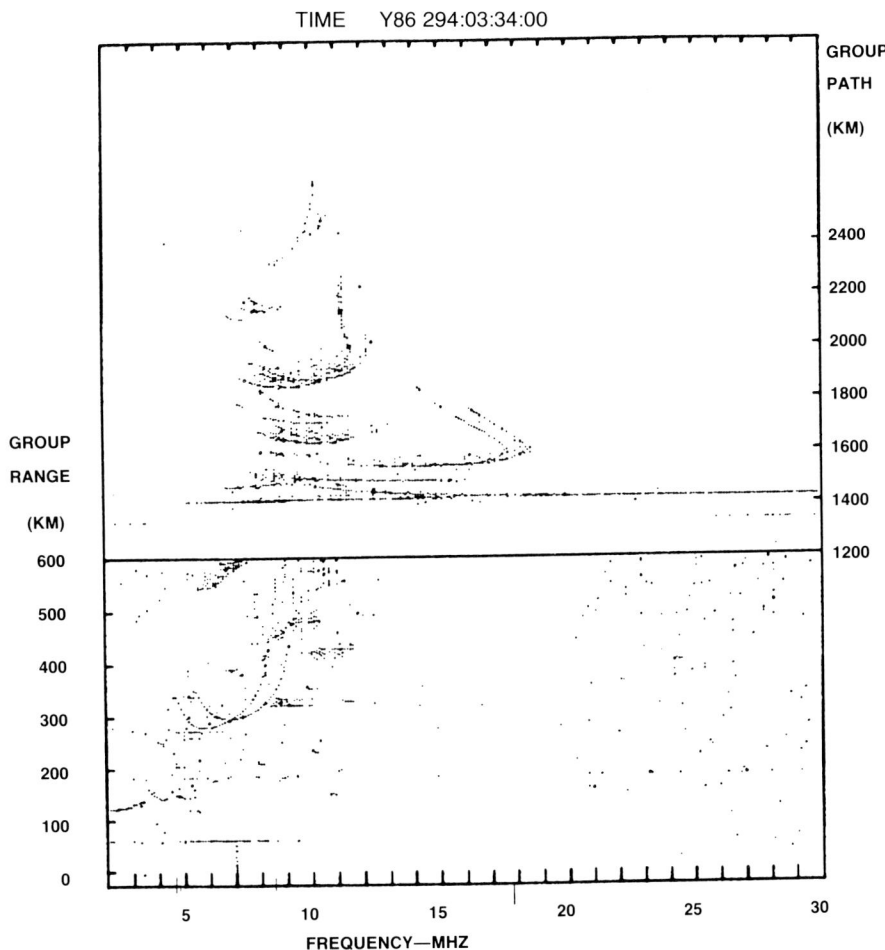

**Figure 5.18** Noon FMCW oblique incidence ionogram for the Alice Springs to Darwin circuit and a simultaneous vertical incidence ionogram for the Alice Springs area at 0330 UT in September 1986. One hop $E_s$ and $F_2$ modes can be clearly seen in both ionograms. The apparent extension of the oblique $E_s$ trace to 30 MHz (at 1,420 km) is spurious. *Courtesy Australian Defence Science and Technology Organization.*

larger antennas. At a fixed frequency, the skip distance decreases as the level of solar activity increases, and closer-in areas can be monitored.

Figure 5.24 shows the effect of moving the BS site towards the equator. Curve 1 had the BS site at 10.0° S, curve 2 at 23.9° S, and curve 3 at 15.0° S. The change in shape for curve 1 is due to the rays hitting the southern crest of the equatorial anomaly, which exhibits large N-S horizontal gradients. Curve 3 is a transition case where some of the frequencies (greater than about 30 MHz) reach the edge of the anomaly. Case 2 is a normal result for a BS sounder at Alice Springs.

Figures 5.25 to 5.27 are "real" BS ionograms obtained with the Jindalee OTH BS system. The ionograms are for the same time, for different azimuths (Figures 5.25 and 5.26), and for different times of the day on the same azimuth (Figures 5.26 and 5.27). Figure 5.27 is for just after sunrise and shows the expected restricted frequency range for which BS occurs. Both the leading and trailing edges for the 1-hop modes can be clearly seen in Figures 5.25 and 5.26. Some fascinating color BS ionograms are given by Earl and Ward [14].

## 5.4.1 Backscatter From Field-Aligned Irregularities

Backscatter from field-aligned irregularities in the E and F layers may also occur at both HF and VHF. For example, BS echoes can appear on normal VI ionograms at high latitudes, as the wall of the mid-latitude trough (Chapter 7) approaches the station. Angle of arrival measurements of the returning ionosonde signals are mandatory if the correct interpretations of the ionograms are to be made. The irregularities may be artificially induced by deliberately heating the ionosphere with large high-power transmitters, and simulation of the BS echoes

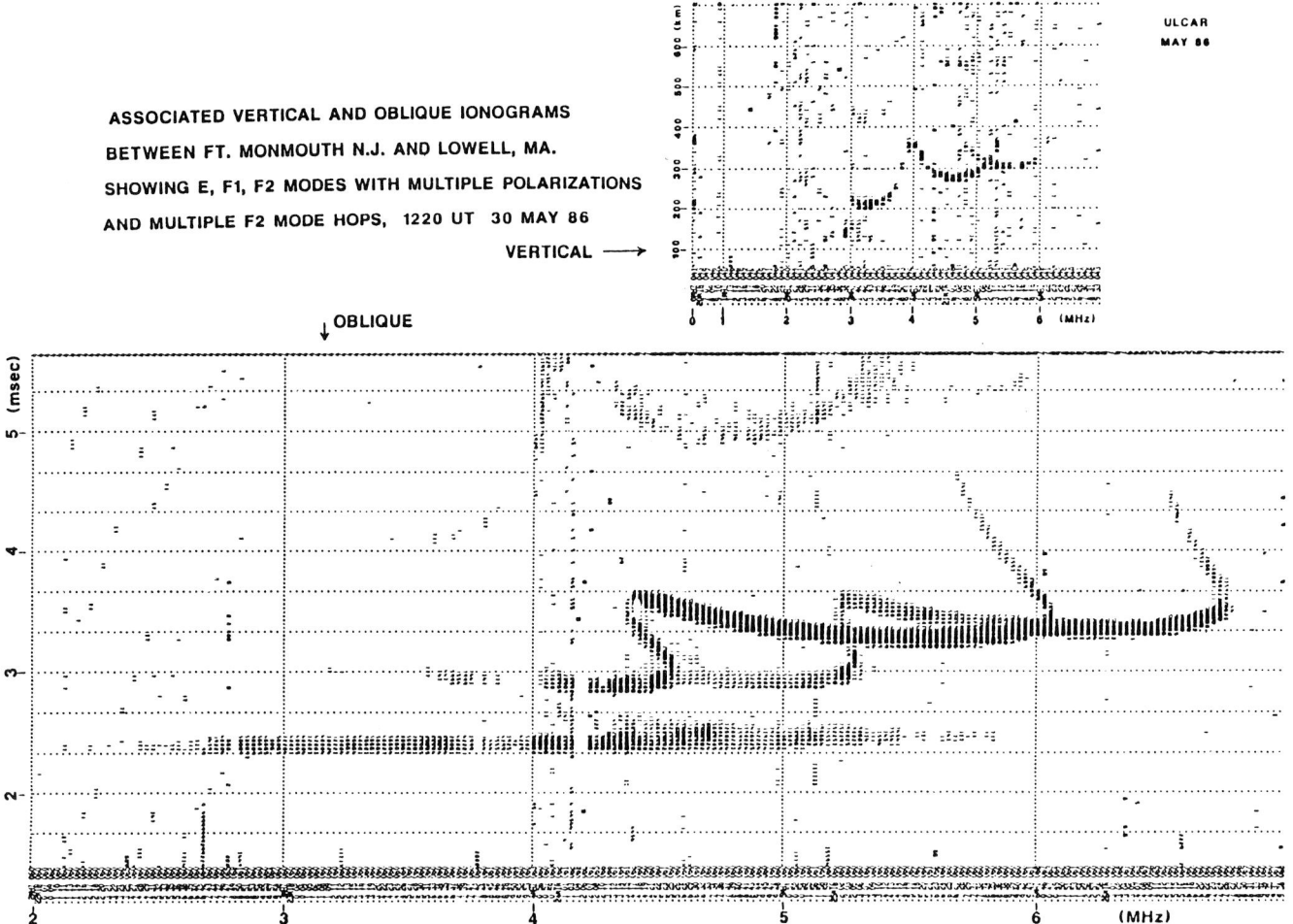

**Figure 5.19** Oblique incidence ionogram for the Lowell, Massachusetts, to Fort Monmouth, New Jersey, circuit for 1220 UT on 30 May 1986. The ionogram shows 1E, $1F_1$, and $1F_2$ modes, as well as a $2F_2$ mode. The two $1F_2$ traces are for the ordinary and extraordinary modes of propagation. *Courtesy University of Lowell Center for Atmospheric Research.*

used to determine the details of the vertical and horizontal distributions of the artificially created irregularities [15, 16].

## 5.5 Real Height Analysis of Ionograms

The real or true height analysis of an ionogram is the inverse of making the ionogram. We start with the ionogram, which is an expression of the relation between the group or apparent path $p'$ and the wave frequency, and attempt to derive the relation between real height and plasma frequency. There is a practical requirement to do this for all four types of ionogram which we have encountered.

### 5.5.1 VI Ionograms

As we have already seen, the group delay $p'$ for a VI ionogram, which we call the virtual height $h'$, can be derived from a knowledge of the electron density profile, $N(h)$. If $h_1'$ is the observed virtual height at a frequency $f_1$ and the base of the ionosphere is at the height $h_0$, we have

$$h_1' = h_0 + \int \mu' \, dh,$$

where the limits of integration are $h_0$ and $h_1$. The height $h_1$ is the real height at which the plasma frequency is equal to $f_1$. Recall that a frequency $f_i$ is reflected at the height where the plasma frequency is equal to $f_i$, if we ignore the Earth's magnetic field or consider just the O ray. We do not know the real height of reflection $h_1$ because we do not know how $f_N$ varies with height. This is what we are trying to determine.

Since the refractive index $\mu'$ is a function of $f_N$, we are obviously going to need to know how $f_N$ varies with height to evaluate the integral. Assume that this variation between frequencies $f_0$ and $f_1$ is

$$f_N = a + b\,h,$$

where $a = f_{N0}$, and b is as yet unknown. We are also going to need to integrate with respect to $f_N$, since by

# Ionograms and Their Interpretation

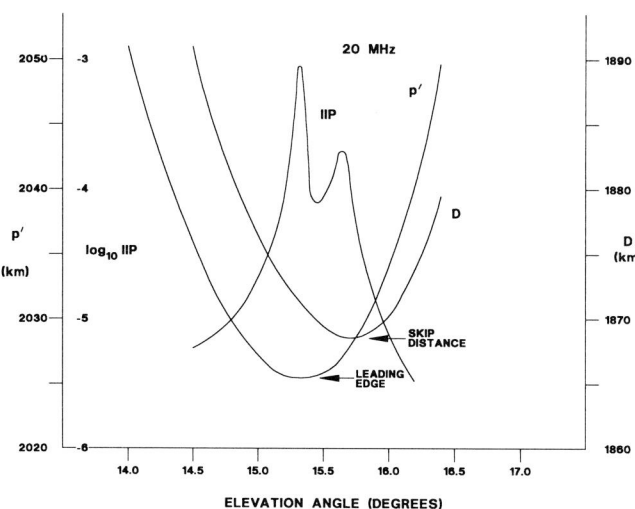

**Figure 5.20** Fixed frequency (20 MHz) cut across the leading edge of a backscatter ionogram, showing how the group path, ground range, and ionospheric illumination power (IIP) vary with elevation angle. The IIP has two sub-maxima which correspond to focussing of the group path (left) and ground range. The leading edge corresponds to an elevation angle of 15.3°, while the familiar skip distance occurs at 15.8°.

doing so the limits of integration will be 0 and $f_1$, which are known.

This change of variable yields

$$h_1' = h_0 + (1/b) \int \mu'(f_N) \, df_N,$$

since $dh/df_N = 1/b$, and is constant over the range of integration. This integral is completely specified and can therefore be evaluated, although with a little difficulty when the Earth's magnetic field is to be included, so there are two unknowns in this equation, $h_0$ and $b$. To proceed any further, we need to make some assumption about the value of $h_0$, so that we can solve for $b$. Suppose, for example, that we assume that a plasma frequency of 0.5 MHz occurs at a height of 90 km. Thus we set $h_0 = 90$ and the lower limit of integration to 0.5 MHz and determine $h_1$. In general, once we know the height $h_i$ at a wave frequency $f_i$, we can determine the height $h_{i+1}$ at the frequency $f_{i+1}$, using

$$h'_{i+1} = h_i + (1/b_i) \int \mu' \, df_N.$$

In this way, we can use the virtual heights $h'$ scaled from an ionogram at arbitrary frequencies to determine the gradients of each linear segment, $b_i$, and the heights $h_i$ corresponding to $f_i$ (and $f_{Ni}$).

We can make life more complicated, but the resultant $N(h)$ profile more accurate for a given number of $(f_i, h'_i)$

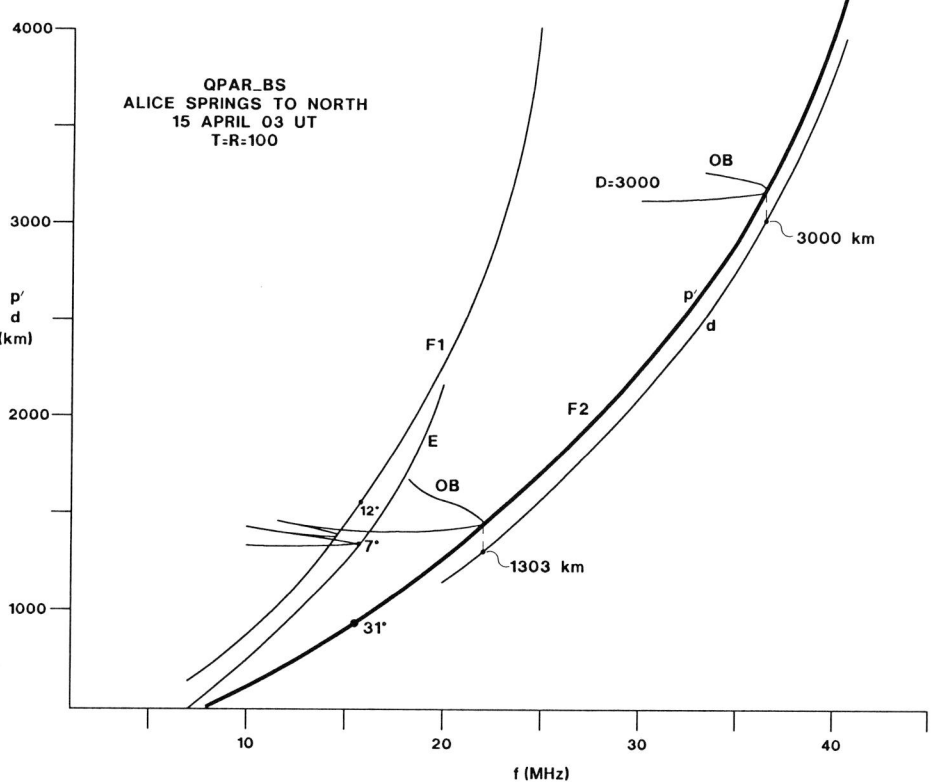

**Figure 5.21** Simulated backscatter leading edge for a backscatter sounder at Alice Springs looking north over Darwin.

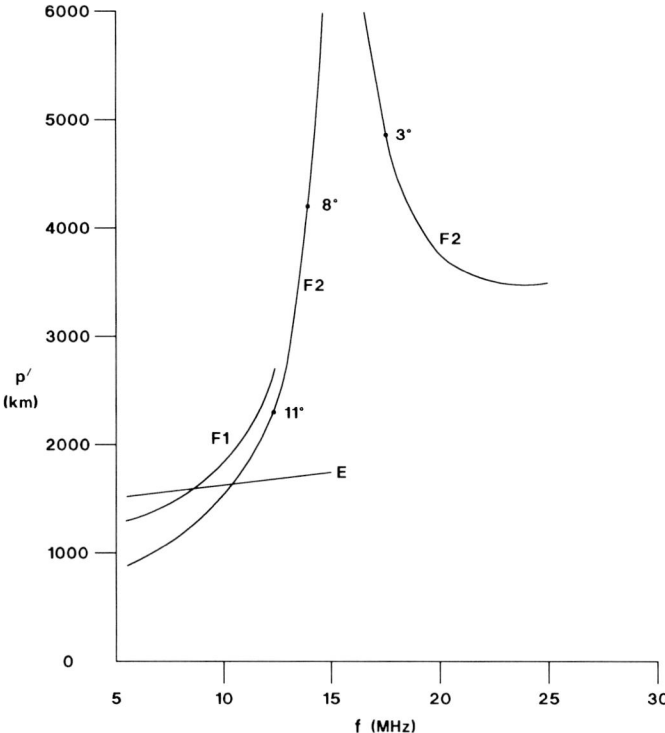

**Figure 5.22** The simulated trailing edge for a backscatter ionogram, with a low elevation cut-off of 3°. The trailing edge is set by the E, $F_1$, and $F_2$ propagation modes in turn.

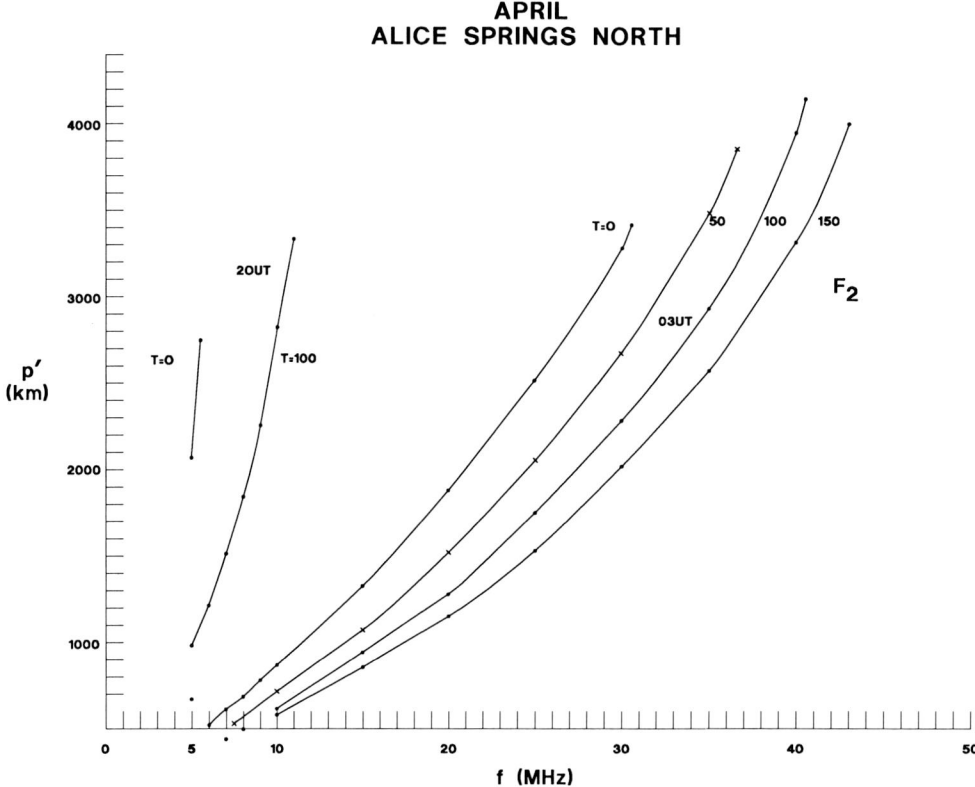

**Figure 5.23** Simulated leading edges for an Alice Springs backscatter ionogram at 03 UT (noon) and 20 UT (dawn) in April at different levels of solar activity. The leading edges do not change shape much with changes in solar activity. In the worst case of dawn at solar minimum, there is a very narrow usable frequency range (5.0 to 5.5 MHz) and a very narrow range of group delays.

# Ionograms and Their Interpretation

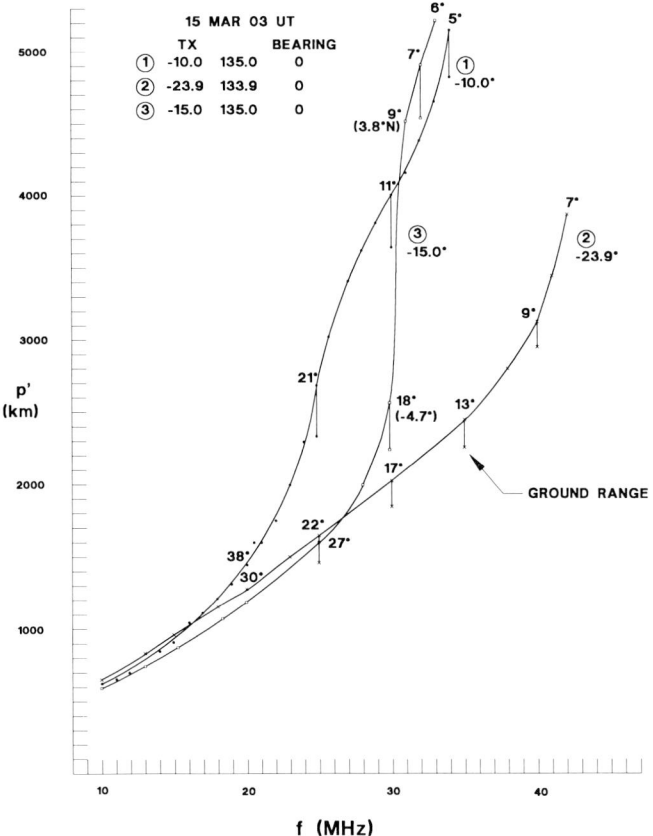

**Figure 5.24** Simulated leading edges for transmitters at different latitudes (March, T = R = 100, 03 UT) and 135° E. As the transmitter approaches the equatorial anomaly, the traces show clear effects of the horizontal gradients associated with the anomaly. The effects are seen at lower frequencies for the most equatorwards transmitter.

pairs, by permitting the variation between points to be parabolic or polynomial in general, rather than linear. We will ignore this complexity here, although practical schemes for real height analysis use at least a parabolic variation. Of more interest here are two points which we skipped over :

1. How valid is our choice of (0.5,90) as our starting point?
2. What happens when $dh/df_N$ goes to infinity (and the change of variable is not valid)?

The first of these points is referred to as the **starting problem** in N(h) analysis. It arises because we do not have any observations below a starting frequency $f_{min}$, whereas we really need them down to zero frequency (if that were possible). The only way to overcome this problem is to make some assumptions about how the plasma frequency varies with height below $f_{min}$. Typically, $f_{min} \approx 1.5$ MHz. We are often better off at night, because the extraordinary ray trace can be used in conjunction with the ordinary ray trace to determine an underlying profile which is consistent with both traces. During the day, the extraordinary trace is not usually visible at low frequencies because of the heavy absorption at those frequencies.

The second difficulty arises when there is a **valley** between the E and $F_1$ or $F_2$ layers, in which the plasma frequency has a local maximum at $f_oE$ and is less than $f_oE$ for a range of altitudes. There are no frequencies which are actually reflected from within the valley. Those below $f_oE$ are reflected by the E layer, and those greater than $f_oE$ penetrate right up to the F layer at the top of the valley. To resolve this "valley" problem, we can assume the general shape for the N(h) profile within the valley and use the observed values of h' to determine the details of the shape. Figure 5.28 shows the form and notation of the standard valley used by John Titheridge in his program POLAN [17, 18]. As with the starting problem, we can get a more realistic solution by using the extraordinary trace as well. One of the more interesting problems of true height analysis is that there is usually not enough information in an ionogram to allow the determination of the detailed shape of the valley. The most that can usually be obtained is the vertical width of the valley.

Titheridge states that there is no purely mathematical procedure for determining the "correct" valley profile from the range of possible solutions, using the ordinary ray only. The curves a, b, and c in Figure 5.29 give real heights for the lower F region which vary over a range of 100 km. All three curves correspond exactly to the given virtual height curve, and so are equally valid mathematical solutions. The full range of possible real height curves extends from the monotonic (no valley) solution to the largest valley which is consistent with a positive value of dN/dh at the base of the upper layer. Selection between different solutions must be based purely on physical grounds, having regard to the reasonableness or physical likelihood of different types of profile. When suitable extraordinary ray data are available (from just above $f_xE$), some information can be obtained about the valley region. At best, two parameters can be estimated, relating basically to the overall width of the valley and to the relative amount of high density ionization. In most cases, only one parameter can be reliably determined. We then use some model for the shape of the valley to give suitable proportions of low and high density ionization and determine the valley width.

There are several practical methods of N(h) analysis which offer sufficiently accurate solutions for a given h' versus f trace, accounting for both the starting and valley problems [19, 20, 21, 22, 23]. Robustness against misinterpretation of the ionogram and against errors in the scaled heights h' is a major requirement for fully

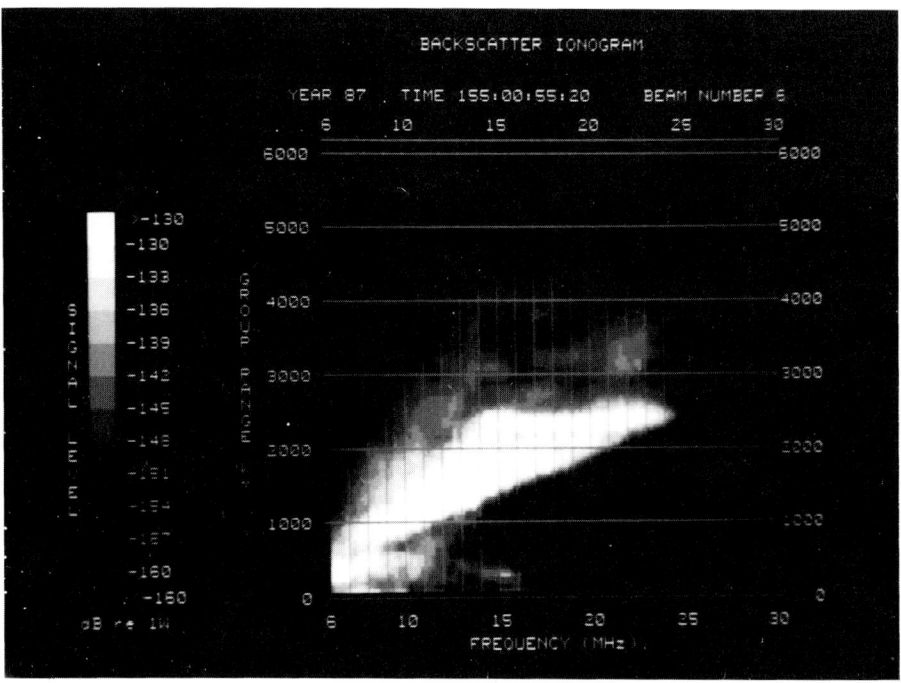

**Figure 5.25** Daytime Jindalee backscatter ionogram for day 155 (4 June) in 1987 at 0055 UT (10 LT). Beam number 6 is on an azimuth of 351°. *Courtesy Australian Defence Science and Technology Organization.*

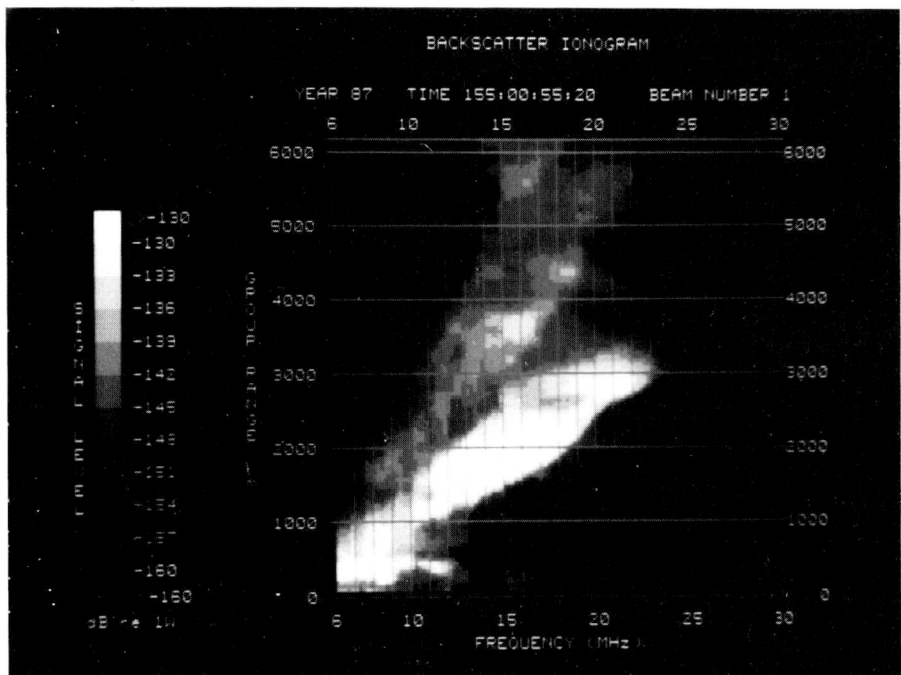

**Figure 5.26** Daytime Jindalee backscatter ionogram for day 155 (4 June) in 1987 at 0055 UT. Beam number 1 is on an azimuth of 296°. *Courtesy Australian Defence Science and Technology Organization.*

automated systems. The method chosen for any implementation is mostly a matter of preference and familiarity, assuming that the method satisfies some basic accuracy criteria. The method most widely used is POLAN. The report UAG-93 by Titheridge is a very extensive treatment of the problem, dealing with the physics as well as the numerical analysis aspects of the problem. It is highly recommended reading.

# Ionograms and Their Interpretation

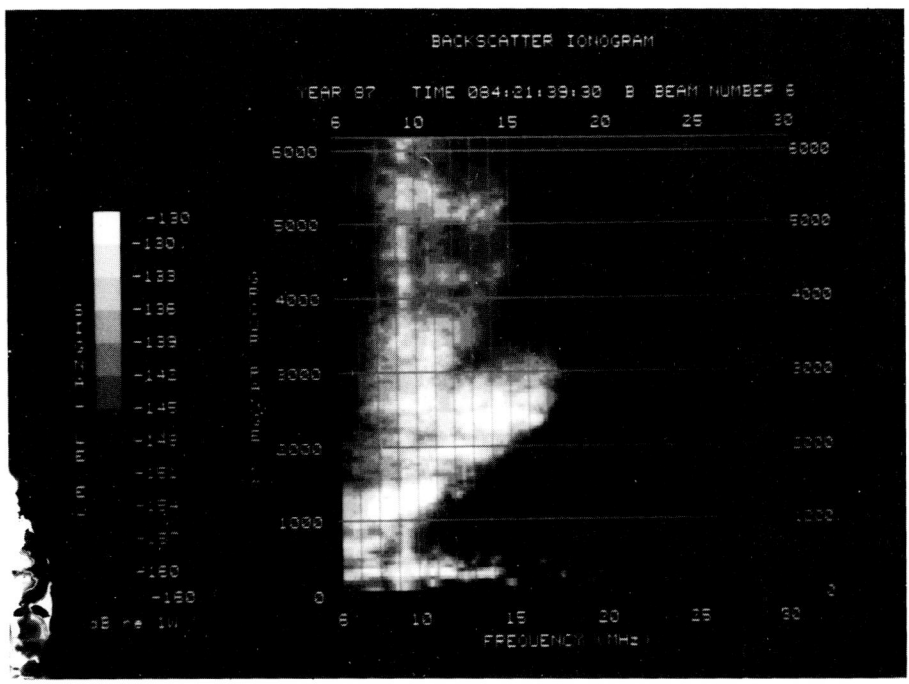

**Figure 5.27** Dawn Jindalee backscatter ionogram for day 84 in 1987 at 2139 UT. Beam number 6 is on an azimuth of 351°. *Courtesy Australian Defence Science and Technology Organization.*

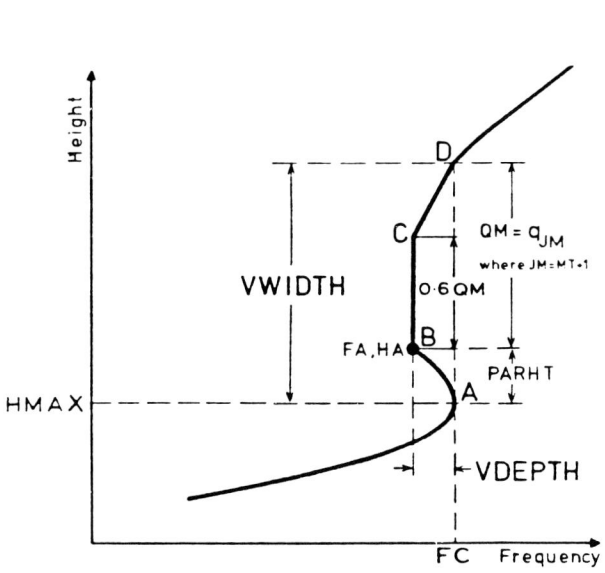

**Figure 5.28** The form and notation of the "standard" valley used in the program POLAN. *After UAG Report 93.*

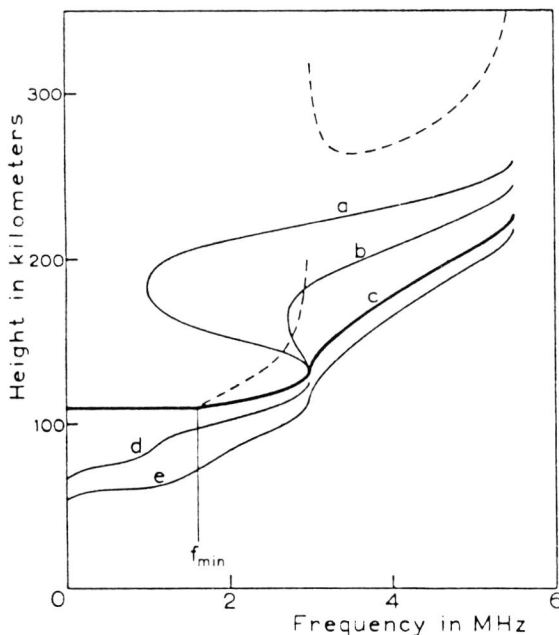

**Figure 5.29** Some of the infinite number of possible real height curves (solid lines) corresponding to the same virtual height curves (dashed lines). *After UAG Report 93.*

### 5.5.2 OI Ionograms

The analysis of oblique ionograms is a somewhat less rewarding task than the analysis of VI ionograms, since they bring with them the following difficulties of their own as well:

1. The *absolute* value of the group delay $p'$ is often not well known.
2. The rays sample the ionospheric layers at different locations along the path.
3. The elevation angles at different frequencies are not known.
4. The calculated profile is not unique, and is rather sensitive to errors in defining the $p'(f)$ trace.

A VI ionogram, of course, samples the layers of the ionosphere at basically the same location, and the elevation angle is 90° and known.

An OI ionogram can be analyzed by using a step-by-step approach similar to that used for VI ionograms or by first converting to the equivalent VI ionogram. The reader is referred to the work of Reilly for the step-by-step approach [24, 25].

### 5.5.3 BS Ionograms

As can be imagined, the inversion of BS ionograms is an order of magnitude more difficult than the inversion of OI ionograms, since the range to the reflection point is not fixed or known [26]. This is true even after we have gone to great lengths just to interpret the ionogram. We shall return to this problem in Chapter 15, when we address the question of how to go from the observed group path at a particular frequency to the required ground range.

## 5.6 Automatic Scaling of Ionograms

Ionospheric physicists spend many a happy hour interpreting their ionograms, seeking to explain the interesting features which appear on them. However not everyone shares this enthusiasm, and some of us would balk at the prospect of interpreting ionograms at the rate of one every 5 minutes forever. Luckily, help is at hand, at least for VI ionograms. The software system ARTIST, which forms an essential part of the Digisonde 256, has proved to be very successful in interpreting many years of ionograms, including some very difficult high latitude ionograms. In these cases, simultaneous angle of arrival observations are essential. ARTIST will determine the leading edge of the ionogram trace, deduce all of the useful critical frequencies and virtual heights, and derive the corresponding true height profile, as illustrated in Figure 5.11. Alternative approaches which have also proved relatively successful have been developed [27, 28], but ARTIST has been much more extensively tested, especially at high latitudes. Automatic scaling (or **autoscaling**) is also being developed for OI and BS ionograms.

## 5.7 References

1. Davies, K. *Ionospheric Radio*. Peter Peregrinus Press, London, 1990.
2. Abramowitz, M., and I. E. Stegun (Eds). *Handbook of Mathematical Functions*. National Bureau of Standards, Applied Mathematics Series, No. 55, Washington, 1964.
3. Davies, K. *Ionospheric Radio. loc. cit.*
4. Hanson, G. H., E. L. Hagg, and D. Fowle. *The Interpretation of Ionospheric Records*. Defence Research Telecommunications Establishment Radio Physics Laboratory Report No. R2, Ottawa, Canada, May 1953.
5. Shapley, A. H. (Ed). *Atlas of Ionograms*. World Data Center A Report UAG-10, Boulder, Colorado, May 1970.
6. Piggott, W. R. *High-Latitude Supplement to the URSI Handbook on Ionogram Interpretation and Reduction*. World Data Center A Report UAG-50, Boulder, Colorado, October 1975.
7. Gonzalez, M. M. de, and L. Kurban. *Atlas of Ionograms for the South American Region*. Centro de Investigaciones Regionales de San Juan, Argentina, 1981.
8. Wakai, N., H. Ohyama, and T. Koizumi. *Manual of Ionogram Scaling*. Radio Research Laboratory, Ministry of Posts and Telecommunications, Japan, October 1987.
9. Bibl, K., and B. W. Reinisch. *The Universal Digital Ionosonde*. Radio Science, 13(13), pp. 519–530, 1978.
10. Reinisch, B. W., and Huang Xeuqin. *Automatic Scaling of Electron Density Profiles from Digital Ionograms, 3, Processing of Bottomside Ionograms*. Radio Science 18(3), pp. 477–492, 1983.
11. Ward, B. D. Private Communication, 1990.
12. Davies, K. *Ionospheric Radio. loc. cit.*
13. Dyson, P. L., and J. A. Bennett. *A Model of the Vertical Distribution of the Electron Concentration in the Ionosphere and its Application to Oblique Propagation Studies*. J. Atmos. Terr. Physics 50(3), pp. 251–262, 1988.

14. Earl, G. F., and B. D. Ward. *Frequency Management Support for Remote Sea-State Sensing Using the JINDALEE Skywave Radar*. IEEE Trans on Oceanic Eng, pp. 164–173, April 1986.

15. Buchau, J. and L. F. McNamara. *Simulation of Ionograms Obtained During the October 1979 Plattevile Heater Experiment*. Radio Science 21(3), pp. 286–296, 1986.

16. McNamara, L. F., B. W. Reinisch and J. Buchau. *Ionosonde Studies of Field-Aligned Irregularities During High-Power HF Heating at Arecibo*. Air Force Geophysics Laboratory Report AFGL-TR-86-0168, Hanscom AFB, Massachusetts, February 1988.

17. Titheridge, J. E. *The Real Height Analysis of Ionograms: A Generalized Formulation*. Radio Science, 23(5), pp. 831–839, 1988.

18. Titheridge, J. E. *Ionogram Analysis with the Generalized Program POLAN*. World Data Center A Report UAG-93, Boulder, Colorado, 1985.

19. McNamara, L. F. (Chairman). *A Comparative Study of Methods of Electron Density Profile Analysis*. World Data Center A Report UAG-68, Boulder, Colorado, September 1978.

20. Paul, A. K. *A Simplified Inversion Procedure for Calculating Electron Density Profiles from Ionograms to Use with Minicomputers*. Radio Science 12, pp. 119–122, 1977.

21. Gulyaeva, T. L. *On a Non-Ambiguous Statement of the Problem of Computing the N(h) Profiles of the Bottomside Ionosphere*. Geomagnetism and Aeronomy 12(3), pp. 551–553, 1972.

22. Titheridge, J. E. *The Real Height Analysis of Ionograms: A Generalized Formulation. loc. cit.*

23. Reinisch, B. W., and Huang Xueqin. *Automatic Calculation of Electron Density Profiles from Digital Ionograms. 3. Processing of Bottomside Ionograms. loc. cit.*

24. Reilly, M. H. *A Method for Real-Height Analysis of Oblique Ionograms*. Radio Science, 24(4), pp. 575–583, 1989.

25. Reilly, M. H. *Ionospheric True Height Profiles from Oblique Ionograms*. Radio Science, 20(3), pp. 280–286, 1985.

26. Dyson, P. L. *A Simple Method of Backscatter Ionogram Inversion*. Accepted for publication in J. Atmos. Terr. Phys.

27. Fox, W. W., and C. Blundell. *Automatic Scaling of Digital Ionograms*. Radio Science, 24(6), pp. 747–761, 1989.

28. McCue, C. G., and J. D. Gilbert. *Automatic Ionogram Scaling*. Bulletin 52, Ionosonde Network Advisory Group, World Data Center, Boulder, Colorado, December 1988.

## 5.8 Problems

1. What is the main advantage of a topside sounder versus a ground-based vertical incidence sounder?

2. What are the two main disadvantages of a topside sounder versus a ground-based vertical incidence sounder?

3. In Figure 5.2, where is the satellite with respect to the ionogram—at the bottom, top, or left hand side?

4. With regard to the "ball and hill" analogy, under what conditions would a ball which has gone over the hill be able to return? Is there an ionospheric counterpart for this situation?

5. Arguing by analogy is acceptable provided the analogy is valid and does not introduce spurious conclusions. Is the ball and hill analogy valid for the daytime equatorial ionosphere?

6. Complete the determination of $h'$, using the transformation to the variable t, as in Section 5.2.2.

7. Is the transformation $t^2 = 1 - (f_N/f_R)^2$ a better one to make than the transformation suggested by W. Becker (see Huang Xueqin and B. W. Reinisch, Radio Science 17(4) pp. 837–844, 1982)? For example, would a computer subroutine be faster or slower with Becker's transformation? (Count the numbers of the different operations and measure the times of each by doing the odd thousand or two.) You might also look at the subroutine GIND in John Titheridge's program POLAN.

8. You have been asked to confirm a model value of the Earth's magnetic field at your site. What do you do? You can assume that you have a pair of compasses and a very good ionosonde at your disposal.

9. The complexity of observed ionograms varies over the surface of the Earth. Which area gives the least complicated ionograms, and which the worst?

10. *The rays corresponding to the MUF never reach the peak of the layer, except at vertical incidence.* Is this statement true in general, only for the ordinary ray, or only when the Earth's magnetic field is ignored?

11. Assuming for the moment that the statement in problem 5.10 is true, why do we go to such effort to observe and model the parameters of the *peak* of the layer? Do not look up the text for this one!

12. Why would anyone want to transform between a VI and OI ionogram?

13. What are the approximations made in the transformation between VI and OI ionograms?

14. Wherein lie the advantages in simulating an oblique ionogram?

15. What is it that stays fixed in the calculation (or experimental making) of an OI ionogram?

16. There are many backscatter sounders currently deployed around the world. What advantages accrue from the use of such sounders? You might like to skip to Chapters 10 and 15 for the answer to this one.

17. What stays fixed in the calculation of a BS ionogram—the frequency, the ground range, the propagation time, the elevation angle, all of the above, or none of the above?

18. What does the leading edge of a BS ionogram correspond to?

19. What does the trailing edge of a BS ionogram correspond to?

20. Which is likely to be simulated more accurately, the leading edge or the trailing edge? Give reasons for your answer.

21. If you had a backscatter sounder at your disposal and were particularly interested in looking at long ranges, would you choose a low or high operating frequency? At what time of day would your coverage area be (a) greatest and (b) smallest?

22. What are the two basic difficulties encountered in the true-height analysis of ionograms? How can these be overcome?

23. Given a very complete O-ray ionogram trace, but no X-ray trace, what are the limits of the height at the top of the valley?

24. Why has so much effort been spent in developing a software system to derive the parameters such as critical frequencies and electron density profile from an ionogram?

25. Ionogram autoscaling systems have become mandatory as labor costs have increased, but have also led to an expanded capability of monitoring the ionosphere on short time scales. They are continually being modified, as ionograms are encountered which they cannot handle. How reliable are these systems in general? You will probably need to do a literature search to answer this one.

26. Referring to Figure 5.13, and the comments in the text about propagation conditions between 10.6 and 13.0 MHz, what important feature of the real world has been ignored?

27. Referring to Section 5.5.1, derive expressions for $b_i$ and $h_{i+1}$, and convince yourself that the whole $N(h)$ profile may be deduced using this step-by-step approach.

# Chapter 6

# Predictions for HF Communications

## 6.1 Introduction

Because HF communications over long distances rely on reflection from the ionosphere, and because the ionosphere is so variable, it is not possible to pick a convenient operating frequency and expect to maintain a communications link using that frequency at all times. Instead it is necessary to choose a frequency which the ionosphere will support at the required time on the circuit being considered. To do this successfully, we must take account of all the variations which the ionosphere along the circuit undergoes—location, season, time of day, altitude, and solar activity. We must also determine the effect of the geometry of the circuit on the range of frequencies available for use on that circuit and consider non-ionospheric restraints such as antennas and transmitter/receiver constraints. This chapter describes the procedures adopted in practice to take account of the ionospheric and propagation factors which determine the choices made to ensure successful communications.

There are four steps which must be followed when taking these factors into account. These are :

1. Predict what the sun will be doing during the time for which the required predictions will apply.

2. Determine what the ionosphere will be doing as a result of the predicted level of solar activity. In other words, set up an ionospheric model.

3. For this ionospheric model, calculate the geometry and propagation modes for the circuit under consideration.

4. Calculate the propagation parameters, namely MUF, LUF, and field strength.

The first step is at best an educated guess, which we will discuss further in Section 6.7, after we have considered steps 2, 3, and 4 in some detail.

## 6.2 Models of the Ionosphere

The development of ionospheric models is a very rich and rewarding task. We do not consider here the various approaches which can be adopted, but rather refer the reader to some excellent papers [1–9]. The status of ionospheric propagation prediction models has been reviewed by Goodman [10]. For our present purposes, a model of the ionosphere is just a recipe that tells us all we need to know about the ionosphere at some point, in particular at the reflection point(s) for a given circuit. For many applications, all we need to know are the critical frequencies for the E and F layers, $f_oE$ and $f_oF_2$, and the height of the peak of the $F_2$ layer, $h_mF_2$. The height of the E layer can safely be set equal to 110 km for most applications. For more detailed studies, we would also need to know how the electron density varied with height, all the way from the D region to the peak of the $F_2$ layer at $h_mF_2$. In other words, we would need to know the electron density profile of the ionosphere.

Ionospheric models exist in various forms, and different applications require models of different levels of complexity. A perfectly adequate model for $f_oE$ is given by equation 3.4. The model for the height of the peak of the E layer, called $h_mE$, simply states that $h_mE = 110$ km. Models for $f_oF_2$ and $h_mF_2$, on the other hand, are not so simple and a large amount of effort has been spent by the international scientific community in perfecting a suitable model. Some of the more accurate models, which will calculate $f_oF_2$ and $h_mF_2$ at any point on the Earth at any time, exist as complex computer programs which require reasonably large computers for their operation. Simplified models have been produced which can be evaluated on hand-held calculators, but the accuracy of the models is severely limited when compared to that of the better computer programs.

Because the $F_2$ region is so complex, models of its variations with time of day, season, position, and solar activity have had to be constructed using enormous numbers of observations made all over the world for many years. About 180 ionospheric stations have been operated throughout the world, although not all 180 were ever operating at the one time. Currently there are about

100 stations in routine operation. Many stations have been operating for more than 40 or 50 years, the first being Slough (U.K.,1930), Watheroo (Western Australia, 1935), Washington (D. C. ,1933), Tomsk (U. S. S. R., 1937), and Tromso (Norway,1932). The number of operational stations is at present undergoing a surge, with the deployment of new generation ionosondes.

The maps of ionospheric parameters such as $f_oF_2$ and $M(3000)F_2$ are based on monthly median values observed at the 180 or so stations for low (R = 0) and high (R = 100) levels of solar activity [11]. They give the variations of $f_oF_2$ and other parameters as a function of position on the Earth; in other words, they take account of the geographic variations of the ionosphere. There is a map for each month and local time, thus allowing for the seasonal and diurnal variations of the ionosphere. The fourth and fifth variations which must be considered are the variations with height and with solar activity. The former is accounted for by mapping the E, $F_1$, and $F_2$ layer parameters separately. As we saw in Chapter 3, the variation with solar activity can be represented by calibration curves (straight lines in fact) which relate monthly median values of $f_oF_2$, say, to monthly average values of the sunspot numbers, for a given station, month, and hour. Each of the calibration curves can be used to derive monthly median values of $f_oF_2$ at two levels of solar activity, R = 0, and R = 100. This is done for each of the 180 stations and contour maps are then drawn. There will be 576 maps of each ionospheric parameter: 12 months × 24 hours × 2 levels of solar activity.

The complexity of the F-region variations as compared to that of the well-behaved E region is illustrated by Figures 6.1 and 6.2. Figure 6.1 shows comparative plots of the solar zenith angle (upper panel) and the observed median values of $f_oE$ for June at solar maximum, as a function of geographic latitude and longitude. Plots such as these can be drawn for each month of the year.

To find the value of $f_oE$ at any time, for a latitude of 30° N for example, we just follow the 30° N horizontal line across the diagram. Thus $f_oE$ increases through 1.5 MHz at 05 LT, passes through 3.0 MHz at 07 LT, and just exceeds 4.0 MHz for the period 11–13 LT. It then decays, passing through 1.5 MHz at around 19 LT.

The important thing to note about the two panels of Figure 6.1 is that the shapes of the sets of contour lines are basically the same. This tells us that the E layer, and $f_oE$ in particular, is strongly controlled by the position of the sun in the sky, which is measured by the solar zenith angle. In fact, the relationship is so close that it is possible to express $f_oE$ as a simple equation in cos Z,

$$f_oE = 0.9 \{(180 + 1.44R) \cos Z\}^{1/4}.$$

This is equation 3.4 again. The sunspot number R takes account of the fact that the ionosphere as a whole has higher critical frequencies at high levels of solar activity.

Figure 6.2 is the F-layer counterpart of the upper panel in Figure 6.1 and we can see immediately that while the contours of $f_oF_2$ follow the shape of the zenith angle curves during the sunrise period in the southern hemisphere (90° E to 120° E at 00 UT), they bear no resemblance to the zenith angle curves in other parts of the world. The shaded areas indicate the crests of the equatorial anomaly which we know to be strongly affected by the Earth's magnetic field and not completely solar controlled. The sunset period (90° W to 60° W) is particularly irregular and is followed by quite large values of $f_oF_2$ which do not decrease to zero after the sun sets. In this case it is winds blowing in the neutral atmosphere (at F-region altitudes) which keep the $F_2$ layer at relatively high altitudes and minimize the loss of free electrons by recombination processes.

Because we need to map $f_oF_2$ in spite of its complexities, it is common practice to draw maps of $f_oF_2$ at a fixed local time, rather than a fixed *universal* time. Although the maps have no physical significance (recall that we get a snapshot of the ionosphere if we specify a fixed universal time), they have simpler shapes than fixed UT maps and are thus easier to represent by formulas or on a computer. Figure 6.3 is a local time map of $f_oF_2$ for 00 LT, June, solar maximum. Note that the edges of the two maps in Figures 6.2 and 6.3 are identical because 00 UT is equivalent to 00 LT at 0° longitude. The contours have much simpler shapes than those in Figure 6.2, because the large diurnal variations caused by the changing zenith angle of the sun do not exist in this type of plot. If the position of the sun were the only factor controlling $f_oF_2$, the contours would be straight lines parallel to the geographic equator. This is true for $f_oE$. However the $F_2$ layer is also affected by winds and by the Earth's magnetic field, both of which vary with geographic longitude, and the contours are not so simple. The effect of the Earth's magnetic field in the equatorial regions is clearly shown in Figure 6.3 by the 6, 7 and 8 MHz contours, which follow the geomagnetic equator plotted in Figure 3.18.

## 6.3 Geometry of the Circuit

The geometry of the circuit has to be calculated because it determines the obliquity factor and thus the MUF, according to equation 4.2. The first thing we have to do is decide what propagation modes we are going to consider. We have previously seen that there are often many modes possible, but it is not feasible or necessary to consider them all every time we want to calculate an MUF. In practice, we need to consider only the lower order modes. As a working rule, a propagation mode may

# Predictions For HF Communications

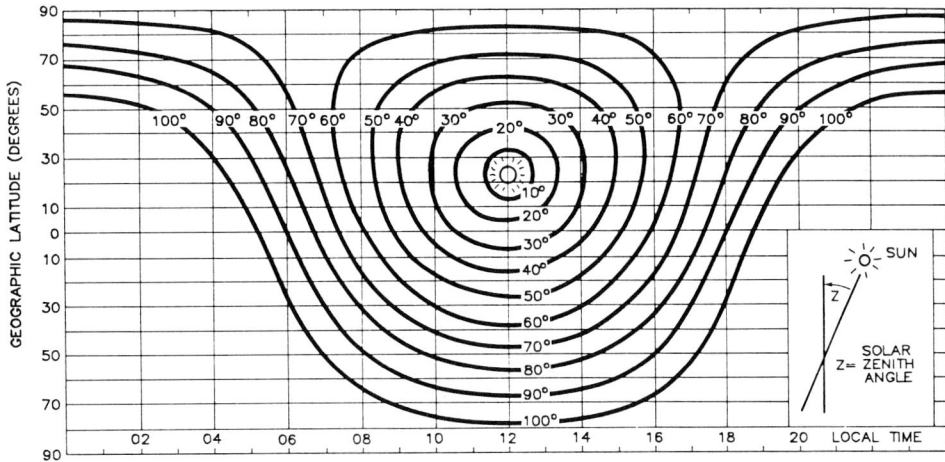

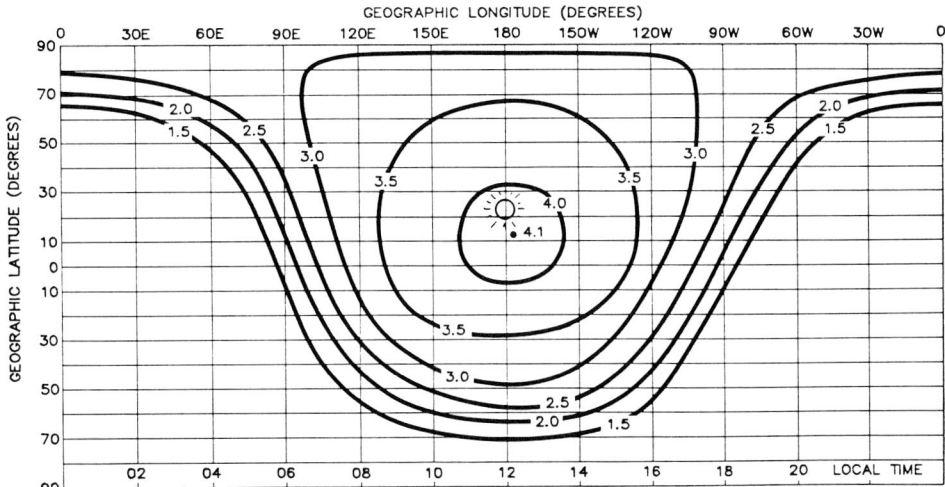

**Figure 6.1** Corresponding plots of the solar zenith angle and the critical frequency of the E layer, $f_oE$, for 00 UT in June at solar maximum. The important thing about the two plots is that the shapes of the contours are almost identical. This means that the value of $f_oE$ is largely determined by the size of the solar zenith angle, making it easy to predict a value of $f_oE$. All we need to know is where the sun is in the sky. *Courtesy IPS Radio and Space Services.*

be ignored if it involves more than two reflections from the E layer. This is because signals reflected from the E layer suffer much heavier absorption than those reflected from the F layer.

F-region hops are normally considered possible on circuits up to about 3,000 km in length at mid-latitudes, and up to about 4,000 km at low latitudes where the F layer is higher, while E-region hops are considered possible on circuits up to about 2,000 km in length. Note that these circuit lengths correspond to very low elevation angles and cannot be attained with simple antennas which radiate very little energy at these angles. If an F-layer circuit is longer than 3,000 km, propagation by a 1F mode would not usually be possible and the lowest order mode would be the 2F mode. For very long circuits, the **control point** technique is used, in which it is assumed that the only parts of the ionosphere relevant to the problem of calculating the MUF are at the first and last reflection points in the F layer, and these are assumed to be 1,500 km in from each end of the circuit.

Once we have decided what propagation modes to consider, it is a simple matter to calculate the obliquity factor for each hop of the propagation modes, and the values of $f_oE$, $f_oF_2$, and $h_mF_2$ at each reflection point. Reflection is assumed to take place at the midpoint of each hop.

## 6.4 Calculation of the MUF and LUF

Calculation of the lowest usable frequency (LUF) is somewhat empirical because it depends on things other than the absorption caused by the ionosphere. However for short circuits and with well-planned equipment, it can

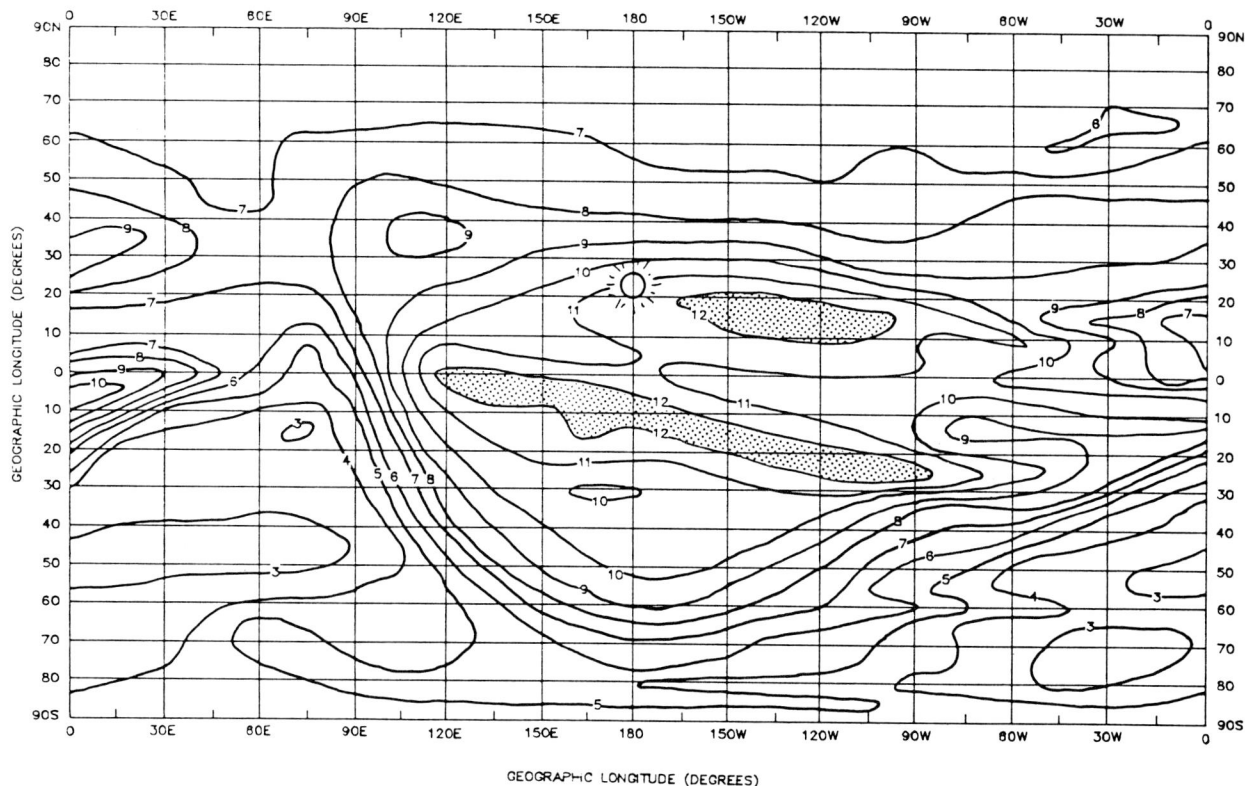

**Figure 6.2** The critical frequency of the $F_2$ layer, $f_oF_2$, at 00 UT in June at solar maximum. Note that 15° longitude is equivalent to one hour in local time. The contours of $f_oF_2$ do not follow the contours of the solar zenith angle in the way that contours of $f_oE$ do. This means that knowing where the sun is in the sky is not sufficient for us to be able to work out the value of $f_oF_2$. The only time that the contours of $f_oF_2$ follow the zenith angle contours in this figure is during the sunrise period in the winter hemisphere (around 20° S, 105° E). The sun is overhead at a longitude of 180°, and at the Tropic of Cancer. *Courtesy IPS Radio and Space Services.*

be related quite simply to $f_oE$ by a simple formula. This is not surprising because the absorption is mostly due to the D region, which depends on the zenith angle in much the same way as $f_oE$. For long circuits, the LUF can be set equal to the first frequency which is just too high to be propagated on a mode involving two E-layer reflections, since any signal reflected twice by the E layer will be severely attenuated. At night, when there is no absorption, the LUF is controlled by the antenna characteristics, the transmitter power, and the noise level, all of which vary with frequency.

We have already discussed the calculation of the MUF in some detail in Chapter 4. The steps involved, for the special case of a 3,000 km circuit, are illustrated in Figure 6.4, and are:

1. Predict the value of the ionospheric index, $T_A$ say, for the month for which predictions are required.

2. Calculate the latitude and longitude of the reflection or control point, C.

3. Determine the local time at the reflection point, given the universal time and longitude.

4. Select world maps of $f_oF_2$ and of the obliquity factor $M(3000)F_2$ for the appropriate month and local time. Two maps are selected for each of $f_oF_2$ and $M(3000)F_2$, corresponding to T = 0 and T = 100.

5. Determine by two-dimensional interpolation in the maps the values of $f_oF_2$ and $M(3000)F_2$ at the location of the reflection point.

6. Interpolate linearly between the values at T = 0 and T = 100 to determine the values at T = $T_A$.

7. Calculate the MUF using the simple formula

$$\text{MUF}(3000) = f_oF_2 \cdot M(3000)F_2.$$

For circuits not 3,000 km in length, the appropriate obliquity factor $M(D)F_2$ for the length D can be obtained from $M(3000)F_2$ via an empirical formula.

## 6.5 GRAFEX Frequency Predictions

This section describes one particular form of HF propagation predictions which contains a wealth of information about the expected propagation conditions as well as the frequencies which will be supported. The form is known as GRAFEX predictions and is the one used by the

# Predictions For HF Communications

Australian Government IPS Radio and Space Services. A slightly different format is provided by the IPS PC-based Advanced Stand-Alone Prediction System [12].

A GRAFEX can be overwhelming at first sight because it is crammed with so much information. Therefore, it is more readily understood if it is first assembled piecemeal. The standard form for a circuit prediction is usually a graph. Panels A and B of Figure 6.5 show the variations of:

1. upper decile MUF for the F layer (UD)
2. median MUF for the F layer (FMUF)
3. lower decile MUF for the F layer (OWF)
4. median MUF for the E layer (EMUF)
5. absorption limiting frequency (ALF)

throughout the day for the required propagation modes. Panel C of the figure shows the superimposed curves for both the first and second orders of propagation mode. It is normally not necessary to consider more than two modes for any one circuit. Indeed, if more than two modes were considered, the graph would become too cluttered for easy use. The space between the curves on the graph, where each pixel represents a time-frequency pair, is taken up with symbols and letters which characterize the propagation conditions for each hour for a number of regularly spaced frequencies; see Figure 6.6.

Figure 6.7 shows a full GRAFEX prediction (without the curves) for a circuit from Perth to Adelaide. The main features of the GRAFEX are:

1. Identification. Usually the names of the terminals of the circuit (normally western terminal first).
2. Circuit length. Great-circle distance between terminals.
3. Time period. Usually a specified month.
4. T index. Predicted value of ionospheric index.

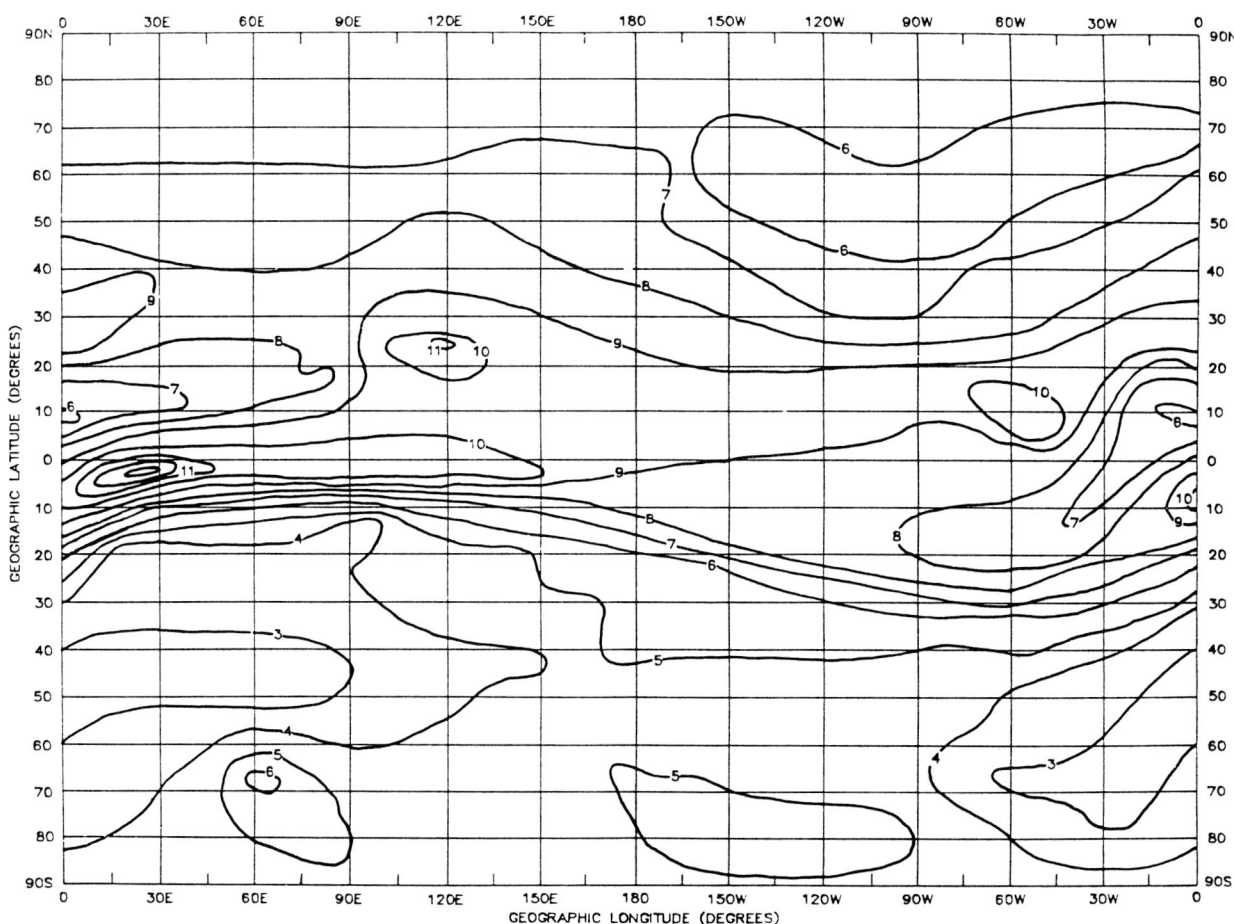

**Figure 6.3** A local time map of $f_oF_2$ at 00 LT in June at solar maximum. The contours of $f_oF_2$ in this map have a much simpler structure than those in Figure 6.2 and are thus easier to represent by formulas. In the equatorial regions, the contours follow the magnetic equator. The sun does not appear in the diagram because it is midnight at all longitudes in this form of presentation of the data. *Courtesy IPS Radio and Space Services.*

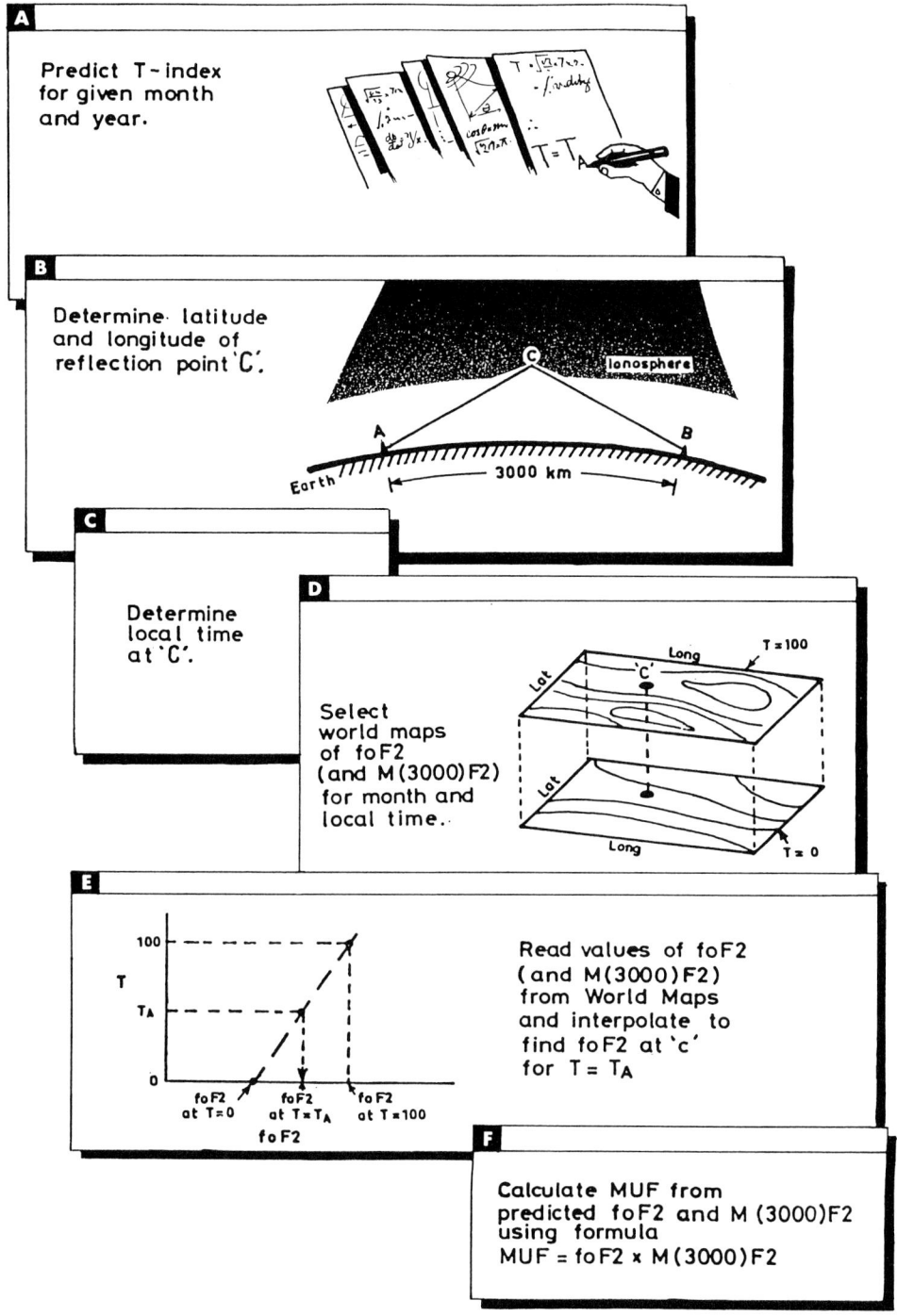

**Figure 6.4** Summary of the prediction technique for HF communications. *Courtesy IPS Radio and Space Services.*

5. Two layers. Predictions are made for both the E and F layers [Height variation of the ionosphere].

6. OWF, EMUF, ALF. Tables of optimum working frequency, E-layer maximum usable frequency, and absorption limiting frequency.

7. Two modes. Predictions are made for the two lowest order modes (least number of hops), except for circuits < 500 or > 12,000 km.

8. Key to symbols. Brief description of all GRAFEX symbols used.

# Predictions For HF Communications

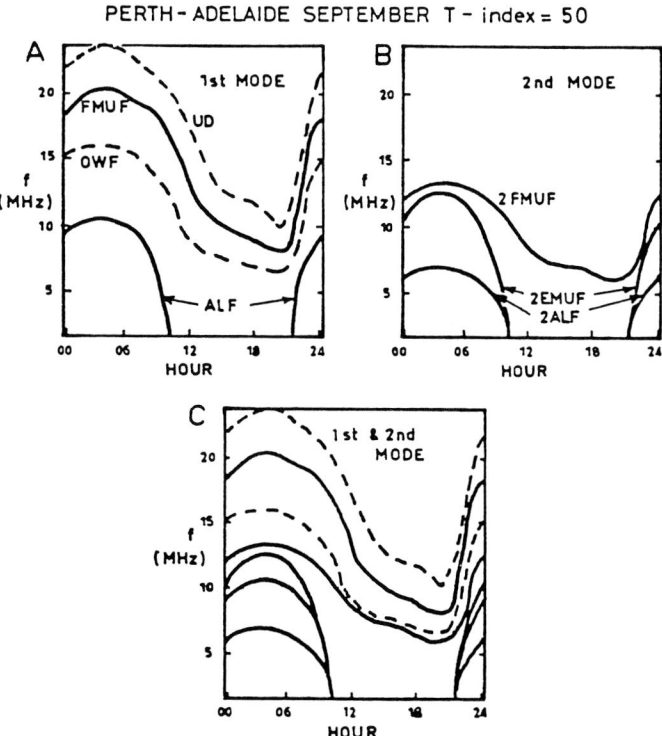

**Figure 6.5** The graphical form of circuit predictions for the first and second propagation modes on the Perth to Adelaide circuit. *Courtesy IPS Radio and Space Services.*

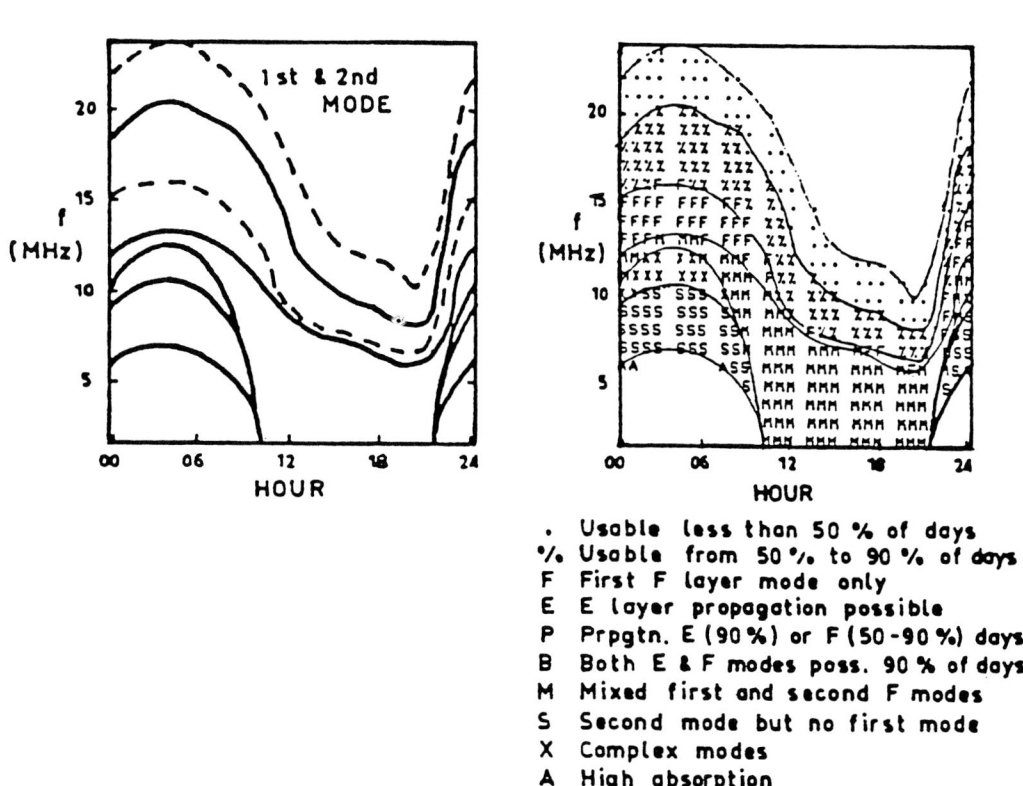

**Figure 6.6** The transformation from the graphical to the GRAFEX form of the frequency predictions for the Perth to Adelaide circuit. *Courtesy IPS Radio and Space Services.*

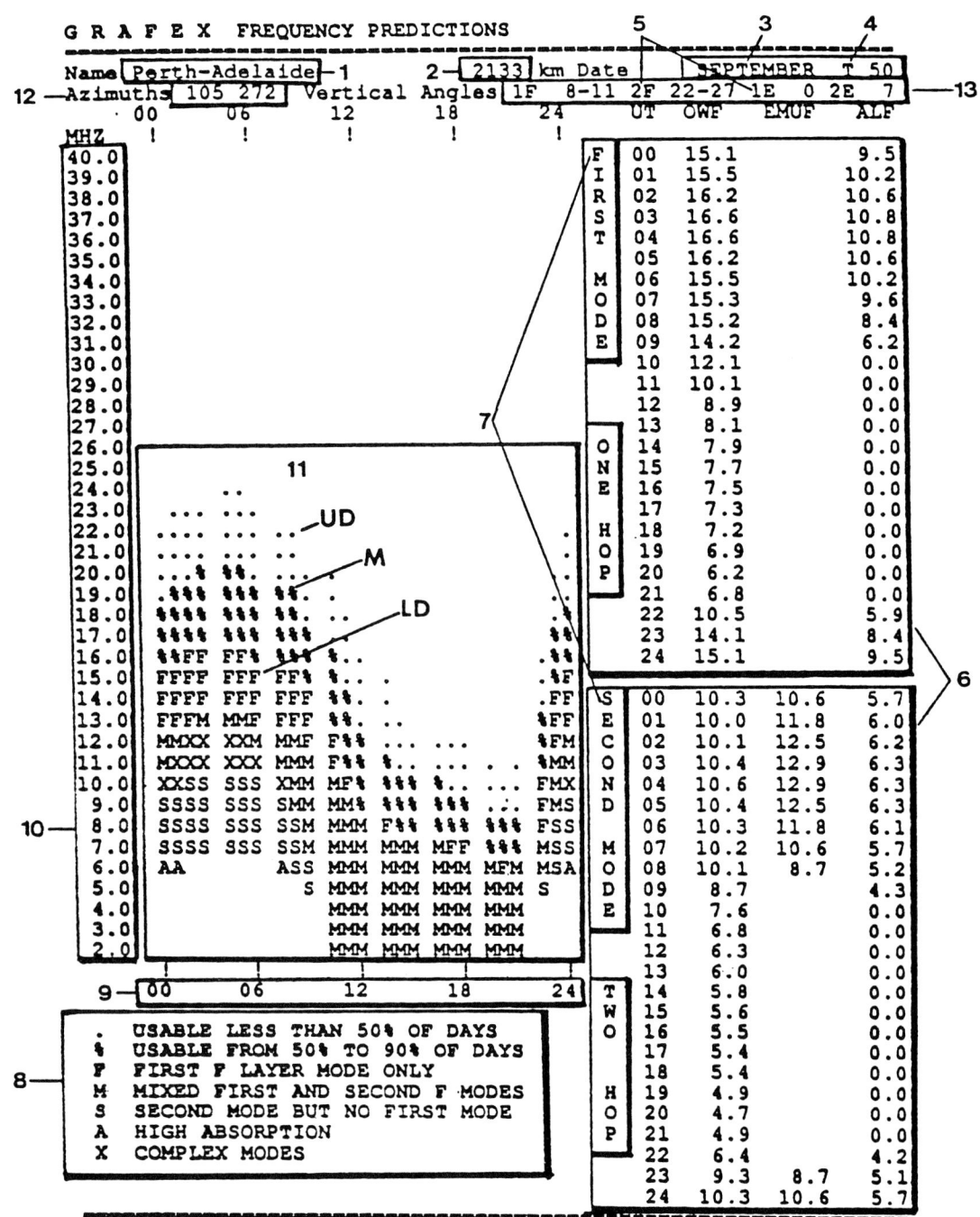

**Figure 6.7** A full GRAFEX circuit prediction for the Perth to Adelaide circuit—see text. The absorption limiting frequency (ALF) is very similar to the LUF, except that it does not take non-ionospheric factors into account. *Courtesy IPS Radio and Space Services.*

9. Horizontal axis. Universal time [Diurnal variation].
10. Vertical axis. Frequency in MHz.
11. Statistics. Upper decile, median, and lower decile of the MUF for the first F mode.
12. Azimuth. Geographic bearings of circuit (west terminal first).
13. Elevation angles. Estimated take-off angles for the radio waves.

The GRAFEX symbols/letters are:

| Symbol | Meaning |
|---|---|
| 'top blank' | frequency above upper decile MUF. Skywave communication possible on fewer than 3 days of the month. |
| . | frequency between upper decile and median MUF. Propagation possible on 3 to 15 days of the month via the first F-layer mode. |
| % | frequency between median and lower decile MUF. Propagation possible on 15 to 27 days via the first F-layer mode. |
| F | first F-layer mode usable for more than 27 days in month. |
| E | first E-layer mode usable for more than 27 days, plus 'top blank' or '.' conditions. |
| P | first E-layer mode usable for more than 27 days, plus '%'. |
| B | both first E- and F-layer modes possible for more than 27 days. |
| M | via first F-layer mode on 27 days, the second F-layer mode on 15–27 days, and possibly the first E-layer mode. |
| S | second mode only on 15–27 days, since the frequency is below the first mode ALF. |
| X | complex modes such as the second E-layer, higher order F-layer, and mixed E- and F-layer modes. |
| A | frequency near ALF, leading to high absorption of signals. |
| 'bottom blank' | frequency below the lowest ALF-skywave is completely absorbed. |

*Suppose that you are required to maintain successful communications on the Perth-Adelaide circuit during a September for which the ionospheric index is 50. Use Figure 6.7 to investigate your chances of doing so with allocated frequencies of 10 and 20 MHz.*

Since you have only two frequencies, we can surmise that you are a "small user," and will probably not have the world's best antennas. So we really cannot expect to use the E modes at all, because the elevation angles are so low. You would probably also have trouble using the 1F mode since the elevation angles are only 8° to 11°. This leaves you with the 2F mode, at an angle of 22° to 27°, for which suitable antennas are easy to come by. The predictions for the second mode show that the allocated frequency of 10 MHz would be supported between 00 and 08 UT. During most of this time, the ALF for the 1F mode is 10 MHz or higher, so only the 2F mode would be usable. This is indicated in the body of the GRAFEX by the letter S. Your 20 MHz frequency may be useful in the middle of the day, but only for less than about half the days of the month. You could not expect to have any communications on any of your allocated frequencies between 09 and 23 UT, which means that they were not a good choice for that month and level of solar activity. They were also not a good choice because these frequencies are used throughout the world for standard time signals, with high power transmitters.

## 6.6 Uses of HF Predictions

Although there is a lot to be said for the experience which HF communicators build up over the years, there is no substitute for timely and accurate frequency predictions. The serious user of HF predictions will use them for planning a communications network, for determining a frequency schedule, or for choosing between allocated frequencies.

### 6.6.1 Planning a Communications Link

Somewhere along the line, every communications link that is in use has been planned in one way or another. In planning a communications link, we must consider all aspects of the problem—the choice of transmitting and receiving sites, the choice of antennas, the choice of appropriate transmitters and receivers, the choice of operating frequencies, and the choice of transmitter power. We shall concentrate here on the choice of operating frequency.

Ideally, we would like to use each frequency as it becomes the best for our purposes, but this would lead to utter chaos in the HF band if we all did that. Every frequency channel occupies a bandwidth of 3 kHz in the HF band, so there are only 30 MHz / 3 kHz = $10^4$ channels available to satisfy everybody. Consequently the use of the HF band (and every other band from ULF to SHF) is heavily regulated both nationally and internationally, and we can use only the few frequencies allocated to us by the appropriate Frequency Allocation Board. One of our first aims must therefore be to ensure that the frequencies allocated to us are the best for our purposes.

We can determine what frequencies we need by the use of what are called **planning predictions.** This is a set of predictions in general containing values of MUF, LUF, angles of elevation, probability of propagation, and so on for four months (March, June, September, and December) to cover the variation of the ionosphere throughout a year, and for several levels of solar activity (typically low, medium, and high) to take account of the variation of the ionosphere with solar activity. We then sit down with a pencil and ruler and work out what frequencies we would require to maintain communications. This is usually a trial and error approach (even if it is done by computer) during which we make our decisions about how important it really is to ensure communications at a particular time. We would normally like to have at least a 90% chance of success, so we would choose a frequency below the OWF. A typical circuit requires a set of four frequencies to cover a whole solar cycle, with a fifth necessary if communications are essential during the predawn period in winter at solar minimum, when the OWFs are very low.

Having decided what frequencies we require, we then approach our national Frequency Allocation Board which does its best to give us what we need, within the constraint of the frequencies not having already been allocated to another user who could interfere with our operations (and vice versa). This is not an easy matter and what we ask for is not always what we get. The HF spectrum is a finite resource, and all frequencies have already been allocated many times over, even in the one country.

In practice, the large governmental organizations get more or less what they need while the smaller user must normally be content with two frequencies, a day frequency and a night frequency. A great deal can be done with just two frequencies, but it is not normally possible to ensure 24-hour communications. However this is often not required by the small user. If only one daytime frequency has been allocated, it may be necessary to have this changed as the solar cycle progresses. A frequency allocated for daytime use at high solar activity will probably not be useful at low solar activity, and vice versa.

#### 6.6.2 Selecting a Working Frequency

Once we have a set of frequencies allocated to us for use on our circuit(s), the next step is to choose which frequency to use at a given time. This is no problem if we have only two frequencies—during the day we use the higher of the two, while during the night we use the lower one. We could even change the frequency automatically in a real-time frequency management system (Chapter 10) just by using a photoelectric cell. If things go wrong during the day, such as an ionospheric storm (Chapter 8) or interference from another user, we could always try the night frequency but since this is probably below the daytime ALFs, we cannot expect much success.

If we have a set of frequencies to choose from, we can adopt several strategies. We could keep working a frequency until it drops out and then change to our next highest or lowest allocated frequency, depending on what time of day it is. Alternatively, we could stick rigidly to HF predictions for our circuit and change to another frequency when the predictions tell us to do so. As well as indicating a faith in the usefulness of HF predictions in general, this strategy could save a lot of hassle because both communicators have a predetermined plan of what frequency to use at a given time.

A third strategy is to throw the HF predictions away and rely solely on our experience to tell us what to do next. This option is not recommended.

### 6.7 Errors in Predictions

Many predictions made for HF communications will be wrong to some extent. This is the nature of the game. However we can and always should attempt to minimize the errors in our predictions. How do the errors arise? To answer this question, all we need to do is review the stages involved in making HF predictions :

1. Predict the general level of solar activity expected to exist at the time when the predictions are to be used.

2. Estimate what the corresponding gross behavior of the ionosphere will be.

3. Calculate the ionosphere and propagation geometry for the given circuit.

4. Calculate the propagation parameters such as MUF and LUF.

There will be errors introduced at all stages of this procedure.

Predicting the general level of solar activity becomes easier and more accurate as the required lead time decreases. If we wish to make predictions for a few hours ahead, we will have no problem predicting the general level of solar activity because we can safely assume that it will not change in that time. This assumes, of course, that we know what the level of activity is at the moment, but that is no problem in principle because there are organizations around the world which cooperate to keep a 24-hour watch on the sun [13]. The problems arise when we want to use longer lead times. Routine predictions are made several months ahead, mainly so that the HF organizations can distribute them to the appropriate sections within their organization and work out a reasonable plan to share the allocated frequencies in an efficient fashion. On this time scale, there is a good chance of introducing a reasonable error in predicting the general level of solar activity because absolutely no one knows yet how to tell what active regions are going to appear on the sun, with corresponding increases in ionizing EUV flux. Predictions are never made more than a few months ahead, except in the form of planning predictions which apply for different months and levels of solar activity but do not tell us to which particular months they apply. Is a March planning prediction for high solar activity any good for March 1990, March 2010, or what?

The errors introduced in the other stages of making HF predictions all depend on how clever we are and how much effort we wish to spend in setting up accurate models of the ionosphere and HF propagation. In general, these can be modelled to only a reasonable precision, and large efforts are still being made to improve the precision further.

One of the more interesting problems to arise is the choice of propagation modes. On many transequatorial circuits, which are those from a transmitter in one hemisphere to a receiver on the other side of the equator in the other hemisphere, it is quite common for the signals to travel by unusual propagation modes which have a higher MUF than the normal modes. Propagation still takes place by the normal modes as well. Figure 6.8 illustrates what happens when a comparison is made between predicted MUFs and **maximum observed frequencies** (MOFs) for a circuit between Yamagawa (southern Japan) and Townsville (northern Australia). There is an MOF for each propagation mode, each MOF corresponding to the observed value of the junction frequency at which the low and high rays merge. The frequency actually scaled from the observed OI Yamagawa to Townsville ionograms was the highest frequency at which propagation was observed to occur, regardless of the propagation mode. This means that the tabulated MOFs are a mixture of the MOFs for different propagation modes.

It is important when analyzing such data to allow for the existence of more than one propagation mode. For example, Figure 6.9 shows normalized histograms of the MOF scaled from OI ionograms for each UT hour on the Guam to N.W.Cape (Western Australia) circuit in June 1970. At each hour, there are 30 values for the MOF, each corresponding to the propagation mode which "carried the MUF" (supported the highest frequency) at the time the MOF was observed. The fact that there are several peaks in some of the histograms immediately tells us that simply calculating the median value of the MOF would give a median value corresponding to *neither* of the modes supported. During the day, the two modes supported were the normal $2F_2$ mode and an $E_sF_2$ mode [14]. During the night, both $2F_2$ and $3F_2$ modes were supported. On some nights, the $2F_2$ mode was not supported, or it had faded so that it was weaker than the $3F_2$ mode at the time the observation was made. This left the $3F_2$ mode as the mode supporting the highest frequency or MOF for the OI ionogram. Similarly, the $E_sF_2$ mode was not supported on some days at a given hour, leaving the $2F_2$ mode as the mode supporting the highest frequency.

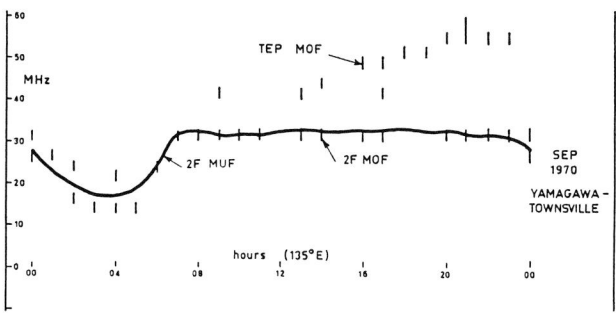

**Figure 6.8** The diurnal variation of the maximum observed frequency (MOF) and predicted maximum usable frequency (MUF) on the Yamagawa to Townsville circuit for September 1970. The figure shows the general agreement between the 2F MUF and 2F MOF, as well as the presence of a second propagation mode in the evening with an MOF about 10 to 20 MHz above the normal MUF. This mode has been identified as a TEP mode, as described in Chapter 9. *After McNamara (1974).*

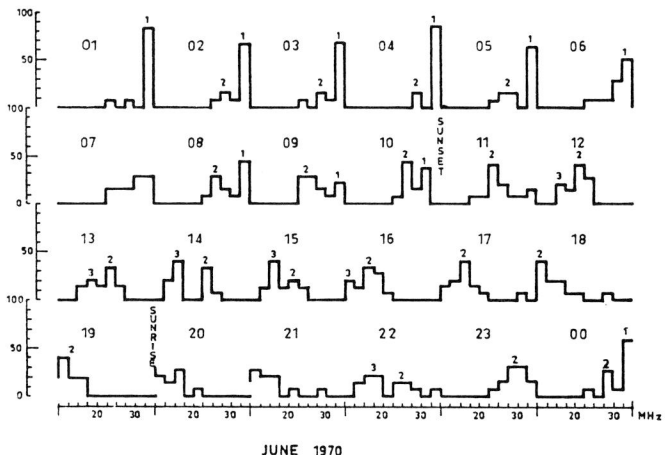

**Figure 6.9** Histograms showing the occurrence rate in 2.5 MHz frequency bins of the MOF for the Guam to North West Cape circuit in June 1970 (normalized to 100 events). *After McNamara (1974).*

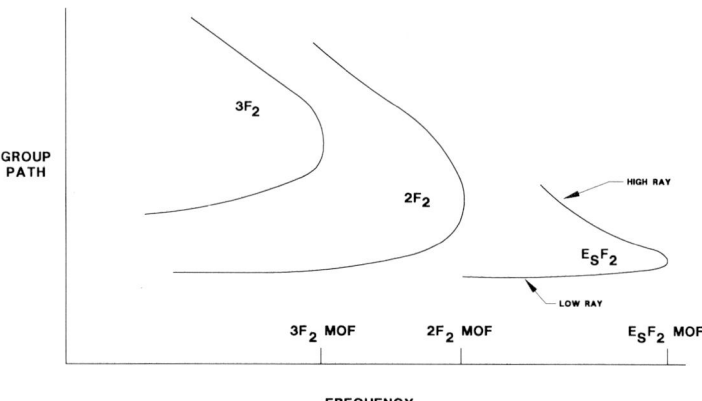

**Figure 6.10** Idealized oblique ionogram for the Guam to North West Cape circuit, illustrating the three propagation modes which appear to have existed on the circuit at different times—$3F_2$, $2F_2$, and $E_sF_2$. The $2F_2$ and $3F_2$ modes could be identified at night, while the $2F_2$ and $E_sF_2$ modes could be identified during the day. When the $E_sF_2$ mode is not supported during the day, the highest frequency observed on the ionogram would correspond to the $2F_2$ mode. Likewise, when the $2F_2$ mode is not supported at night, the MOF for the ionogram would correspond to the $3F_2$ mode.

Figure 6.10 shows an idealized ionogram with $3F_2$, $2F_2$, and $E_sF_2$ propagation traces, serving to illustrate how the ionogram MOF can correspond to different propagation modes as the higher frequency modes are successively disallowed.

The vertical bars in Figure 6.8 represent channels for which isolated peaks occurred in the Yamagawa-Townsville histograms. There is quite good agreement between the predicted 2F MUF and a set of peaks in the histograms which can be identified as corresponding to the 2F mode. Figure 6.8 also shows the existence of a second propagation mode in the afternoon and evening. This mode(s) is referred to as a TEP mode, and is discussed in Chapter 9. The comparison between predicted MUFs and MOFs can be a rather dull business, except when it leads to the discovery of hitherto unsuspected propagation modes [15, 16].

## 6.8 Field Strength of the Received Signal

We have concentrated so far on the problem of choosing the correct operating frequency and antenna (Chapter 4) for communications on a given circuit. While these are very necessary preconditions for successful communications, there remains the problem of ensuring that the signal reaching the receiver can be detected above the level of the background noise. The various types of noise are discussed in Chapter 7 and will not be discussed further here. Rather we shall take a brief look at the various factors which contribute to the strength of the signal reaching the receiver.

The strength of the signal at the receiver is controlled by four things—the transmitter power, the gain of the transmitting antenna at the particular frequency and elevation angle, the attenuation of the signal between the transmitting and receiving antennas, and the gain of the receiving antenna. Our main interest here is the loss of signal strength between the two antennas, which is called the pathloss.

The **pathloss,** L, is the total loss of signal strength along the path [17], and is made up of six main contributing factors, which are listed in Table 6.1, along with typical values. By far the largest loss is the free-space loss, the signal strength falling off as $1/D^2$, where D is the distance along the propagation path (note—not the ground range). The next most important is usually the absorption loss.

The **free-space loss,** $L_b$, is equal to $20\log_{10}(4\pi D/w)$, where w is the wavelength of the signals, measured in the same units as D.

The **absorption loss,** $L_a$, is made up of deviative and non-deviative components, as described in Chapter 4. The former is important for vertical incidence reflection from the E layer, especially near $f_oE$, but the latter is the important component for long distance F-region circuits. Non-deviative absorption depends inversely on the square of the operating frequency and has other variations described in Chapter 4. In practice, it is derived from empirical maps of vertical incidence absorption measured at a standard frequency of around 2.2 MHz, since calculations based on models of the neutral particle and electron densities lead to very unreliable estimates [18].

The **ground reflection loss,** $L_g$, depends on the conductivity and dielectric constant of the Earth at the reflection point. It is less for sea reflections than for dry ground reflections (typically 0.5 and 3 dB).

The **polarization coupling loss,** $L_p$, is the loss of power pickup because of the difference between the polarizations of the received wave and the receiving antenna [19]. It depends on the orientation of the incident wave at the ionosphere relative to the Earth's magnetic field, and the polarization of the antennas (Chapter 4). In the worst case, $L_p$ can actually be the controlling influence on the received signal strength.

The **sporadic-E obscuration loss,** $L_q$, depends very much on the value of the critical frequency of any $E_s$ layer at the reflection point, $f_oE_s$, relative to the operating frequency. The higher the value of $f_oE_s$, the higher the probability that the signals will be prevented from reaching the F layer.

# Predictions For HF Communications

**Table 6.1** The components of the total pathloss between transmitting and receiving antennas for an HF circuit. *After IPS Explanatory Leaflet No. 4.*

| Symbol | Meaning | Typical Range of Values (dB) | Comments |
|---|---|---|---|
| $L_a$ | Absorption Loss | 0-20 per hop | Non deviative and deviative<br>$L_a \propto 1/f^2$<br>f = frequency |
| $L_b$ | Basic free space transmission loss | 50-120 | $20 \log(4\pi d/\lambda)$<br>d = slant path length<br>$\lambda$ = signal wavelength |
| $L_g$ | Multihop ground reflection loss | 1-3 per reflection | Function of conductivity and dielectric constant of the earth |
| $L_p$ | Polarization coupling loss | 3-6 | Ionosphere splits initial wave into two oppositely polarized waves |
| $L_q$ | Sporadic E obscuration loss | 0-1 | Function of sporadic E critical frequency $f_oE_s$ |
| $G_f$ | Horizon focus gain | 0-9 | Large for small angles |

The **horizon focus gain,** $G_f$, is a gain (we do not get many of these!), which is largest for low elevation angles. It is due to the arrival at the receiver of rays from a cone of elevation angles at the transmitter, rather than from just one elevation angle.

The calculation of L requires an accurate model of the ionosphere and an iterative search for the propagation modes which link the transmitter and receiver, without many of the simplifying assumptions made when calculating just the MUF. As such, it is a time-consuming and fairly expensive procedure. However the calculations are necessary when a circuit is first being engineered, since the results tell us what antennas and transmitter power are required. Once the pathloss and elevation angles have been calculated, it is a simple matter to choose a suitable antenna. The calculation of the required transmitter power involves a consideration of the noise levels which must be overcome, and the required fading margin, both of which are described in Chapter 7.

The pathloss can be used to determine the strength of a signal at the receiver, if the transmitter power is known, in terms of

1. Power
$$P = P_t + G_t(\beta) + G_r(\beta) - L$$

2. Amplitude or field strength (independent of the effective area of the receiving antenna)
$$E = 107.2 + 20 \log_{10} f + P_t + G_t(\beta) - L,$$

where

$P$ = available power at the output of the receiving antenna (dBW)

dBW = dB above 1 watt

$E$ = field strength at the input to the receiving antenna (dB$\mu$)

dB$\mu$ = dB above $1\mu$V/m

$f$ = transmitter frequency in MHz

$P_t$ = transmitter power in dBW

$G_t(\beta)$ = transmitting antenna gain in dB at an elevation angle $\beta$

$G_r(\beta)$ = receiving antenna gain

$\beta$ = elevation angle of the propagation mode

$L$ = predicted total pathloss in dB.

The **signal-to-noise ratio** (S/N) at the receiver can be computed from the predicted pathloss and estimated HF noise level. The noise taken into account in S/N predictions includes atmospheric and galactic noise, but usually not man-made noise, since this is so site-specific. The noise data is obtained from world maps of radio noise for a given location, season, local time, and frequency [20]. A typical map is given as Figure 7.8.

The **noise/pathloss parameter** (N/P) seems an unlikely parameter at first sight, but is sometimes used because it is independent of the system configuration [21]. The N/P is defined as the negative of the sum of the noise power (dB) and pathloss: N/P = − (Noise + Pathloss). It is positive when the absolute magnitude of the noise power is greater than the magnitude of the pathloss. The N/P figures refer to a bandwidth of 1 Hz. For optimum propagation conditions, the N/P values are large. The relationship between transmitter power, S/N ratio, and the N/P parameter is

$$P_t + G_a = (S/N)_b - (N/P)_1 + B \text{ (dBW)},$$

where

$G_a$ = total antenna gain in the direction of propagation

$(S/N)_b$ = S/N ratio for a receiver bandwidth of b Hz

$(N/P)_1$ = noise/pathloss for a receiver bandwidth of 1 Hz

$B = 10 \log_{10} b$, where b is the receiver bandwidth in Hz.

Once the right-hand side of this equation is given, the antenna gain and transmitter power can be traded against each other. Alternatively, if the left-hand side is fixed, then the expected S/N can be estimated and, by comparison with the standard CCIR required S/N ratios [22], the expected grade of service or its availability over the time considered can be determined. The following sections give two examples of the use of N/P predictions. Further examples are available elsewhere [23, 24].

### 6.8.1 Estimation of Required Transmitter Power

*Estimate the required transmitter power to ensure 50 baud F1B telegraphy, with 1 in 1000 character error rate, over a 24-hour period. The transmission frequency is 10 MHz.*

Suppose for simplicity that the elevation angle for the circuit is 10° for the whole 24-hour period, and that the N/P figures have been calculated for the appropriate circuit, season, and ionospheric index. The required S/N is given by CCIR Recommendation 339-5 as 63 dB into a 1 Hz bandwidth under fading conditions with non-diversity (see Chapter 7) reception. Hence

$$P_t + G_a = 63 - N/P \text{ (dBW)}.$$

This means that the transmitter power plus antenna gain must exceed all predicted values of (63 − N/P) throughout the 24-hour period in order to maintain the specified grade of circuit. The minimum transmitter power required can best be illustrated by plotting the predicted values of (S/N − N/P + B), as in Figure 6.11. If the possible transmitter powers are marked on the same diagram, the lowest power that exceeds all of the (S/N − N/P + B) values can be chosen. Thus in the present example, a transmitter of 300 W would be sufficient. Note also that if the antenna at each end had a gain of g dB at the 10° elevation angle, then 2g dB could be subtracted from the (S/N − N/P + B) values, and a (25 − 2g) dBW transmitter would be adequate.

### 6.8.2 Estimation of Available S/N

*Assuming the availability of appropriate N/P figures, estimate the S/N available over a given circuit for the period 0500–0800 UT during December 1983, at a frequency of 10 MHz. The effective radiated power is 300 W. Suppose that the antenna gains at the predicted elevation angles are 6 dB for the transmitting antenna and 3 dB for the receiving antenna.*

We have $G_a = 9$ dB and $P_t = 25$ dBW (10 $\log_{10} 300$), so that for a 100 Hz bandwidth,

$$S/N = N/P + 25 + 9 - 20 = N/P + 14 \text{ dB}$$
$$= 34 + N/P - B \text{ dB}.$$

By comparison with the CCIR recommended S/N ratios, it is found that this circuit will support 50 baud FSK telegraphy (F1 transmission) with character error rates less than 1 in $10^5$ under stable conditions, provided N/P > 22 dB (recommended S/N is 36 dB). However under fading conditions, the error rate could reach 1 in

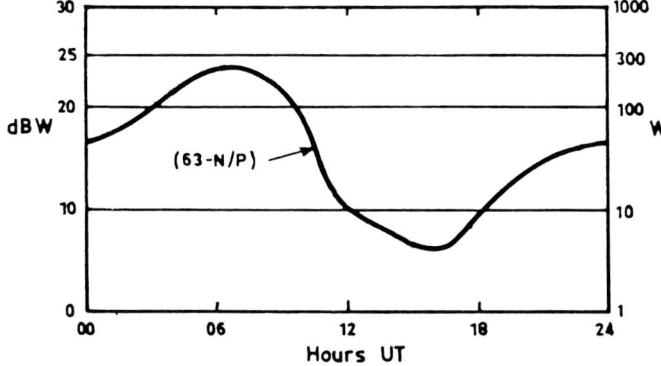

**Figure 6.11** Diurnal variation of the predicted noise/pathloss figure for example 1 in the text. *Courtesy IPS Radio and Space Services.*

$10^2$ for the same N/P (recommended S/N to achieve this error rate is 33 dB).

## 6.9 "Now-casting" the Ionosphere

As we saw in Chapter 3, the ionosphere is monitored routinely on a grand scale. Traditionally, the plot of virtual height versus frequency was displayed on a cathode ray tube and photographed, giving an ionogram. The film was processed and the ionogram analyzed by hand to determine features such as $f_oE$, $f_oF_1$ and $f_oF_2$, the presence of $E_s$, and so on. This "scaled" data was then entered into a computer and made available for later studies of the ionosphere.

The time scale for this chain of events is far too long for the serious communicator who wants to know what is going on *now* [25]. However recent advances in modern technology have permitted the manufacture of ionosondes so reliable and versatile that it is now possible to monitor the ionosphere in real time at a useful number of stations. A picture can thus be built up of what the ionosphere is doing when we want to use it, rather than having to rely on predictions made of the average behavior of the ionosphere a few days or weeks in advance. In other words, instead of forecasting the state of the ionosphere, we *now-cast* it—hence the name. The purist may prefer the term "real-time specification of the ionosphere."

In the simplest situation, a base station wishing to maintain the best possible communications with field stations out to a few hundred km away could use an ionosonde at the base station to obtain ionograms. Using the rule of thumb [26] that the layers of the ionosphere do not change much over a horizontal distance equal to the height of the layer, the overhead echoes can be used to characterize the E layer out to a radius of about 100 km and the F layer out to about 300 km.

The use of an ionosonde at the base station has several advantages:

1. In conjunction with HF predictions, it will indicate whether or not the critical frequency $f_oF_2$ is at the expected level or depressed because of an ionospheric storm (described in Chapter 8) or a general lowering of the level of EUV flux from the sun. If the $f_oF_2$ is clearly lower than the expected value, it can safely be assumed that it will be depressed over a region of at least 300 km radius.

2. It can assure us that the ionosphere is still there for us to use, and has not been rendered unusable by large increases in absorption due to solar flares (see Chapter 8 for SWF and PCA).

3. The ionosonde can be used to help select the operating frequency. Multipath interference (described in Chapter 7) is detrimental to HF communications, so it is best to choose a frequency that is supported by only one propagation mode.

4. An ionosonde will also indicate the presence of an $E_s$ layer. When a very dense $E_s$ layer is present, propagation will be via $E_s$ modes which have better propagation characteristics than the normal F layer (no fading, no Doppler shift, strong signal strength) and offer the possibility of faster and more reliable transmission of data.

For long distance communications, the ideal place to have an ionosonde is at each of the expected ionospheric reflection points. This fortunate state of affairs does not happen very often, so we are reduced to the less attractive option of using an ionosonde at one location to infer what the ionosphere is doing at some other location. One technique which can be adopted for this purpose is to use calibration graphs for that location of $f_oF_2$ versus sunspot number, or ionospheric index, to determine what sunspot number would have been expected to lead to the observed value of $f_oF_2$. This "effective" sunspot number or index is then used [27, 28], in conjunction with calibration graphs for the desired location, to derive the expected value of $f_oF_2$ at that location. There are many variations of this technique, which we shall return to in Chapters 13 and 15. The ideal situation is one in which a network of ionosondes controlled by a central coordinator covers the whole part of the ionosphere in which reflections take place. The problem then becomes similar to that of weather mapping and forecasting.

## 6.10 Sources of HF Predictions and Solar-terrestrial Forecasts

CCIR Recommendation 313 lists organizations which provide routine HF propagation predictions, and real-time specifications of solar activity and its resulting effects on the Earth's magnetic field, the ionosphere, and HF communications [29]. The international watch on the sun is coordinated by the International Ursigram and World Days Service (IUWDS) of the International Union for Radio Science (URSI).

## 6.11 References

1. Szuszczewicz, E. P. *The Modelling, Measurement and Predictability of the Global-Scale Ionosphere: Current Perspectives and Future Needs.* Solar-Terrestrial Physics Workshop, Leura, 1989.

2. Daniell, R. D., D. T. Decker, D. N. Anderson, J. R. Jasperse, J. J. Sojka, and R. W. Schunk, *A Global Ionospheric Conductivity and Electron Density (ICED) Model.* Ionospheric Effects Symposium IES-90, Washington, D.C., May 1990.

3. Davies, K. *Review of Recent Progress in Ionospheric Prediction.* Radio Science 16(6), pp. 1407–1430, 1981.

4. Jursa, A.(Ed). *Handbook of Geophysics and the Space Environment,* Air Force Geophysics Laboratory, Air Force Systems Command. NTIA Accession Number ADA 167000, 1985.

5. Rawer, K., and D. Bilitza. *Electron Density Profile Description in the International Reference Ionosphere.* J. Atmos. Terr. Phys., 51(9/10), pp. 781–790, 1989.

6. Rush, C. M. *Ionospheric Radio Propagation Models and Predictions—A Mini-Review.* IEEE Trans. Ant. & Prop., AP-34(9), pp. 1163–1170, September 1986.

7. Sojka, J. J. *Global Scale, Physical Models of the F region Ionosphere.* Rev. Geophysics, 27(3), pp. 371–403, August 1989.

8. Schunk, R. W., and E. P. Sczeczewicz. *First-Principle and Empirical Modelling of the Global-Scale Ionosphere.* Ann. Geophysicae 6(1), pp. 19–38, 1988.

9. Tascione, T. F., H. W. Kroehl, R. Creiger, J. W. Freeman, Jr, R. A. Wolf, R. W. Spiro, R. V. Hilmer, J. W. Shade, and B. A. Hausman. *New Ionospheric and Magnetospheric Specification Models,* Radio Science 23(3), pp. 211–222, 1988.

10. Goodman, J. M. *Shortwave Propagation Prediction Methodologies.* IEEE Trans. Broadcasting, 34(2), pp. 230–237, June 1988.

11. Fox, M. W., and L. F. McNamara. *Improved World-Wide Maps of Monthly Median foF2.* J. Atmos. Terr. Phys. 50(12), pp. 1077–1086, 1988.

12. ASAPS. *Advanced Stand-Alone Prediction System.* IPS Radio and Space Services, Sydney, 1989.

13. CCIR Recommendation 313-5. *Exchange of Information for Short-Term Forecasts and Transmission of Ionospheric Disturbance Warnings.* Dubrovnik, 1986.

14. McNamara, L. F. *Ionospheric Predictions on Transequatorial Circuits.* Proc. IREE (Aust), pp. 117–126, May 1974.

15. McNamara, L. F. *The Accuracy of MUF Predictions within Australia,* Ionospheric Prediction Service Report IPS-R28, Sydney, November 1975.

16. McNamara, L. F. *Ionospheric Predictions on Transequatorial Circuits. loc. cit.*

17. CCIR Report 252-2. *CCIR Interim Method for Estimating Sky-Wave Field Strength and Transmission Loss at Frequencies between the Approximate Limits of 2 and 30 MHz.* New Delhi, 1970.

18. Ferguson, B. G., and L. F. McNamara. *Calculation of HF Absorption using the International Reference Ionosphere.* J. Atmos. Terr. Physics, 48(1), pp. 41–49, 1986.

19. Davies, K. *Ionospheric Radio.* Peter Peregrinus Ltd., London, 1990.

20. CCIR Report 322-3. *Characteristics and Applications of Atmospheric Radio Noise Data.* Dubrovnik, 1986.

21. Explanatory Leaflet No. 4. *Pathloss, Noise and Noise/Pathloss Predictions.* IPS Radio and Space Services, Sydney, June 1989.

22. CCIR Recommendation 339-5. *Bandwidth, Signal-to-Noise Ratios and Fading Allowances in Complete Systems.* Geneva, 1974.

23. Braun, G. *Planning and Engineering of Shortwave Links,* Siemens Aktiengessellschaft, Heyden & Son Ltd., London, 1982.

24. Kirby, R. C. *Radio-Wave Propagation.* Section 18 in *Electronics Engineer's Handbook,* D. G. Fink and D. Christiansen (Eds.), pp. 18–50 to 18–126, McGraw Hill Book Company, New York, 1982.

25. Kelly, T. D. *The Need for Real-Time Remote Control of Vertical Incidence Ionosondes.* Proceedings of the Workshop on Solar Terrestrial Physics, Meudon, June 1984.

26. It is believed that the first thumb to be used in this respect belonged to W. R. Piggott, a leading figure in ionospheric physics for over 40 years.

27. Reilly, M. H., F. F. Rhoades, J. M. Goodman, and M. Singh, *Updated Climatological Model Predictions of Ionospheric and HF Propagation Parameters.* Ionospheric Effects Symposium IES-90, Washington, D.C., May 1990.

28. McNamara, L. F. *The Use of Ionospheric Indices to Make Real and Near Real Time Forecasts of $f_oF_2$ Around Australia.* Solar-Terrestrial Predictions Proceedings. Vol. 1: Prediction Group Reports. R. F. Donnelly (Ed.), Space Environment Laboratory, Boulder, Colorado, pp. 249–258, 1979.

29. CCIR Recommendation 313-5. *Exchange of Information for Short-Term Forecasts and Transmission of Ionospheric Disturbance Warnings. loc. cit.*

## 6.12 Problems

1. List briefly the four main steps followed in making predictions for HF communications.

2. What propagation modes should be considered for circuits of 5,000 km, 7,000 km, and 12,000 km ?

# Predictions For HF Communications

3. If you receive HF predictions, how would you go about selecting the best frequency to use?

4. What are the factors affecting the $F_2$ region which are not important for the E region?

5. Why would one use maps of $f_oF_2$ plotted at a fixed local time? Are such maps "real"?

6. What sets the lower and upper limits of HF propagation?

7. Assume your task is to maintain HF communications on a fixed point-to-point circuit "under all conditions." How many frequencies would you require to have allocated to you?

8. What are the two phenomena contributing most to path loss on an HF circuit?

9. Explain briefly the advantages of "now-casting" the ionosphere.

10. What are the advantages to an HF communicator in having a VI ionosonde?

11. Use the IPS GRAFEX predictions in Figure 6.12 to determine the minimum number of frequencies you would require to maintain communications between Boulder and St. Louis during December 1990, assuming that you want a success rate of 90% or more. What frequency and what antenna would you use to ensure that only one propagation mode will be supported during the day? What is the nighttime value used for $f_oE$?

12. Use the IPS field-strength predictions (from the program ASAPS) given in Figure 6.13 to answer the following questions:

    a. The best usable frequency (BUF) out of the frequency set (the amateur frequencies) is given as 21 MHz between 17 and 22 UT. Is this consistent with your answer to question 6.11?
    b. If you always work at the OWF, what is the minimum power you would require to maintain a fading margin of 10 dB?

13. Why are the noise levels lower during the day than during the night, as indicated by Figure 6.13?

```
=================================================================================
IPS GRAFEX PREDICTIONS ----------------- ADVANCED STAND ALONE PREDICTION SYSTEM
=================================================================================
Circuit: boulder-st louis        Distance: 1299 KM     Date:    December,1990
Tx: boulder         40.03  254.73  Bear: 92   281      T-index: 152
Rx: st louis        38.67  269.75  Path: Short Path    Circuit#: 1
=================================================================================
First Mode                                                          Second Mode
1F 17-19 1E  5    |--------F r e q u e n c y  (MHz)---------|  2F 35-38 2E 14
UT  OWF EMUF  ALF  1...5...10 ...15...20 ...25...30 ...35...40  OWF EMUF  ALF UT
00 17.6  2.6  0.0  XMMMMMMMM MMFFFF%%  .                        11.4  1.6  0.0 00
01 15.0  2.6  0.0  XMMMMMMMM FFFF%%..                            9.8  1.6  0.0 01
02 12.0  2.6  0.0  XMMMMMMFF F%%.                                7.8  1.6  0.0 02
03  9.1  2.6  0.0  XMMMMFFF%                                     6.3  1.6  0.0 03
04  7.7  2.6  0.0  XMMMMMF%.                                     5.6  1.6  0.0 04
05  7.0  2.6  0.0  XMMMMF%.                                      5.2  1.6  0.0 05
06  6.7  2.6  0.0  XMMMMF%.                                      5.1  1.6  0.0 06
07  6.7  2.6  0.0  XMMMMF%.                                      5.0  1.6  0.0 07
08  6.5  2.6  0.0  XMMMMF%.                                      4.8  1.6  0.0 08
09  6.2  2.6  0.0  XMMMMF%.                                      4.6  1.6  0.0 09
10  5.8  2.6  0.0  XMMMM%%.                                      4.4  1.6  0.0 10
11  5.6  2.6  0.0  XMMMM%%.                                      4.2  1.6  0.0 11
12  5.6  2.6  0.0  XMMMM%%.                                      4.2  1.6  0.0 12
13  6.7  2.6  0.0  XMMMMF%%.                                     4.6  1.6  0.0 13
14 12.2  5.3  3.3   AMMMMMFF FF%..                               7.1  0.0  2.6 14
15 18.5 10.7  6.0    ASMMMMM MMFFFFFF%% %...                    11.2  6.0  3.6 15
16 20.9 12.6  6.9    ASSXMMM MMMMMFFFF %%%....                  13.2  7.4  4.0 16
17 22.8 13.6  7.3     SSSXMM MMMMMFFFF FF%%..                   14.4  8.1  4.3 17
18 23.2 14.1  7.5     SSSXMM MMMMMFFFF FFF%%.                   14.9  8.5  4.3 18
19 23.0 14.1  7.5     SSSXMM MMMMMFFFF FF%%..                   14.8  8.5  4.3 19
20 22.8 13.7  7.3     SSSXMM MMMMMFFFF FF%%..                   14.2  8.2  4.2 20
21 21.8 12.6  6.9    ASSXMMM MMMMMFFFF F%%%..                   13.8  7.5  4.0 21
22 21.2 10.7  6.0    ASXMMMM MMMMMFFFF F%%%..                   13.6  6.1  3.6 22
23 20.5  5.6  3.5   AMMMMMMM MMMMFFFFF %%%..                    12.8  0.0  2.6 23
UT  OWF EMUF  ALF  1...5...10 ...15...20 ...25...30 ...35...40  OWF EMUF  ALF UT
=================================================================================
```

**Figure 6.12** IPS GRAFEX predictions for the Boulder to St. Louis circuit, for use with question 6.11.

```
========================================================================
Circuit: boulder-st louis         Distance: 1299 KM     Date: 15 December,1990
Tx: boulder          40.02   254.72   Bear: 91   281    T-index: 152
Rx: st louis         38.66   269.75   Path: Short Path  Circuit#: 1
FSet: amateur                         Antenna: ISOTROPIC  RxNoise: -148 dBW/Hz
Required S/N: 0 dB                    %Days: 90         Min. Angle: 3 deg.
Modes: 1F 1E 2F 2E       Best Usable Freq.   TxPwr: 1000W  BandWidth: 3.0kHz
========================================================================
UT  MODE  PROB  ANGLE  NOISE  S/N  SN@MUF  SN@OWF    BUF     MUF    OWF
00   1F    99    17    -33    39    43      42     14.000   20.3   17.6
01   1F    99    18    -33    39    41      40     14.000   16.9   15.0
02   1F    99    19    -30    35    38      37     10.000   13.3   12.0
03   1F    99    19    -27    31    35      33      7.000   10.1    9.1
04   1F    99    20    -25    30    31      30      7.000    8.5    7.7
05   1F    99    21    -25    29    31      29      7.000    8.0    7.0
06   1F    99    19    -22    23    30      29      3.000    7.8    6.7
07   1F    99    19    -21    22    32      30      3.000    7.8    6.7
08   1F    99    19    -20    21    33      30      3.000    7.5    6.5
09   1F    99    19    -20    21    33      30      3.000    7.5    6.2
10   1F    99    19    -20    21    32      28      3.000    7.3    5.8
11   1F    99    18    -23    24    32      29      3.000    7.1    5.6
12   1F    99    18    -25    26    33      31      3.000    7.1    5.6
13   1F    99    18    -25    26    35      32      3.000    8.1    6.7
14   1F    99    18    -33    36    41      39     10.000   14.0   12.2
15   1F    99    18    -36    39    43      42     14.000   21.1   18.5
16   1F    99    18    -34    35    43      42     14.000   23.9   20.9
17   1F    99    18    -37    41    43      42     21.000   25.4   22.8
18   1F    99    19    -37    41    42      42     21.000   25.2   23.1
19   1F    99    19    -37    41    42      42     21.000   25.0   23.0
20   1F    99    19    -36    40    43      42     21.000   24.9   22.8
21   1F    99    19    -36    41    43      41     21.000   24.6   21.8
22   1F    99    19    -36    42    43      42     21.000   24.4   21.1
23   1F    99    18    -33    38    44      43     14.000   23.7   20.5
========================================================================
```

**Figure 6.13** IPS field strength predictions for the Boulder to St. Louis circuit, for use with question 6.12.

# Chapter 7

# Communication Problems under Normal Conditions

## 7.1 Introduction

Even when things are going well, and we have done everything correctly, we can still encounter problems in the use of the ionosphere for communications. Some of these arise from basic properties of the normal undisturbed ionosphere which act as fundamental limitations to HF propagation and it is these problems which we shall consider here. Other problems arise from disturbances to the ionosphere caused by events on the sun, but we shall defer discussion of these to Chapter 8.

## 7.2 Fading

Fading is a repetitive rise and fall in signal level, the signal getting stronger and weaker in turn. It is described by its depth—whether it is shallow or deep—and by its rate—whether it is fast or slow. The rate usually increases with frequency. In general, deep and fast fading is a disaster, but slow and shallow fading can be tolerated. Under conditions of deep fading, the signal level can drop below the noise level and thus be lost. To avoid loss of signal, we need to ensure that the signal level is sufficiently above the noise on the average that even a deep fade will not cause loss of signal. In other words, we have to ensure an adequate **fading margin.** When an adequate fading margin is available, the amplitude changes may be eliminated by using an automatic gain control (AGC) circuit. Examples of fading signals are given in Figure 7.1 [1].

Fading may usually be attributed to one of four effects:

1. Movement of the ionosphere and changes of the propagation path length
2. Rotation of the plane of polarization of the wave
3. Variations of ionospheric absorption with time
4. Focussing and temporary disappearance of the signal due to "MUF failure" (i.e., the MUF drops below the operating frequency)

### 7.2.1 Multipath Fading

On any particular HF circuit, and for a given elevation angle, the transmitting antenna will effectively illuminate an area of the F region (called the Fresnel zone) about a few km wide, and rays (or waves, if you prefer to think in terms of waves rather than rays) reflected from all parts of that region will arrive at the receiving antenna. These rays will have random phases with respect to each other because the path lengths will all be slightly different, and the phases will change in response to the ever present fluctuations of the ionosphere. Thus

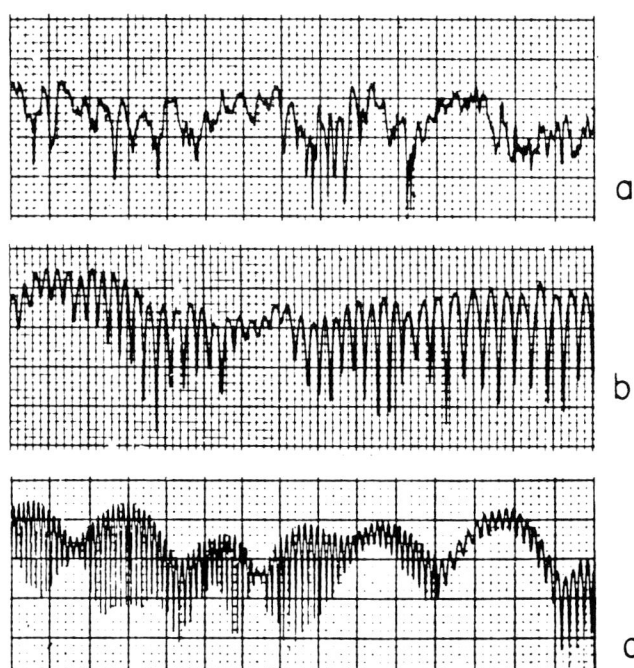

**Figure 7.1** Some examples of fading signals. The fading in panel (a) is random. The periodic fading in panel (b) is the resultant of two waves of almost equal amplitude whose phase difference is changing at a constant rate. Panel (c) illustrates combined slow and fast fades. The fast (short period) fading could be due to interference between the high and low rays, while the slow (long period) fading superimposed on the fast fading could be due to beating between the ordinary and extraordinary waves.

when the rays combine at the receiver, the resulting signal will fluctuate in intensity as the individual rays go in and out of phase. In other words, the signal will fade. This fading will not be too severe because the phase differences are small and there is little chance that the multitude of signals will cancel each other out at any time. There will therefore be a bundle of rays, with changing overall amplitude, which can be considered as *the* ray for the given elevation angle.

The situation is, however, quite different when we have only two rays, travelling by two different propagation modes. It is then possible for the rays to interfere with each other to such an extent that they will occasionally cancel each other out completely, leaving no signal to be detected by the receiver.

If the ionosphere were a rock-steady reflector, the phase difference between any two propagation modes would remain fixed and the net result of combining two signals at the receiving antenna would remain constant. However the ionosphere is a moving and irregular medium, so the path lengths on any propagation mode are continually changing by small amounts. The difference between the two path lengths can change by half a wavelength or more. Thus when the two signals combine, the resulting signal will fade as the two signals interfere with each other, alternatively canceling and reinforcing each other. This is called **interference fading.** The period of this type of fading is usually relatively short, about a second or two. Fast interference fading is called **flutter fading.** If the fading rate changes rapidly with frequency, it is known as **selective fading** and can produce distortion of the modulation envelope and "garbling" due to loss of information, if the carrier fades to a level below that of the side-bands.

Multipath fading can be expected to be severe whenever 1F and 2F modes, or 2E and 1F modes, exist on the same circuit, because the ray paths will be very nearly, but not quite, equal in length. Every time the difference in path length becomes equal to half a wavelength, the two signals would be exactly out of phase, and would cancel each other out to a large extent. The cancellation would be complete if the amplitudes of the two rays were equal.

Different signal modulation techniques are affected in different ways by multipath fading. Analogue voice transmissions are relatively insensitive to selective fading, whereas digital data transmission is strongly affected. Selective fading is very severe on circuits for which the ground and sky waves are of comparable intensity.

The slow and deep fading observed at night on signals from distant MF broadcast stations is due to interference between signals travelling along slightly different raypaths. The extraordinary ray is much weaker than the ordinary ray at these frequencies, because they are near the electron gyro-frequency, so there can be no O-X interference. Instead, the fading can be traced to the nature of the nighttime E region. The nighttime E region does not completely disappear, but drops to about $f_oE = 0.6$ MHz. This is a sufficiently high plasma frequency to reflect MF waves up to 1.8 MHz on circuits of 600 km or greater, and the lower parts of the broadcast band on still shorter circuits. The nighttime E region is also believed to be much more irregular then the daytime E region, leading to a family of reflected waves for a single incident wave, rather than just a single wave. The irregularities in the E region tend to be aligned parallel to each other, so the reflected rays interfere with each other, giving rise to an interference pattern which moves slowly over the ground as the irregularities move bodily in a horizontal direction, placing a receiver in a null or signal maximum alternately. Figure 7.2 shows the fading observed on three circuits to Jervis Bay in Australia [2]. The circuits to the MF broadcast stations 2BL, 4QG, and 3LO (frequency around 800 kHz) were 143, 870, and 590 km respectively.

### 7.2.2 Polarization Fading

We saw in Chapter 4 that every plane polarized wave which enters the ionosphere can be considered as splitting into two components, called the ordinary and extraordinary waves, which are circularly polarized in opposite directions and propagate independently. These waves keep the same polarization as they propagate through the ionosphere, but the phase difference between them gradually increases as the waves propagate. Thus when the two waves recombine on leaving the ionosphere, the plane of polarization of the plane polarized wave will have rotated with respect to the original plane of polarization of the transmitted wave. The amount by which the plane of polarization has rotated depends on the total number of electrons along the path between the transmitter and receiver and will thus change during the day. This phenomenon is known as **Faraday rotation,** after Michael Faraday, who discovered the effect in optical experiments.

Polarization fading is the result of changes in the state of polarization of the composite (ordinary and extraordinary) wave relative to the orientation of the receiving antenna. When the electric field of the composite wave points in the same direction as the antenna, which we can consider for simplicity to be a straight length of wire, the induced voltage will be a maximum. On the other hand, when the electric field is at right angles to the antenna, the induced voltage will drop to zero. If the incoming wave at the receiver kept its same polarization, the

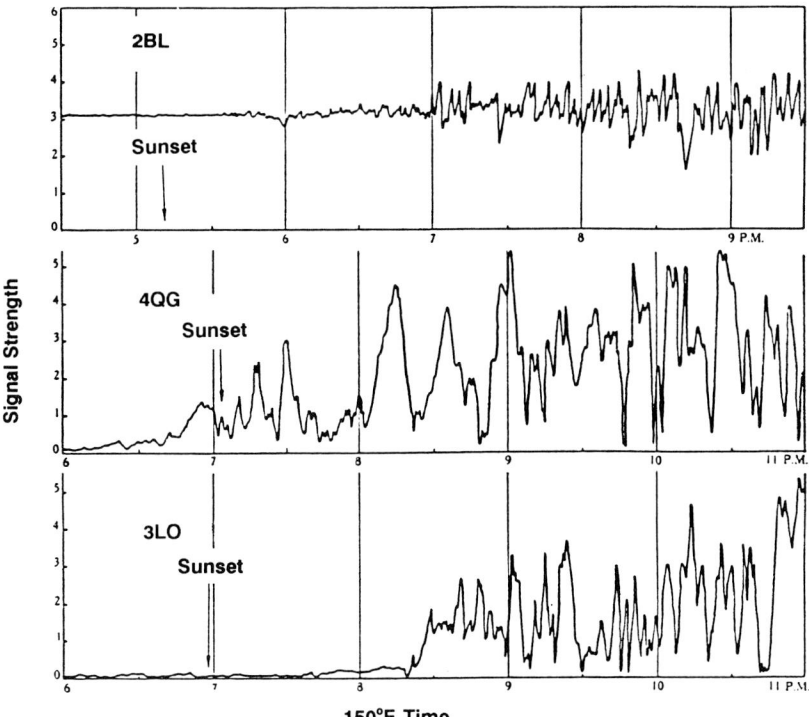

**Figure 7.2** Medium frequency broadcast signals showing the lack of signal during the day and a slowly fading signal at night. From top to bottom, the circuit lengths are 143, 870, and 590 km. This is Figure 6 in Green's work (1946), which had the caption "Natural fading of medium-frequency broadcast signals showing the sunset effect at both short and long distances from the transmitter."

induced voltage would remain fixed. However as the number of electrons along the path between the transmitter and receiver changes, the plane of polarization will rotate and the induced voltage will alternatively increase and decrease, passing through zero whenever the planes of polarization of the wave and antenna are at right angles.

Polarization fading can occur whenever there are changes in the total number of electrons along the path. This includes irregular changes, as well as the expected diurnal changes. An example of polarization fading is given in Figure 7.3 [3]. Deep fading can be avoided by the use of two antennas polarized at right angles. In this way, if a wave is inducing zero voltage in one antenna, it is inducing maximum voltage in the other.

### 7.2.3 Skip Fading

Skip fading, or MUF fading, occurs when the receiver is on the edge of the skip zone for propagation from a given transmitter. This means that the operating frequency is exactly equal to the MUF for the circuit. However as the ionosphere fluctuates, so does the MUF, and at times the actual MUF will drop just below the operating frequency and the signal will disappear. Another way of looking at this is that the skip distance will lengthen and shorten as the ionosphere changes, alternatively covering and not covering the receiver. The signal strength does not in fact drop immediately to zero as the skip zone covers the receiver. Just outside the skip zone, the amplitude oscillates because of interference between the four rays which arrive at the receiver—the high and low angle ordinary and extraordinary rays. Right on the edge of the skip zone, **skip focussing** occurs, increasing the signal amplitude by typically 6 to 9 dB. If we were to increase our operating frequency from well below the MUF up to the MUF itself, we could expect to see the following sequence of events as the frequency increases [4]:

1. First, there is long-period fading as the ordinary and extraordinary rays interfere, the rate of fading increasing (the period decreasing) with frequency.

2. A rapid fading follows, due to interference between the low and high angle rays, superimposed on the slow fading.

3. As the frequency increases, the slow fading gets faster, and the fast fading gets slower.

4. The fading rate decreases as the operating frequency passes through the MUF.

### 7.2.4 Absorption Fading

Absorption fading is due to inhomogeneities in the lower ionosphere, from where the HF absorption arises,

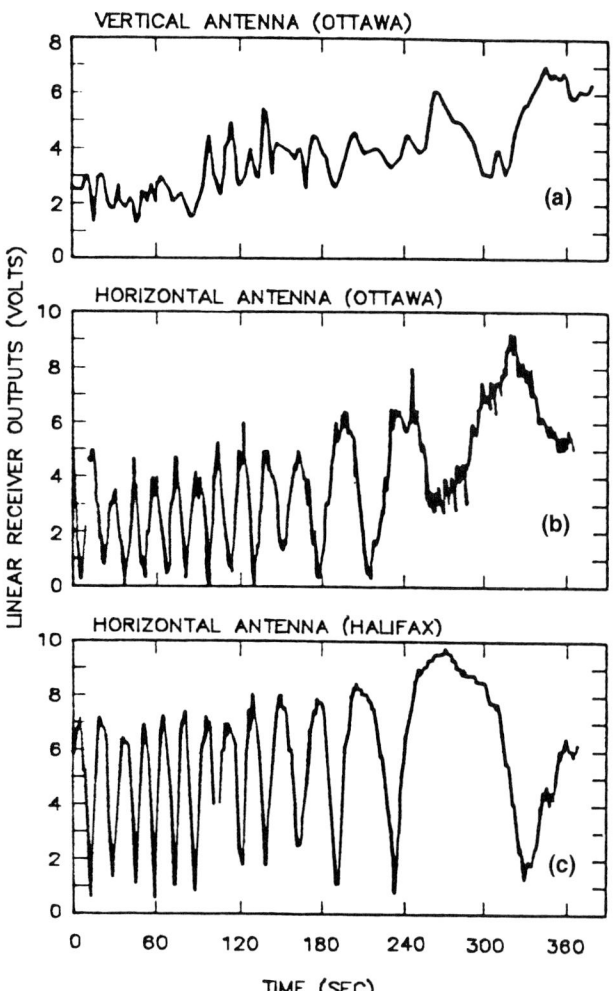

**Figure 7.3** The polarization fading observed at the vertical and horizontal antennas on a 1000 km path, Ottawa to Halifax. Panels (a) and (b) illustrate a much more severe fading problem on the horizontal antenna than on the vertical one. Panels (b) and (c) illustrate the fact that polarization fading is not usually the same at both ends of a circuit. In other words, polarization fading is usually nonreciprocal. In this case, the fades at Ottawa and Halifax are almost antiphase.

because of changes in the background atmosphere in which the ionosphere is embedded. The period is very irregular, and can be an hour or more.

### 7.2.5 Scintillations

Scintillations are rapid fluctuations in amplitude, phase, and angle of arrival imposed by irregularities in the ionosphere on signals passing through it, and are the radio wave equivalent of "Twinkle, twinkle, little star." They are most intense in the equatorial and polar regions, occurring only at night in the former. The fading has adverse, and often severe, effects on satellite communication systems. It is usually fairly slow and deep, with a few fades per minute. It can also cause increases in the signal strength, but these are not important on an operational circuit. The amplitude of the scintillations decreases with frequency, but is still significant in the GHz band currently used for satellite communications systems. Figure 7.4 illustrates the "worst case" fading at 1.2 GHz [5]. Fading depths can exceed 20 dB in the early evening for signals traversing the peaks of the equatorial anomaly at solar maximum.

### 7.2.6 Some Solutions

Even when a signal fades severely, no great problems will be encountered provided the fading does not drop the received signal level below that of the local noise. One way to be certain that this does not happen is to ensure that the transmitter power is large enough to keep the received signal above the noise level even during a deep fade. The fading margin required will vary with the nature of the fading, the modulation system in use, and the desired quality of reception.

However increasing the transmitter power is not always the best solution. An alternative is to use techniques which are less vulnerable to fading. These techniques rely on the fact that the fading is not quite the same on two closely spaced frequencies or on two very similar, but not quite identical, propagation paths. In **diversity techniques,** the signal is received in two or more ways and added together in such a fashion as to enhance the signal and cancel out the fading.

There are at least five types of diversity reception which may be employed. The choice of type depends on the particular application.

1. Space diversity
2. Frequency diversity
3. Angle-of-arrival diversity
4. Polarization diversity
5. Time (signal repetition) diversity

**Space diversity** is achieved through the use of two receiving antennas several wavelengths apart, which receive signals travelling by slightly different propagation paths. **Frequency diversity** involves the sending of the same signal on two slightly different frequencies, but is not recommended because it is wasteful of the usable frequency spectrum and twice as expensive as the use of a single frequency. **Angle-of-arrival diversity** requires the use of large, expensive, directional antennas which can select out each of the several propagation modes. **Polarization diversity** is useful when the ordinary and extraordinary modes of polarization have about equal amplitudes, and therefore a high probability of deep fading; it involves the use of two antennas arranged

# Communication Problems under Normal Conditions

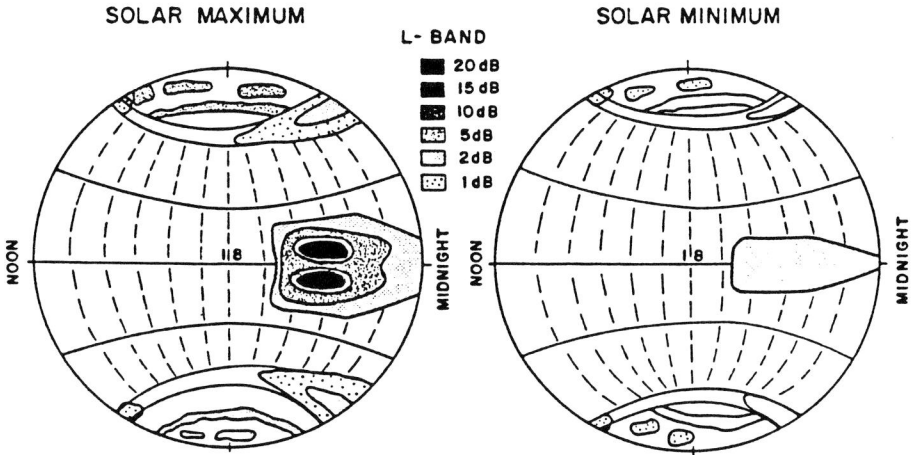

**Figure 7.4** The intensity of scintillations on trans-ionospheric signals at 1.2 GHz for solar maximum and minimum. The scintillations reach very high levels during the evenings at equinox and solar maximum.

at right angles to each other. **Time diversity,** in which the signal is sent more than once, has been found useful for the transmission of digital data.

## 7.3 Sporadic E

We have already encountered sporadic E (or $E_s$) in Chapter 3, where we described its properties and occurrence. The $E_s$ layer is a problem in HF communications because of its irregular and unpredictable behavior. If we were able to predict its presence and its critical frequency reliably, we could take it into account. $E_s$ is often a very good reflector at HF and when it is there, it is well worth using. However, we still know too little about what causes $E_s$ to be able to predict it.

One of the most important effects of $E_s$ on HF propagation is the **screening effect,** in which an $E_s$ layer reflects a signal which would otherwise have gone up to the F layer. As illustrated in Figure 7.5, the $E_s$ layer will limit the range of transmission and the signals will not usually reach the intended receiver. Even if they do, they will probably arrive at a very low angle not allowed for in the design of the circuit.

## 7.4 Problems at Low Latitudes

In many ways the low-latitude ionosphere could be expected to be an exceptionally good medium for HF communication—the critical frequencies are normally quite high, it is rarely affected by ionospheric storms (Chapter 8), and it does not suffer the irregular perturbations found at high latitudes (Section 7.5). However, the region has large horizontal gradients of critical frequency, in which the critical frequency can change rapidly over a few hundred kilometers. Rapid increases in critical frequencies during sunrise also cause difficulties.

### 7.4.1 Temporal Gradients in $f_oF_2$

Here we shall be concerned with rapid changes of $f_oF_2$ with time, the sunrise period being the most important. We saw in Chapter 3 that $f_oF_2$ at Manila in January at high solar activity increases from around 6 MHz at 06 LT, to 14 MHz at 07 LT, only one hour later. See Figure 3.11. Most operators receive only two allocated frequencies from their national Frequency Allocation Board, a day frequency and a night frequency. For short distance communications around Manila in January at high solar activity, suitable values for these frequencies would be 5 MHz (night, with possible failure of communication around dawn) and 12 MHz (day). What frequency do we use between 06 and 07 LT? We cannot use 12 MHz, because it is too high until almost 07 LT, so we would be forced to

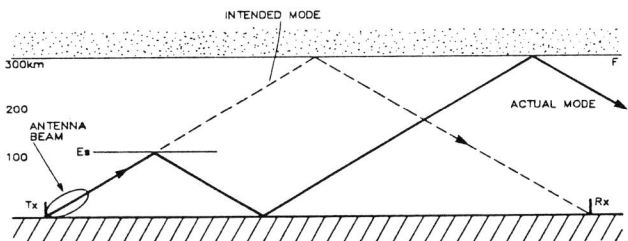

**Figure 7.5** The screening effect of a dense sporadic-E layer. The $E_s$ layer will reflect waves at all frequencies up to the MUF for the $E_s$ layer. For example, if $f_oE_s$ is 5 MHz and the $E_s$ layer is 500 km down range, the obliquity factor will be about 4.4 (the path length would be 1,000 km) and the $E_s$ layer will reflect all frequencies up to a value of $5 \times 4.4 = 22$ MHz. After reflection by the $E_s$ layer, the signals will hit the ground, then the F layer, and finally the ground again, but not at the desired location. The only way to avoid screening by an $E_s$ layer is to use a propagation mode which will either miss the $E_s$ layer or hit it at such a steep angle that the lower obliquity factor will ensure that the signal is not reflected by the $E_s$ layer. The normal E layer also screens the F layer, but not to as high a frequency as an $E_s$ layer, since $f_oE$ is normally only 2 to 3 MHz.

use 5 MHz. However, we then have the problem that 5 MHz signals are heavily absorbed during the latter part of this transition period, so we are in a no-win situation. Even if we had a large number of allocated frequencies, we would still have problems because we would have to change the antennas whenever we changed frequency. The factor of two in operating frequencies (5 to 12 MHz) would prohibit the use of the one antenna, especially if it is tuned to the operating frequency, as is usually the case. The situation is especially difficult for HF broadcasters, and a problem for all, if the predictions of the MUF do not give an accurate picture of when this sudden rise of $f_oF_2$ will occur.

For long-distance circuits in the east-west direction, which would involve multi-hop F-region propagation, the problem of the sunrise period will extend to several hours, as sunrise affects each hop of the circuit in turn.

### 7.4.2 Horizontal Gradients in $f_oF_2$ and $h_mF_2$

The steep gradients in $f_oF_2$ and $h_mF_2$ associated with the equatorial anomaly cause problems with north-south circuits, especially from the point of view of predicting the correct MUFs. The maps of $f_oF_2$ and $h_mF_2$ in the equatorial region are not as reliable as the maps for mid-latitude regions; the simple methods normally used for calculating MUFs do not allow for hops of unequal lengths (equal length hops happen only when $h_mF_2$ is constant along the circuit); and $f_oF_2$ and $h_mF_2$ change rapidly with latitude. All three of these facts lead to discrepancies between predicted and observed MUFs, which can make it difficult to use HF efficiently.

### 7.4.3 Scattering by Irregularities in the F Region

We encountered F-region irregularities in Chapter 3. Field-aligned irregularities in the equatorial F region have both advantages and disadvantages for HF communications. Satisfactory communications via modes which rely on scattering by irregularities is sometimes possible at frequencies well above the predicted MUF. On the other hand, irregularities can result in interference signals. The main disadvantage in using scatter signals at HF is that movements of the irregularities cause deep and rapid fading (flutter fading). Fading rates of up to 20 Hz occur, depending on the frequency used. Changes in the azimuth of arrival also occur, deviations of up to 50° having been observed on transequatorial paths associated with spread F. The systematic movement of a group of irregularities results in a Doppler shift of the transmitted frequency (up to about 30 Hz at 20 MHz), while random motions of individual irregularities lead to Doppler spreading (up to about 20 Hz). The observed fading rates are different for different wave frequencies.

## 7.5 Problems at High Latitudes

The high latitudes, that is the regions near the north and south poles, are by far the worst regions to be in from the point of view of HF communications [6]. This is because the high-latitude ionosphere is exceedingly variable both in space and time. Most of us live in the mid-latitude regions, where the ionosphere is relatively benign, but at high latitudes we really have our work cut out for us when we try to use the ionosphere. Worse still, we often have no other choice. The same problems occur for circuits between mid-latitude stations which go over the polar regions, i.e., on transpolar circuits.

Many of the HF problems at high latitudes arise because the Earth's magnetic field is almost vertical there. This means that any particles from the sun, arriving in either the quiet solar wind or in gusts associated with events on the sun (see Chapter 8), can penetrate right down into the low levels of the ionosphere. The charged particles spiral around the lines of force of the Earth's magnetic field, down to low altitudes where they collide with components of the neutral atmosphere causing increased ionization in a process called **collisional ionization.** The electron density in the high-latitude ionosphere therefore depends not only on where the sun is in the sky, but also on what particles are arriving into the region directly from the sun. As we shall see further in Chapter 8, any event on the sun can therefore affect the high-latitude ionosphere and hence HF communications.

The most important feature of the high-latitude ionosphere is the **auroral oval,** which is an oval-shaped annulus centered on the Earth's magnetic pole. It is the region in which the aurora can be seen most often—the aurora borealis (northern lights) or aurora australis (southern lights). The auroras are produced at E-region heights (about 100 km) during the process of collisional ionization. Some of the energy carried by the high speed charged particles manifests itself as visible light during collisions with the atoms of the neutral atmosphere.

The auroral oval can be considered as being fixed with respect to the sun, with the Earth rotating about the geographic pole beneath it. The minimum distance of the oval from the magnetic pole occurs during the daytime, while its maximum distance occurs at night. The auroral oval expands and contracts in step with the changes in the solar wind. When all is quiet (as described in Chapter 8), the auroral oval will contract to a narrow band about 20° from the magnetic pole and only a few degrees wide. Under disturbed conditions, however, the oval gets wider and expands towards the equator. In the southern hemisphere, the oval will occasionally pass as far north as Tasmania, while in the northern hemisphere it will occasionally pass as far south as New York City.

# Communication Problems under Normal Conditions

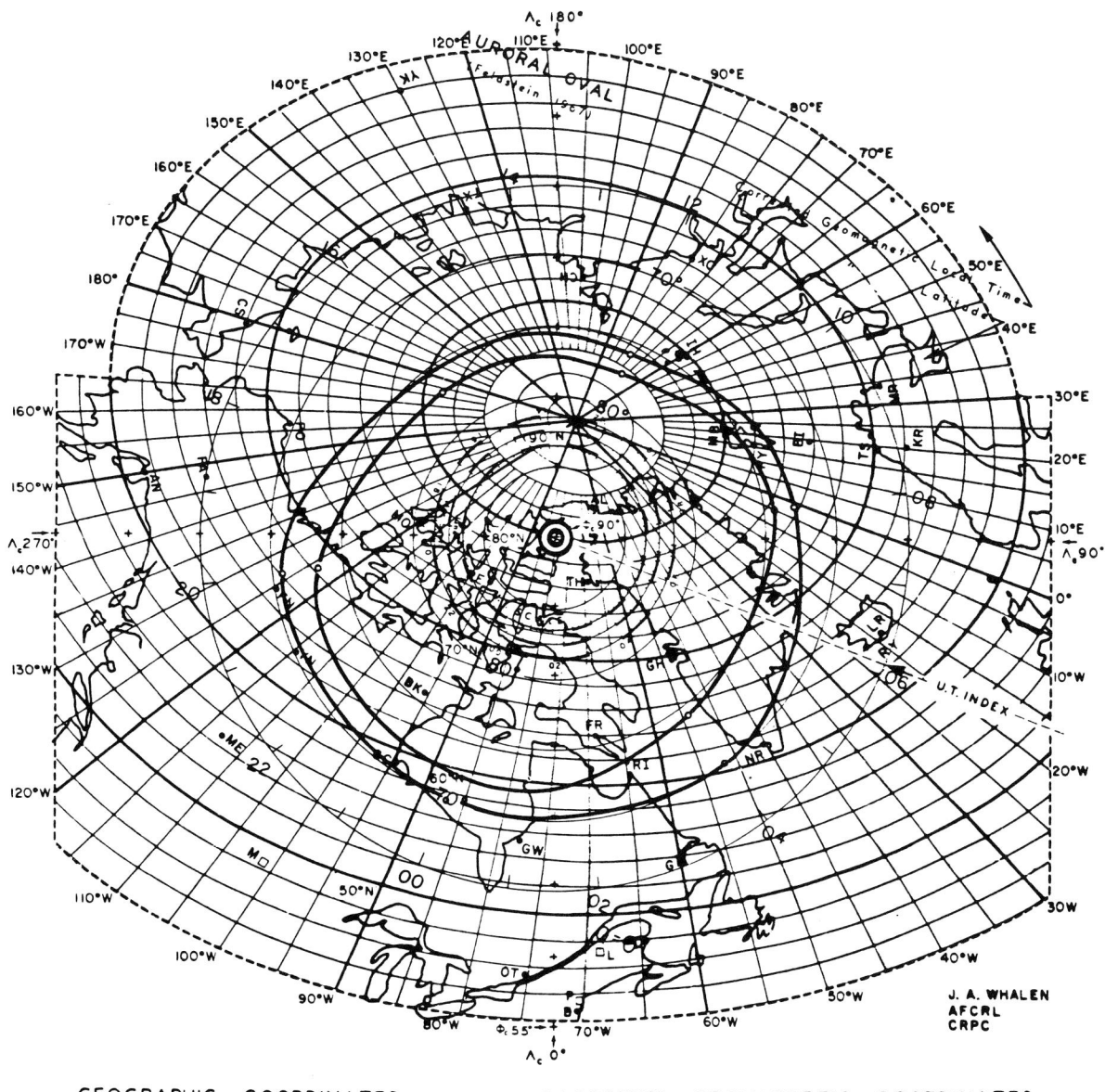

GEOGRAPHIC COORDINATES plotted in CORRECTED GEOMAGNETIC COORDINATES

**Figure 7.6** Approximate position of the northern auroral oval under magnetically quiet conditions. 04 UT.

Typical positions of the northern oval are shown in Figure 7.6 (quiet conditions) and Figure 7.7 (disturbed conditions) [7].

Small scale field-aligned irregularities in the E and F regions associated with the auroral oval and high-latitude spread F may seriously degrade the performance of a high-latitude HF circuit. Severe "garbling" of AM transmissions is very common in aurorally affected signals because of destructive interference between the two sidebands. A considerable improvement can be achieved using the SSB and CW modes of operation, but even then the signals are sometimes unintelligible during severe auroral disturbances.

Many of the difficulties associated with HF communications at high latitudes can be put into some sort of order by considering the location of the circuit relative to the auroral oval. The ionosphere at the latitudes covered by the auroral oval itself is disturbed at all heights, so any HF signal reflected from it will be very difficult to use. The increased ionization in the D region causes increased absorption, while the irregularities in the E and F regions lead to heavy scattering of HF signals. Increased ionization in the E region, which is called auroral sporadic E, can also lead to E-layer screening which we met earlier in this chapter.

On the equatorwards edge of the auroral oval lies the **mid-latitude trough** [8], which is a region of the

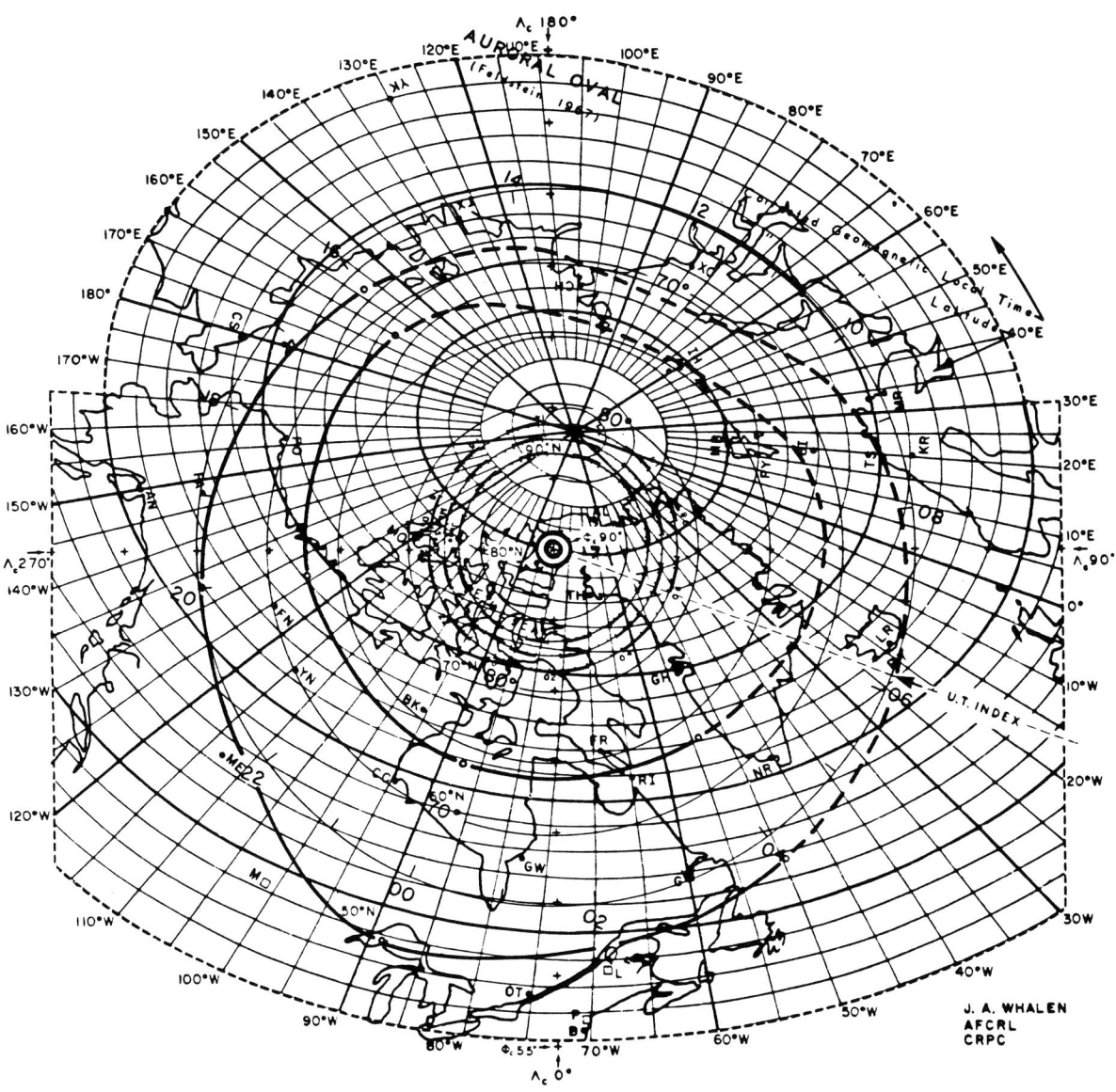

**Figure 7.7** Approximate position of the northern auroral oval under magnetically disturbed conditions. The oval has expanded equatorwards and become much wider. 06 UT.

ionosphere only a few degrees wide in which the critical frequencies suddenly drop by a factor of two or more, and the altitude of the peak suddenly rises by 100 km or more as we go across it in latitude. For short circuits with a reflection point near the trough, the MUFs can suddenly drop by a factor of two or more when reflection occurs within the trough rather than outside it. Long circuits can also be affected if reflection takes place within the trough, because it is possible to get off-great-circle propagation, as the signals bounce off the walls of the trough. The mid-latitude trough is mainly a nighttime phenomenon. On a given circuit, the trough will have different effects at different times of day, as it moves across the circuit.

The second region of importance which can be related to the auroral oval is the **auroral absorption zone,** which is a zone of particularly severe absorption in which all HF signals can be absorbed. The increased absorption is important from about midnight until late morning, and is restricted to a band of latitudes a few degrees wide, approximately at the latitude of the auroral oval.

## 7.6 Polarization Mismatch

We saw earlier in this chapter that we need to exercise some caution to ensure that the wave incident on our receiving antenna is not plane polarized in a direction at right angles to the polarization of the antenna itself. If it is, the voltage induced in the antenna will be zero, even though a strong incident wave actually reaches the antenna. In some cases, however, we have very little control over what the ionosphere does to our signal as far as its polarization is concerned. Under some conditions of propagation, depending particularly on the wave frequency and the orientation of the raypath relative to the Earth's magnetic field, only one of the characteristic waves (the ordinary and extraordinary waves) will propagate through the ionosphere. For example, it is possible for a plane polarized incident wave to be converted completely into the ordinary wave, no energy going into the extraordinary wave. The wave which reaches the receiver will be circularly polarized and a plane polarized antenna will pick up only half the incident energy. If the polarization conditions are such that all of the energy is converted into the extraordinary wave, we would encounter problems if we are using low frequencies (say 0.3 to 3.0 MHz), because at these frequencies the extraordinary wave is heavily absorbed and we could end up with no signal arriving at the receiver.

As a general rule, if polarization mismatch is found to be, or expected to be a problem, appropriate antennas should be used:

1. For east-west propagation along the magnetic equator, horizontally polarized antennas should be used.

2. At high latitudes, vertically polarized antennas should be used.

3. In mid-latitudes, there are several sets of conditions under which polarization mismatch can be a problem. In these cases, which will not be described here, vertical polarization is preferable to horizontal [9].

Polarization effects become less important as the frequency increases, because absorption (especially of the extraordinary wave) decreases.

## 7.7 Noise and Interference

No matter how great a signal level that a transmitter provides at a receiver, it will be to no avail if the level of the wanted signal is not greater than the level of signals already existing at the receiving site—signals which we do not want but are forced to contend with. A loud shout would doubtless be effective in a church, but at a football match even a loud shout may not be heard above the background noise. Radio noise is similar in many respects to ordinary noise and arises from four main sources—atmospheric, galactic, local, and interference.

**Atmospheric noise** is caused by distant thunderstorms, lightning strokes being profuse emitters of radio waves. Spark transmitters, which are very similar in operation to lightning discharges, were in fact the forerunners of modern HF communication techniques. The radio waves from thunderstorms propagate over the Earth just like any other radio waves and arrive unwanted at all receivers. Atmospheric noise cannot be eliminated and must always be allowed for in the design of an HF link. Some respite is gained during the day when the D region absorbs a large part of the noise, especially at the lower frequencies.

The noise level at any point on the Earth's surface will depend on the geographic location, time of day, season, operating frequency, and bandwidth of the receiver. Atmospheric noise is well understood and has been mapped the same way as the ionosphere has been mapped. Reliable estimates of it can also be calculated using the known distribution of thunderstorms, a model of the ionosphere, and simple HF propagation theory. Atmospheric noise levels decrease with increasing frequency and are greatest at low latitudes, where thunderstorm activity is greatest. Figure 7.8 shows the expected levels of atmospheric noise at 1 MHz, for 08 - 12 UT in summer [10]. The noise level is clearly greatest over tropical land masses. The frequency dependence of the noise for the same conditions is illustrated in Figure 7.9.

**Galactic noise** is radio emission from our own galaxy. It is usually not as important as atmospheric noise and is important only at frequencies above $f_oF_2$ since lower frequency signals cannot penetrate the ionosphere. In sparsely populated countries such as Australia, galactic noise can be the major daytime noise source outside of urban regions, at frequencies greater than about 10 MHz.

Local noise includes noise due to local thunderstorms and **man-made noise.** We have all experienced the effects of local thunderstorms, which also affect the MF broadcast band. The effect is the same as that of distant thunderstorms but is much greater and more obviously related to a thunderstorm. Man-made noise is another source of noise which is all too familiar, being caused by electrical equipment such as motor vehicle ignition systems, diathermy machines, welding machines, and so on.

Interference is a particular kind of man-made noise, being caused by another transmitter working on our frequency. Although this is not supposed to happen in the well-regulated world of HF it does, sometimes because the regulations have been ignored, and sometimes because propagation conditions exist which were not allowed for by the regulatory body when it allocated the frequency to more than one communicator.

The best solution to the problem of noise normally involves locating the receiver well away from centers of population, thus decreasing the local radio noise level. Narrower receiver bandwidths are also invoked, since the

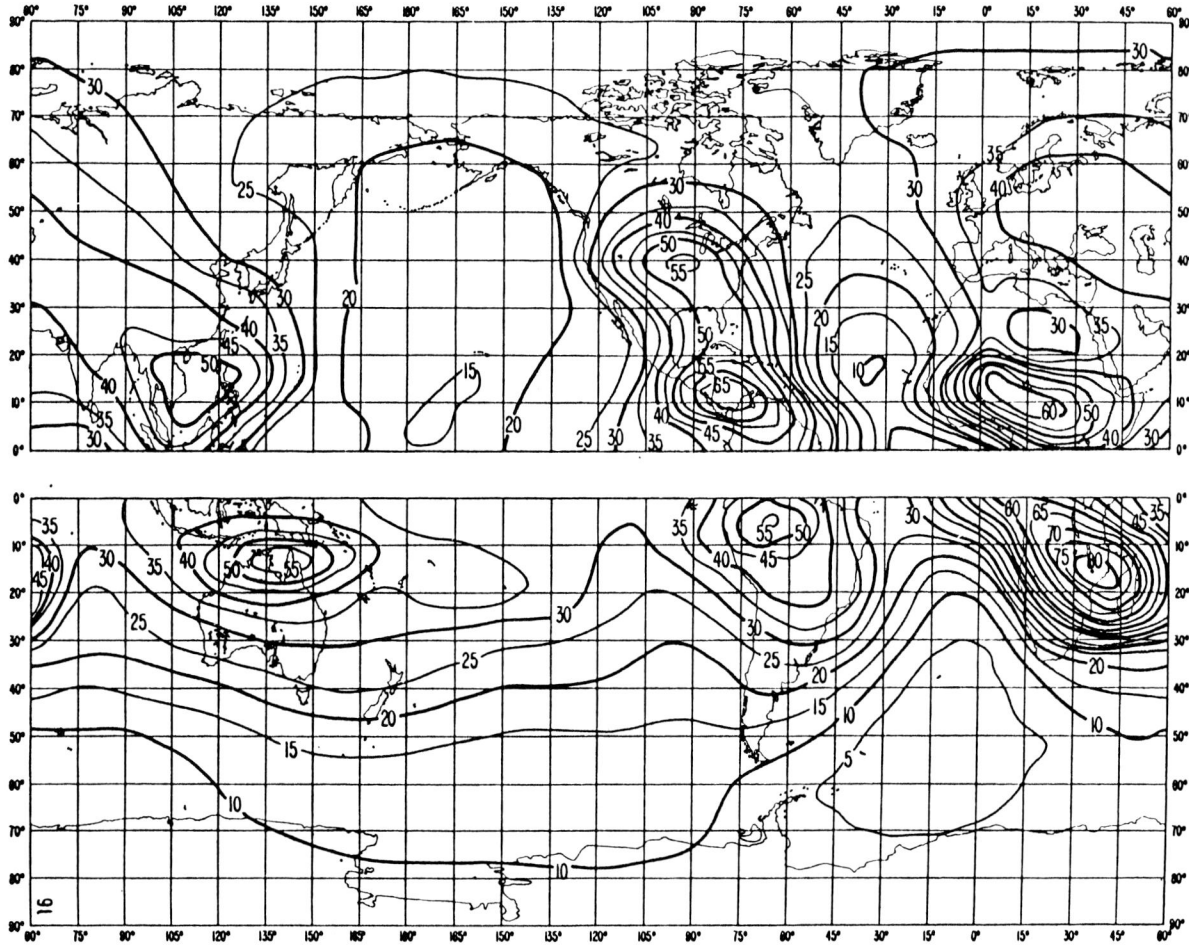

**Figure 7.8** Expected values of atmospheric radio noise, in units of dB above $kT_0b$ at 1 MHz, for summer, 08 - 12 LT.

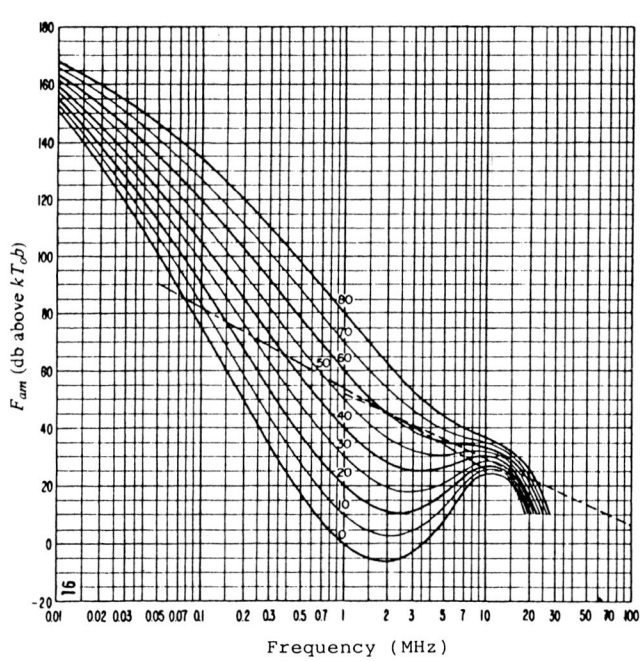

**Figure 7.9** Variation of radio noise with frequency during summer, 08 - 12 LT.

wider the bandwidth, the greater the amount of noise being added to the signal. Signal coding, which allows the signal to be pulled out of the noise, is also a good but somewhat expensive option. One simple solution is to use a horizontally polarized antenna, since a lot of man-made noise is vertically polarized. In some cases, it is possible to use a directional antenna which can be pointed at the transmitter while having a null in the direction of the source of noise.

It is essential when designing an HF circuit to know the likely noise level and what phenomenon will set that level. In lightly populated areas, the noise level is often set by atmospheric noise, whereas in more densely populated regions such as Europe, the noise level is set by man-made noise and interference.

## 7.8 References

1. Auterman, J. See Davies (1965), Figure 5.17.
2. Green, A. L. *Early History of the Ionosphere*. A.W.A. Technical Review, 7, p. 177, Sydney, 1946.

3. Jull, G. W., and G. W. E. Pettersen, Quoted by Davies (1965), Figure 5.21.

4. Davies, K. *Ionospheric Radio Waves*. Blaisdell Publishing Company, Waltham, Mass., 1969.

5. Basu, Santimay. Quoted by Aarons, J., *Forecasting Morphology and Dynamics of F-layer Irregularities*. Ionospheric Effects Symposium IES-90, May 1990.

6. CCIR Report 1012, *Operational Modelling of HF Radio Propagation Conditions at High Latitudes*. Propagation in Ionized Media, Green Book, Volume 6, Geneva.

7. Whalen, J. A. *Auroral Oval Plotter and Nomograph for Determining Corrected Geomagnetic Local time, Latitude, and Longitude for High Latitudes in the Northern Hemisphere*. Air Force Cambridge Research Laboratories Environmental Research Papers, No. 327, July 1970.

8. Whalen, J. A. *The Trough in the Daytime F Layer: A Macroscopic Effect of Ionospheric-Magnetospheric Convection*. Ionospheric Effects Symposium IES-90, Washington, D. C., May 1990.

9. Davies, K. *Ionospheric Radio*. Peter Peregrinus Press, London, 1990.

10. CCIR Report 322. *World Distribution and Characteristics of Atmospheric Radio Noise*. Geneva, 1964.

## 7.9 Problems

1. Explain what is meant by the "fading" of a radio signal, and describe its characteristics.

2. To what four phenomena can fading be mostly attributed?

3. Under what conditions would you expect multipath fading? What could you do to minimize it?

4. What is the difference between multipath fading, interference fading, flutter fading, and selective fading? What sort of fading should you anticipate on circuits which also support a ground wave? At what times of day would this fading arise?

5. What is the cause of polarization fading, and how may it be avoided?

6. You have been working a circuit on 15 MHz all afternoon since lunchtime when your signals start to fade severely, sometimes dropping right out. What should you do?

7. Describe briefly the five types of diversity techniques. Why do/don't we adopt these techniques?

8. Is sporadic-E screening a good thing or a bad thing?

9. Describe the three main problems encountered with HF communications in the equatorial zone.

10. The high-latitude ionosphere is a very difficult propagation medium to use at HF. Describe the main features of the high-latitude ionosphere and the ways in which they affect HF propagation conditions.

11. What polarization would you select for antennas for the following circuits? Why?

    a. East-west propagation along the magnetic equator.
    b. High latitudes.
    c. Mid-latitudes.

12. What are the four types of radio noise with which we must contend at HF? Which is the most important in your area?

13. How would you overcome each of the four types of radio noise? Are there any solutions which are preferable to others?

14. The E layer does not quite disappear at night, as evidenced by the reception of remote broadcast stations in the AM band. What would the E-layer skip distance typically be at night (see also Figure 4.6)?

15. Figure 7.8 gives world maps of the expected radio noise for summer mornings. Why aren't the top and bottom panels of the figure joined?

# Chapter 8

# Disturbances to Normal Communications

## 8.1 Introduction

*If* the HF predictions of usable frequencies at a particular hour are correct, *if* we are then using a frequency between the optimum working frequency and the lowest usable frequency, *if* our equipment is in good condition, *if* we have chosen appropriate transmitting and receiving antennas and these are in good order, *if* the level of radio noise at our receiving site is sufficiently low, and *if* there is no one else using our frequency, then we would expect to have reliable communications on 90%, or 27, of the days of the month at that hour. However, no matter how well we do our job, there will be three or so days of the month when, on the average, communications at predicted usable frequencies will fail. These are the days of the month when the ionosphere reacts to the sun as it changes from being "quiet" to "disturbed," with these changes leading indirectly to disturbances to HF communications.

Some of the effects of these disturbances are controlled to a large extent by the Earth's magnetic field, and in turn modify this field. Even the quiet solar wind has dramatic effects on the Earth's magnetic field. We saw in Chapter 3 that the field is similar to that of a large bar magnet, and we showed what the magnetic lines of force of this magnet would look like. However that was an old fashioned view of what things are like, which prevailed before the solar wind was discovered. In fact, because of the flow of the solar wind over the Earth, the magnetic field lines look more like those in Figure 8.1. The field lines on the day side of the Earth (the side facing the sun) are compressed, whereas the field lines on the night side are stretched out into a long tail.

Figure 8.1 also indicates the position of the **Van Allen belts,** which are belts of charged particles trapped by the Earth's magnetic field, and the **ring currents.** The latter arise when particles of opposite electric charge in the solar wind flow in opposite directions around the Earth at the equator. It is the ring current which causes geomagnetic storms (Section 8.5). Features of the sun-Earth environment such as the Van Allen belts and the ring current were among the first discoveries made by scientific satellites. Note that Figure 8.1 ignores the tilt of the N-S axis of the Earth's magnetic field with respect to the sun-Earth line.

There seem to be three types of disturbance on the sun which affect the ionosphere and HF communications. These are solar flares, high speed solar wind streams from coronal holes, and sudden disappearing filaments. The occurrence rates of each are shown in Figure 8.2 for the 1979 solar cycle. The details of the interactions are still under debate in the solar-terrestrial community [1, 2, 3]. Hewish and his co-workers conjecture that interplanetary shock fronts and disturbances, and sporadic storms, arise from "eruptive streams" emanating from coronal holes [4]. Others believe that coronal mass ejections provide the necessary physical connection between the sun and the Earth's magnetosphere. Since we are more concerned with what happens to the ionosphere, we will stick to the simple picture of three types of disturbance.

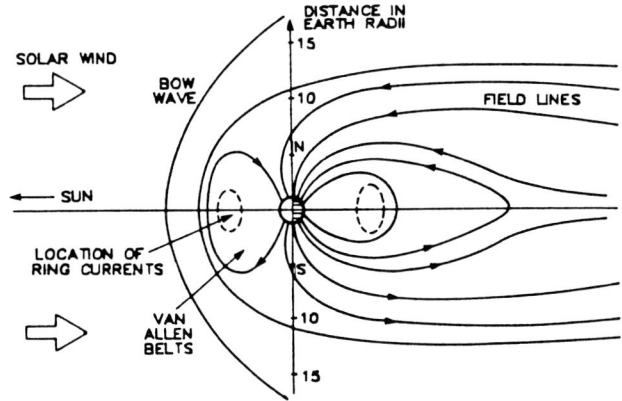

**Figure 8.1** The solar wind distorts the Earth's magnetic field so that it is compressed on the day side and stretched out into a long tail on the night side. A few details of the features shown in this highly simplified diagram are given in the text. *Courtesy IPS Radio and Space Services.*

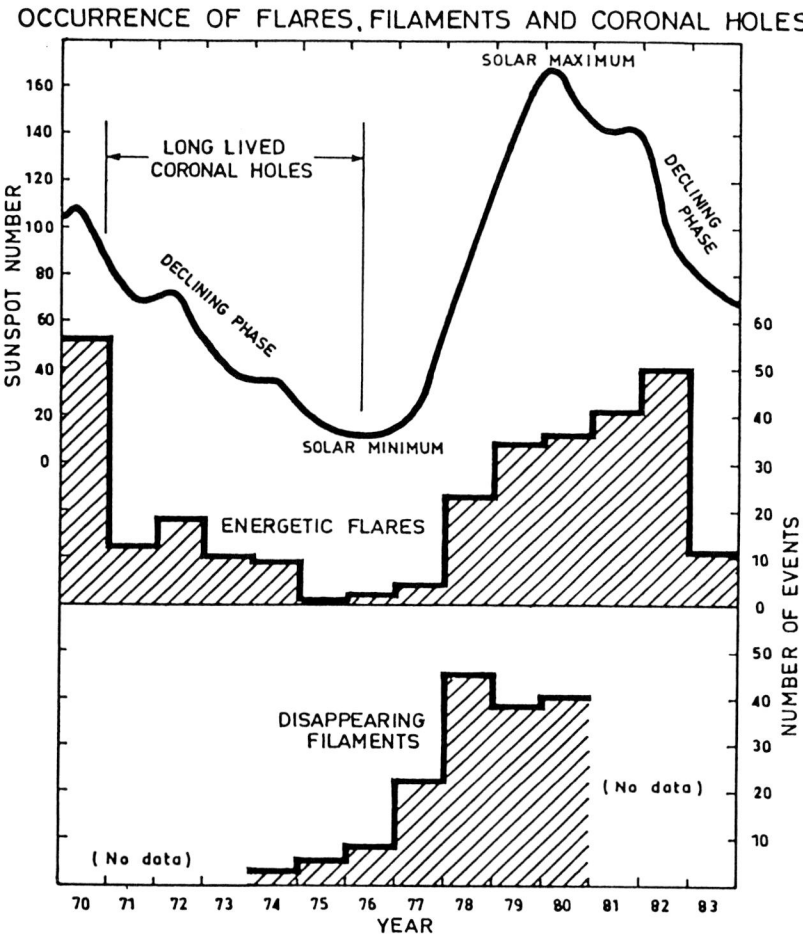

**Figure 8.2** The occurrence rates from 1970 to 1983 of the three solar phenomena which affect the ionosphere and therefore HF communications—coronal holes, energetic solar flares, and disappearing filaments. The sunspot number is also shown, from an average maximum of around 100 in 1970, through a minimum near zero in 1976, through another (high) maximum of around 160 in 1980, and then declining towards a minimum in 1986. *Courtesy IPS Radio and Space Services.*

## 8.2 Solar Flares and Their Effects

A solar flare is a large explosion on the sun, apparently caused by a sudden release of magnetic energy when the magnetic field at some position in an active region gets so contorted that it takes less energy to blow out the offending material and settle back to a less contorted state than it does to stay contorted. The determination of just which active region will spawn flares large enough to have an effect on the Earth is still not an exact science [5]. A solar flare can be seen by any of the techniques used to observe the sun, but a flare must be exceedingly large and bright before it can be seen in white light. Figure 8.3 shows Region 5395, just before and just after it produced two optically bright flares. Figure 8.4 shows convincingly that matter is indeed ejected from the sun, and would hit the Earth if the geometry were favorable (which it is not in this case—why not?).

Flares have a large range of sizes and can last for different lengths of time. A large flare can occupy 500 to 1,000 millionths of the sun's disc and can last for several hours. Flares have traditionally been classified according to their area and brightness as seen in $H_\alpha$, but a more useful classification is the intensity of the X rays emitted. This intensity can be measured by satellites orbiting above the atmosphere of the Earth (which absorbs X rays) and is one of the best indicators of how much energy a flare has released.

Flares occur most often during periods of high solar activity, as illustrated in Figures 8.2 and 8.5. Most are only small, or minor, and only a few percent have any effect on the ionosphere. Figure 8.6 shows the occurrence rates for the strongest class of flares in the X-ray classification, which are called X flares. It can be seen that the occurrence rates are much lower than those given in Figure 8.5 for flares as a whole.

Flares have three major effects on the ionosphere and HF communications, each caused by different types of

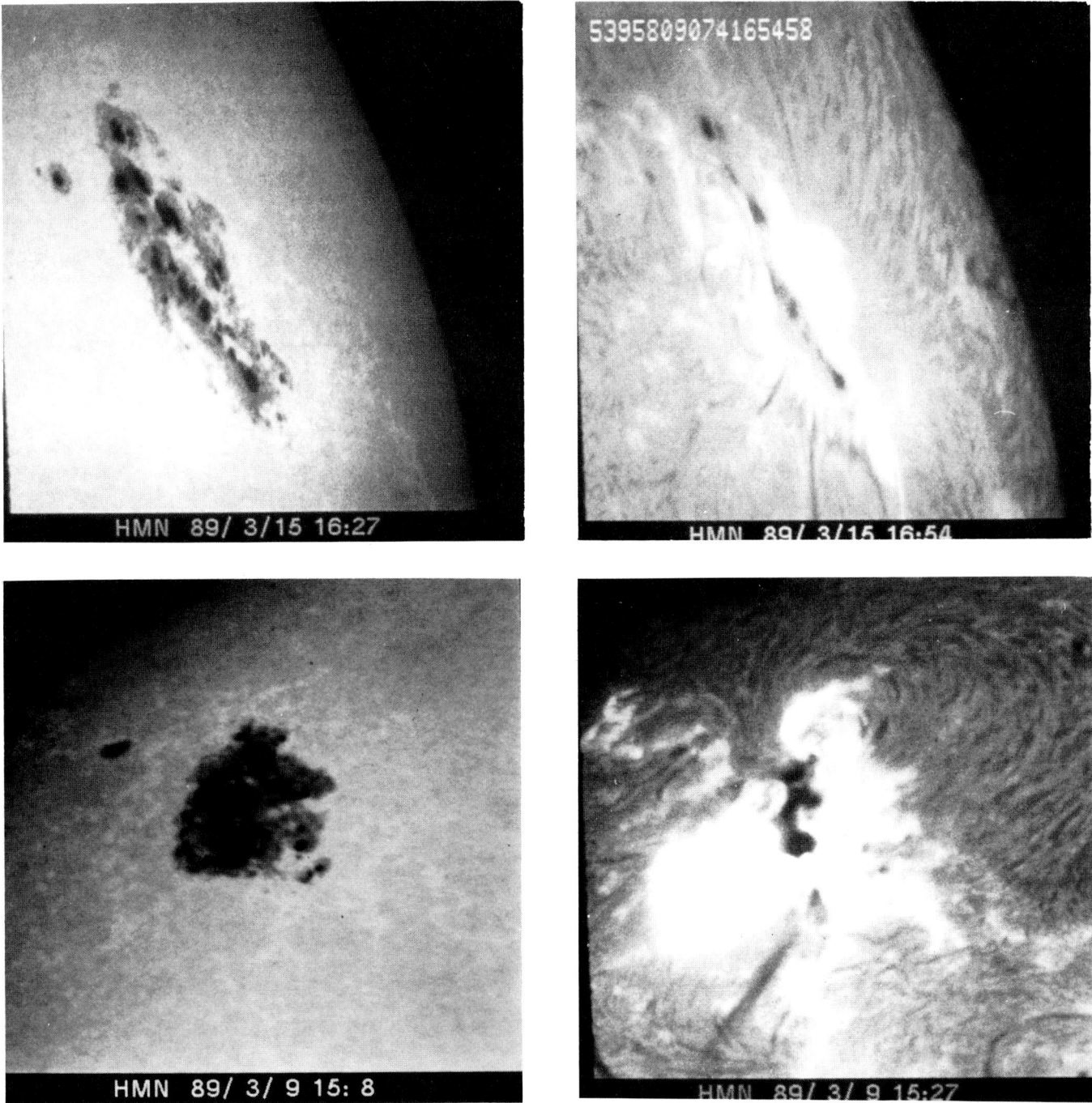

**Figure 8.3** "Before and after" pictures of an active region which produced large flares. Region 5395 produced an optical 4B flare on 9 March 1989, and a 2B flare on 15 March. These and other flares in March 1989 led to some rather trying conditions for the SSL operators during the March 1989 field trials of SKYLOC™ (see Chapter 11). *Courtesy Space Environment Laboratory.*

emissions or ejections from the site of the flare—X rays, protons, and a plasma cloud.

### 8.2.1 Effects of X Rays—The Shortwave Fadeout

Provided a flare is sufficiently energetic, and it occurs on that part of the sun facing the Earth, some of the X rays which it emits will hit the Earth's atmosphere, penetrate down into the D region and cause increased ionization by the process of photoionization. A large flare can increase the electron density of the D region by a factor of 10. This increase does not help us at HF. What it means is that there are now ten times more electrons to take energy from the radio waves and lose it in collisions with the countless millions of neutral atoms. In practice the effect is often disastrous, *all* of the energy of a radio wave being absorbed by the D region, leaving none to continue on to

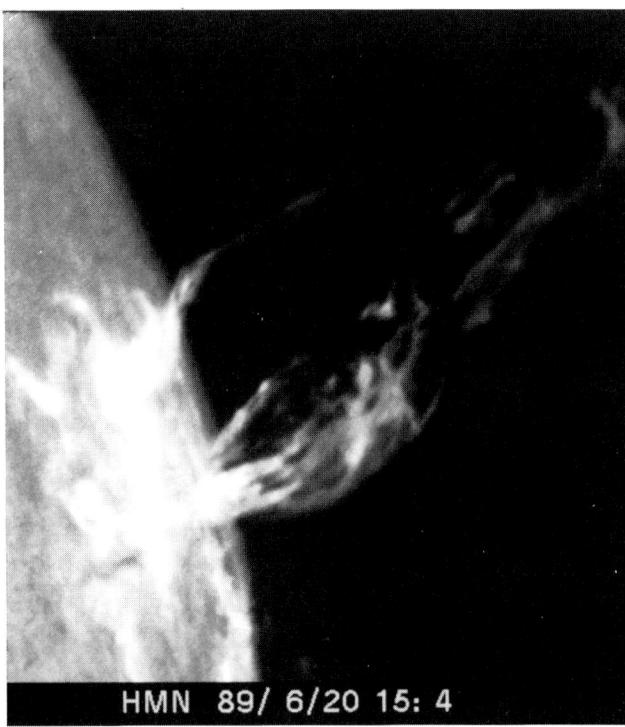

**Figure 8.4** Clear evidence that matter is ejected from the sun. The photograph shows hot solar plasma projected against dark interplanetary space, on 20 June 1989. *Courtesy Space Environment Laboratory.*

the receiver. This phenomenon is known as a shortwave fadeout (called SWF for short). It is also called the Dellinger fade, after J. H. Dellinger who in 1937 first explained the cause of fadeouts.

A SWF can last as long as the flare which is causing it (up to an hour or so) and the size of its effect on HF propagation will depend on the size of the flare. A small flare will have only a small effect and will affect only the low frequency end of the HF band. Absorption increases as the frequency decreases, which means that SWFs are more detrimental to lower frequencies. It also means that during a SWF we should attempt to communicate using the highest allocated frequency which lies below the normal OWF.

Because a SWF is caused by X rays which always travel in straight lines, it can be observed only on the side of the Earth facing the sun, i.e., on the part of the Earth which is in daylight. For this reason, a SWF is also sometimes called a **daylight fadeout.** A SWF is more effective for locations with a smaller solar zenith angle, since these would receive a higher flux (energy per unit area) of X rays. The main features of a SWF are illustrated in Figure 8.7.

Being basically an absorption effect, a SWF is more effective when normal absorption has its greatest values, i.e., at low or equatorial latitudes and in the middle of the day. This means that a given flare can have a very severe effect on one circuit, but only a minor effect on another. As far as a SWF is concerned, the important parts of the circuit are where the raypath cuts through (or attempts to do so) the D region on its way up to the E or F layers and on the way down again. If either of these areas is heavily affected by increased absorption due to a flare, a SWF will result, even though the ionosphere at the reflection point is not affected.

### 8.2.2 Effects of Solar Protons—The PCA

Some of the very energetic flares also eject a stream of protons which can hit the Earth if they are ejected in the right direction. The protons are produced from hydrogen atoms which have been ionized (stripped of their single electrons) by processes within the flare site. On their way to the Earth, the protons can cause severe damage to unshielded satellites or astronauts, since they travel at speeds up to about 0.8 times the velocity of light, or about $2.5 \times 10^8$ meters/second, and are thus highly penetrating. The occurrence of these **proton events** is highly correlated with the occurrence of large solar flares [6]. Figure 8.8 shows the occurrence rate of proton events for the last three solar cycles. Studies over the last few decades have led to some potential for forecasting whether or not any particular flare will eject a stream of protons [7].

The stream of protons can arrive at the Earth anything from ten minutes to a few days after the start of the flare, depending on how large the flare is and where on the sun it is located. When they arrive at the Earth, the protons encounter the Earth's magnetic field. Being a charged particle, a proton cannot cross the lines of force of the field, but must gyrate or revolve around them. The magnetic lines of force are horizontal near the equator and vertical near the poles. This means that any protons on a path towards the equatorial ionosphere cannot penetrate directly into the ionosphere, and the equatorial ionosphere is thus spared from their disruptive effects.

The situation at higher latitudes towards either pole is, however, somewhat different. Here the field lines are almost vertical and electrons gyrating around them can penetrate right down into the ionosphere. Once they have penetrated into the D region, they cause a dramatic increase in the electron density by ionizing atoms of the neutral atmosphere in a process known as **collisional ionization.** What happens is that the very energetic and fast protons just knock electrons off atoms with which they collide. As we saw in the previous section, increased ionization of the D region causes increased absorption, not enhanced reflection. In the case of ionization by solar protons, the absorption is very severe but is confined to

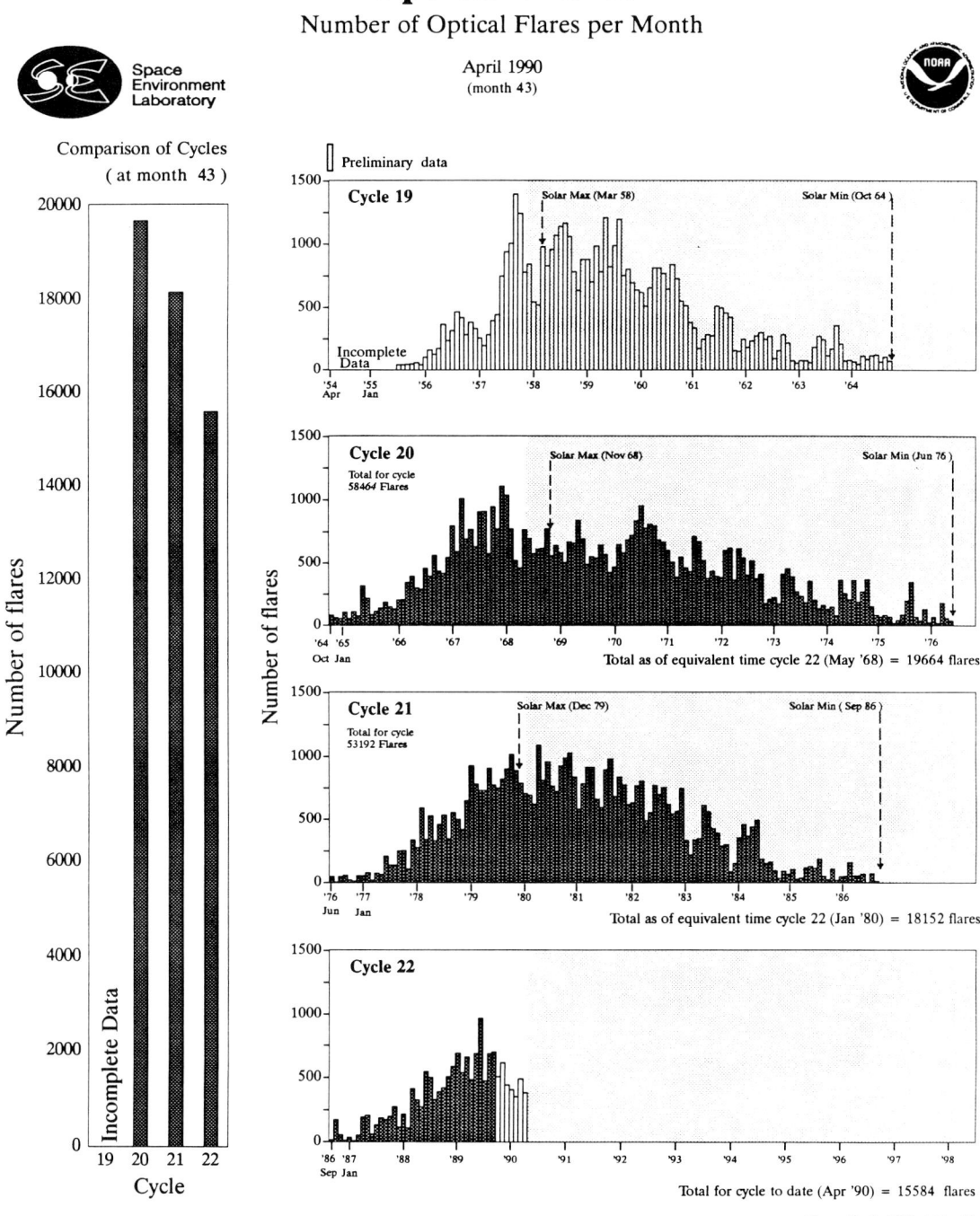

**Figure 8.5** The number of optical flares per month from 1954 (cycle 19) to April 1990. The number of flares varies in step with the sunspot number. *Courtesy Space Environment Laboratory.*

the polar caps or within about 20° of the poles themselves. The event is known as a PCA or **polar cap absorption event** and sometimes as a polar blackout, and is illustrated in Figure 8.9. The effects of the PCA can last for several days, depending on the size of the flare, and usually completely disrupt HF communications within and to the polar regions. The PCA will also prevent communications on any circuit which has one of its reflection points within the polar cap. On these circuits, communications can be maintained by the use of relay stations and a dog-leg circuit which bypasses the disturbed area. For circuits contained within the polar cap, the only choice for an HF communicator is to wait until the stream of protons has declined and the D region has

110                                                                                                          The Ionosphere

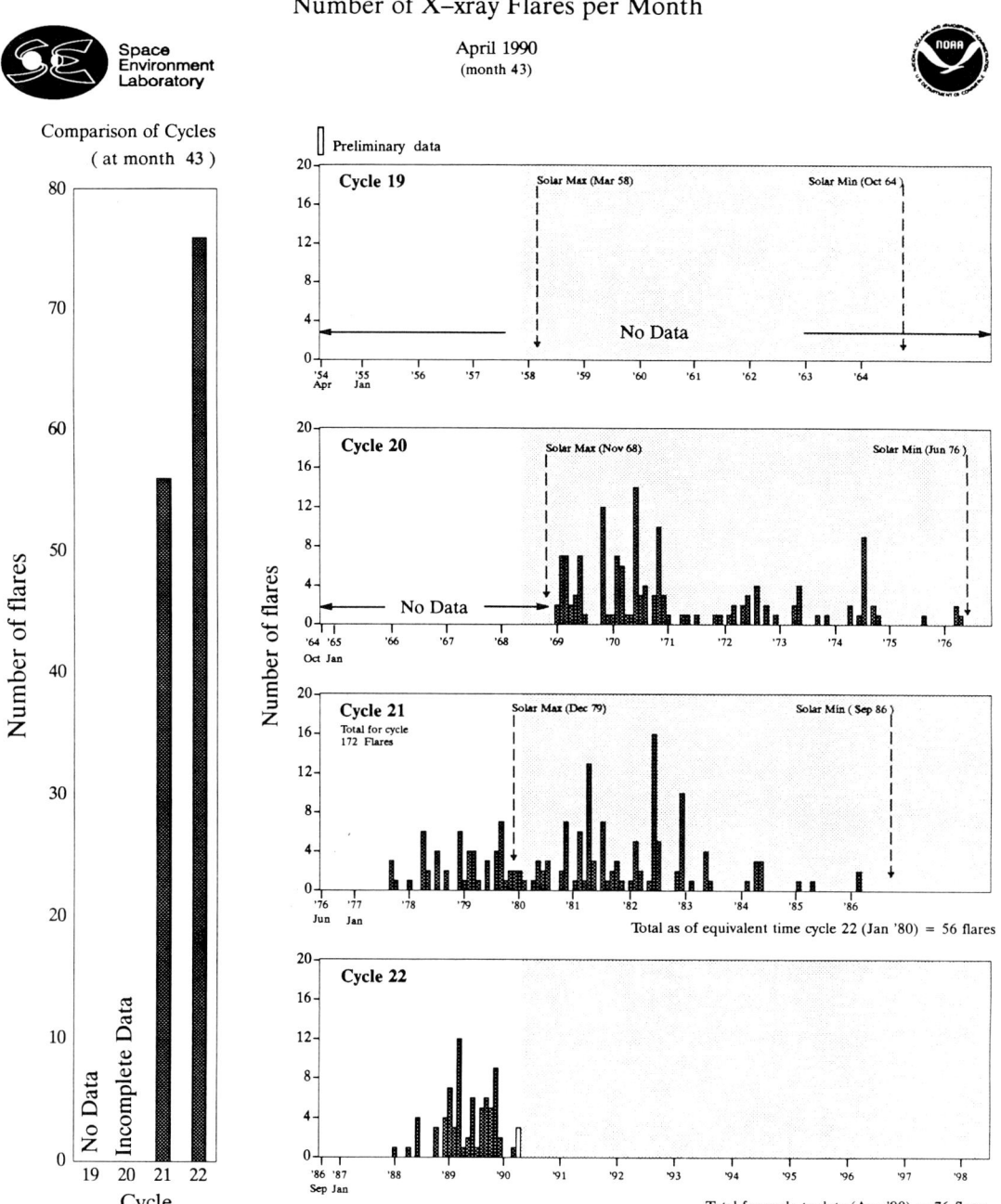

**Figure 8.6** The number of X-class X ray flares from 1969 onwards. The occurrence rate of these very large flares is much less than for all flares. These are the flares which are most likely to affect the ionosphere. *Courtesy Space Environment Laboratory.*

recovered. This can be anything up to a week or so for a large flare. If reliable communications are vital, consideration should be given to the use of communication satellites. The very high frequencies used in satellite communications are much less affected by the absorption of the PCA than frequencies in the HF band. There are, however, other things that can go wrong at the very high frequencies (Section 7.2.5 of Chapter 7).

Fortunately, PCAs are rare events, and large ones are even rarer. They occur most often at solar maximum when there are more flares on the sun, and about seven or eight can then be expected each year. Table 8.1 gives the number of principal PCA events (events with peak absorption greater than 20 dB) which occurred during the 1957 to 1963 solar cycle. This is the highest cycle yet recorded. Also included is the sunspot number for each

# Disturbances to Normal Communications

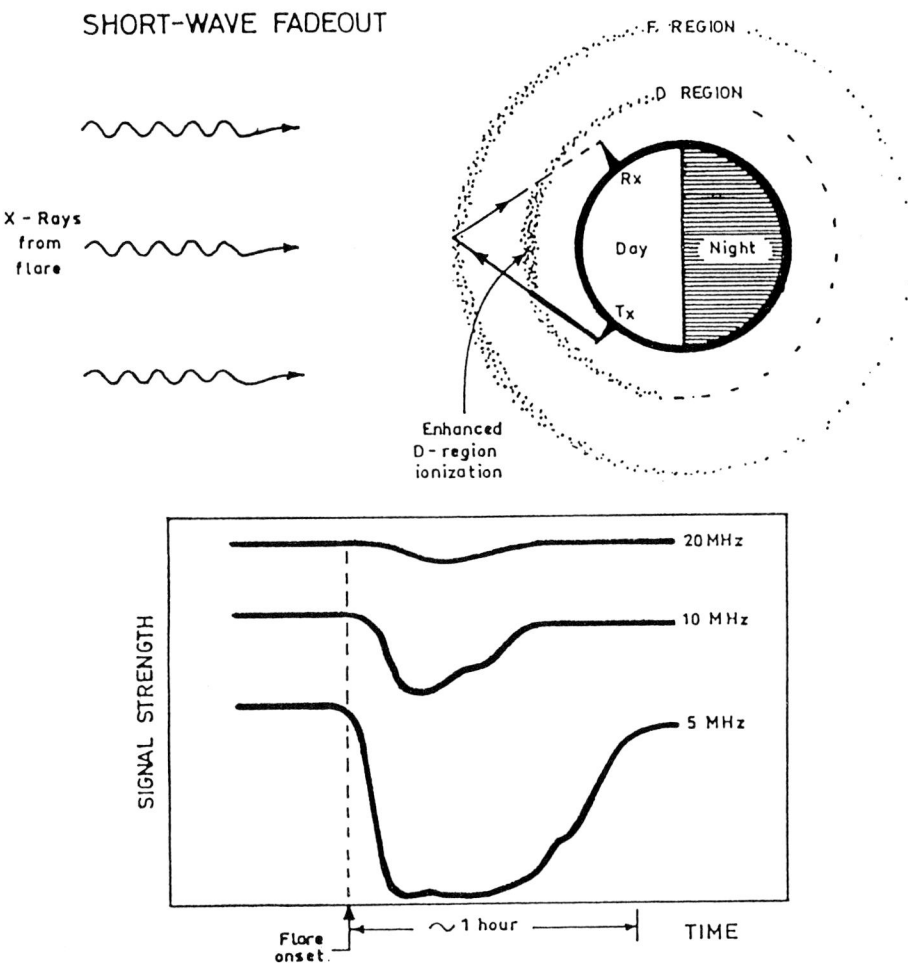

**Figure 8.7** The effect of a short wave fade out (SWF) on the day-side ionosphere and on the signal strengths at frequencies of 5, 10, and 20 MHz on a circuit passing through the day-side ionosphere. X rays from the flare cause a large and sudden increase in the density of electrons in the D region, which causes a corresponding increase in the absorption suffered by an HF signal passing through it. The low frequencies on an HF circuit are affected most and are the last to recover from the SWF. The night-time side of the ionosphere is not affected by X rays. *Courtesy IPS Radio and Space Services.*

year, the average duration of each event, the average peak absorption at a reference frequency near 30 MHz with respect to a reference path not suffering any increased absorption, and the maximum peak absorption with respect to the reference path. The absorption during a PCA is usually much less during hours of darkness than it is during the daylight.

## 8.2.3 Effects of a Plasma Cloud—The Ionospheric Storm

The third effect of a large solar flare on the ionosphere and on HF communications, which can in many cases be more important than the SWF or PCA, is called an ionospheric storm. An ionospheric storm is analogous to the familiar atmospheric storm which brings rain and wind, and the term is used to describe the condition of the ionosphere when unusual things are happening to it. In the case of an ionospheric storm, the ionosphere is changed, sometimes severely, especially as far as the critical frequency $f_oF_2$ is concerned. As a direct consequence of the ionospheric storm, conditions for HF propagation are changed, with resulting effects on HF communications. The effects are, naturally enough, usually detrimental.

Ionospheric storms are caused when a cloud of plasma ejected from a large flare hits the Earth. For a flare to cause a storm, the flare must first eject such a cloud of plasma, and that cloud must then hit the Earth. This normally happens only for large, energetic flares situated near the center of the face of the sun as seen from the Earth—in other words, near the central meridian (CM) of the sun. When a plasma cloud hits the Earth, it causes changes to the electric fields in which the ionosphere is embedded and also to the chemistry and large-scale movements of the $F_2$ region. The result of all these changes is that the critical frequency in the $F_2$ layer can

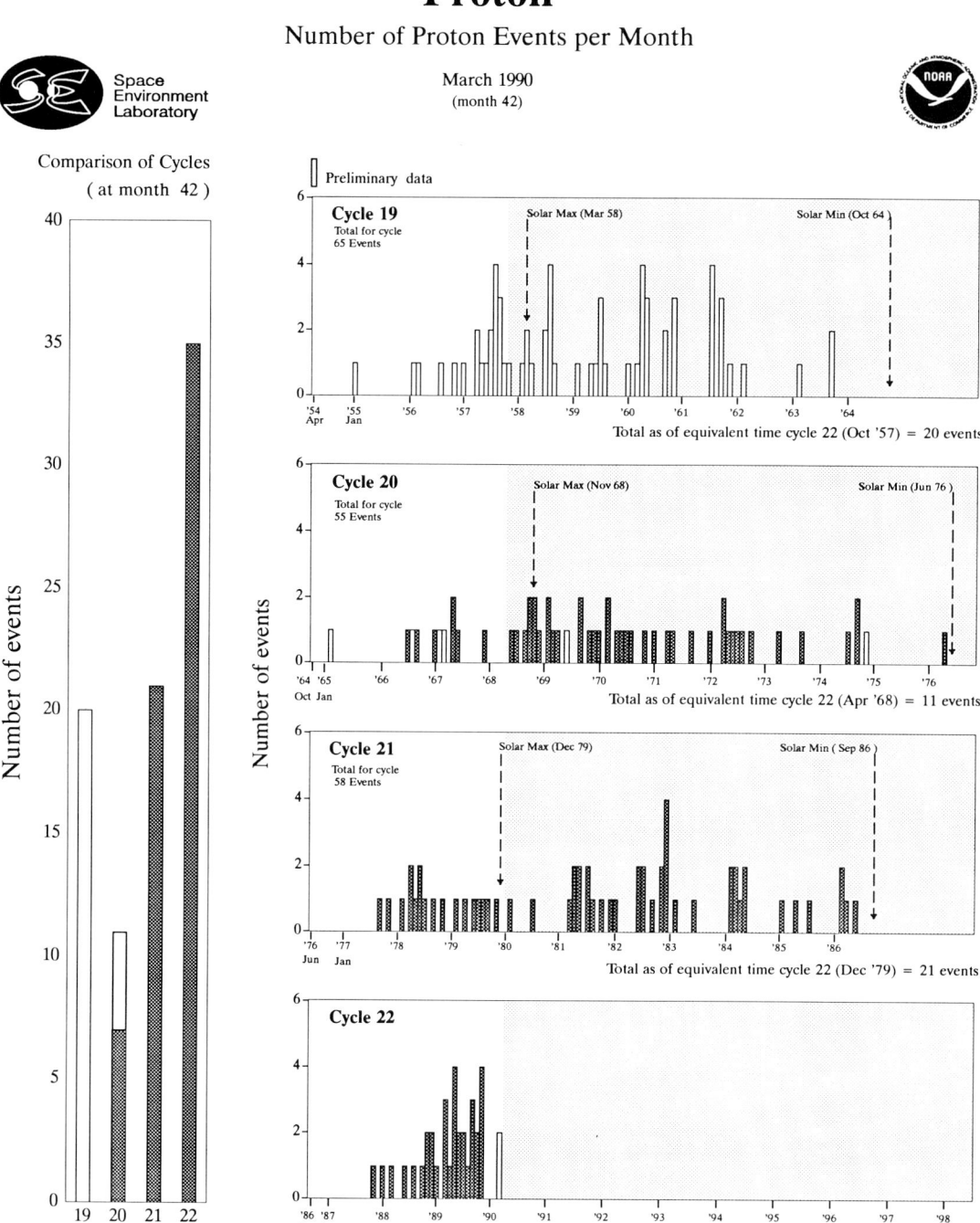

**Figure 8.8** The number of proton events per month from 1969 onwards. Proton events are most likely to be caused by X-class flares, so they have similar occurrence rates. *Courtesy Space Environment Laboratory.*

be either increased or decreased. Whether the critical frequency is increased or decreased at a particular location depends on such things as the time of day when the plasma cloud hits the Earth, the local time, the season and the latitude at the point in question, and how long the storm has been going on.

An ionospheric storm will normally commence about two to four days after the flare which caused it. This means that it is possible to get ample warning that the storm will take place. Any increases in critical frequency usually go unnoticed by a communicator since they just make things a little easier for him. For the direction-

# Disturbances to Normal Communications

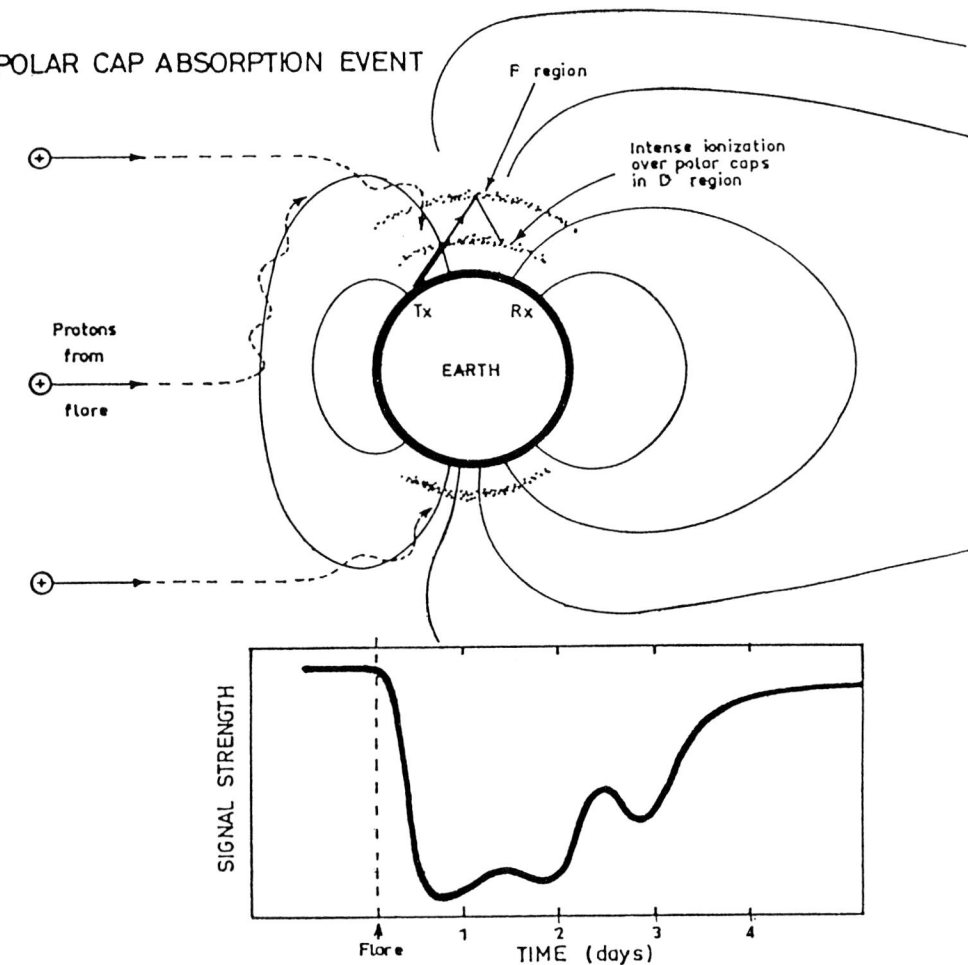

**Figure 8.9** A polar cap absorption event (PCA) is caused by high-energy protons from large flares which penetrate to lower altitudes of the atmosphere over the polar regions of the Earth, thus causing large increases in the electron density in the polar D region. This increase in density leads to a large increase in HF absorption, which is known as a PCA. A PCA can last for days, depending on the output of protons from the flare. There is usually some recovery of the signal strength during the hours of darkness. *Courtesy IPS Radio and Space Services*.

**Table 8.1** The number of principal polar cap absorption events (events with peak absorption greater than 20 dB) occurring during solar cycle 19. Included also are the smoothed sunspot number for each year, the average duration of each event, the average peak absorption at 32.2 MHz with respect to an unaffected reference path, and the maximum peak absorption with respect to the reference path.

| Year | 1952–5 | 1956 | 1957 | 1958 | 1959 | 1960 | 1961 | 1962 | 1963 |
|---|---|---|---|---|---|---|---|---|---|
| Smoothed Zurich sunspot number | 6 to 38 | 142 | 190 | 188 | 160 | 114 | 55 | 38 | 28 |
| Number of principal polar cap events | 0 | 4 | 13 | 9 | 5 | 12 | 3 | 0 | 2 |
| Average duration (hours) | — | 104 | 60 | 76 | 149 | 54 | 69 | — | 72 |
| Average peak absorption with respect to the reference path (dB) | — | 54 | 41 | 79 | 143 | 57 | 76 | — | 31 |
| Maximum peak absorption with respect to the reference path (dB) | — | 104 | 74 | 190 | 190 | 160 | 136 | — | 37 |

finding operator trying to determine the location of an HF transmitter (chapter 11 et seq.), and for the over-the-horizon radar operator trying to locate a ship or aircraft (Chapter 15), life becomes more difficult because the ionosphere they are actually using may be affected differently by the storm from the ionosphere that they are able to observe. It is the decreases in critical frequency which are important to the HF communicator since these can lower the MUF for his circuit below the frequency at which he would normally be operating, but the DF and OTHR operators are also affected by any increases.

During a large ionospheric storm, the critical frequency of the $F_2$ layer can drop by a factor of two, causing a corresponding drop in MUFs on a circuit passing through the disturbed region. In general, the D, E, and $F_1$ regions are not affected by ionospheric storms. When the $F_2$ region is severely depleted of electrons during major storms, the critical frequency of the $F_2$ layer can drop below that of the $F_1$ layer, and the $F_2$ layer cannot be observed from the ground. The ionosphere is then said to be in a **G condition**, and the highest frequency propagated is then supported by the $F_1$ layer, rather than by the usual $F_2$ layer.

The effects of ionospheric storms are greater in the equinoxes and in summer than in winter and are greater at higher latitudes. That is, the equatorial regions are less affected than the polar regions. During winter, the main storm effect is an increase in $f_oF_2$, although a severe storm can result in a following decrease of $f_oF_2$ (and MUFs). Figure 8.10 shows what happened to the ionosphere over Perth (Western Australia) following the large solar flare of 2 August 1972.

Increased absorption will occasionally occur at midlatitudes during a storm because of ionization of the D region by charged particles (mainly electrons). Recall that absorption at high latitudes will be very severe on most occasions.

### 8.2.4 Summary—Effects of Solar Flares

Figure 8.11 illustrates the three ways in which a major flare can affect the F region and thereby affect HF communications: by X rays, protons, and a plasma cloud in turn. The effects on a communications circuit of a SWF and an ionospheric storm are illustrated in Figure 8.12. The flare which caused the earlier SWF could have been the one which led to the ionospheric storm a day or so later.

## 8.3 Coronal Holes and HSSWS

We saw in Chapter 2 that coronal holes are relatively cool "open" structures in the solar corona which can be observed using special techniques. By "open" we mean that the lines of force of the sun's magnetic field stretch out into space, rather than folding back down to the surface of the sun. Because the field lines stretch out into space, and because ionized material can travel easily along field lines, ionized material pours out of a hole into interplanetary space in what is called a **high speed solar wind stream** (HSSWS for short—this can be read as hiss-wiss). Solar wind flows out from the sun over the whole surface, but above coronal holes the streams are faster. Typical speeds are 300 km/sec for a slow speed solar wind and 500 km/sec for a HSSWS. Material in a HSSWS therefore takes about four days to travel from the sun to the Earth. This may be compared with the two days (roughly) that it takes for material ejected from a solar flare to reach the Earth.

Figure 8.13 illustrates a "plan view" of a HSSWS, which is best understood by comparing it to the action of

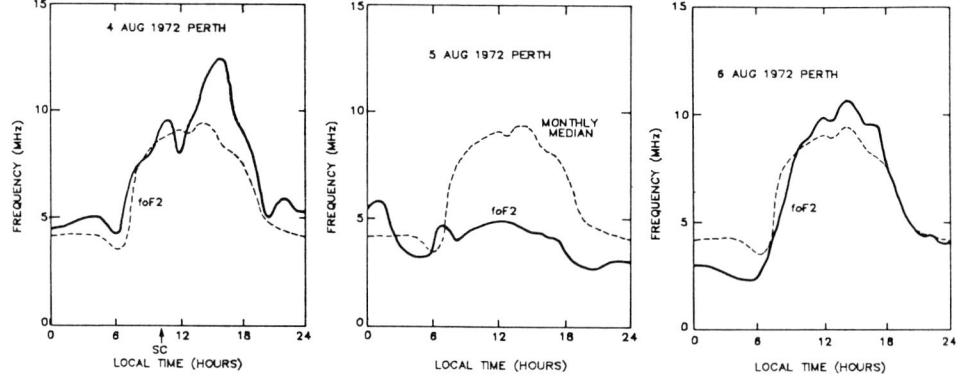

**Figure 8.10** The effects of a large ionospheric storm on the $F_2$ region over Perth, caused by several large flares which occurred two days earlier on 2 August. The shock front from the flares hit the Earth at about 1100 local time on 4 August, the plasma cloud itself following a few hours later. The effect on the ionosphere that day was an increase of $f_oF_2$ to a value about 50% higher than the normal monthly median value. This increase lasted about six hours. For the whole of the next day, $f_oF_2$ remained below 5 MHz, which represented a decrease of $f_oF_2$ by up to 50%, relative to the expected median behavior. The ionosphere completely recovered following sunrise on 6 August. *Courtesy IPS Radio and Space Services.*

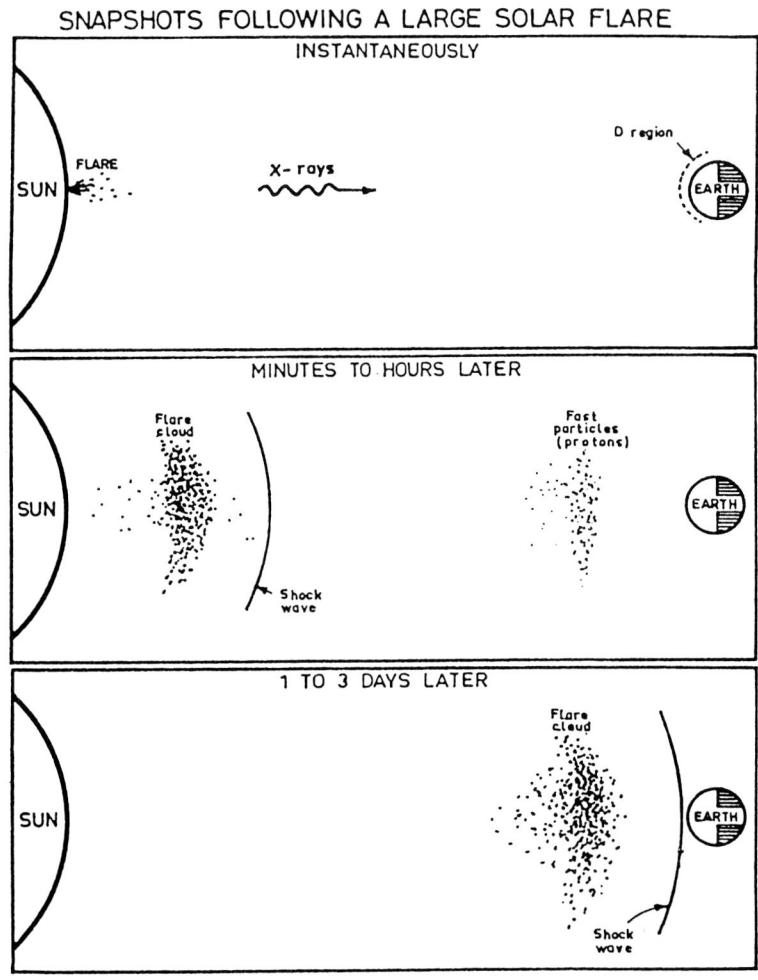

**Figure 8.11** The three events in interplanetary space which follow a large solar flare :
(1.) Top. X rays are one form of radiation emitted by the flare. These travel in straight lines at the velocity of light, taking about 8 minutes to reach the Earth.
(2.) Middle. Fast protons reach the Earth after a delay which can be as short as a few minutes. These are followed by the more slowly moving flare cloud, which is preceded by a shock at the position where the clouds hits the quiet solar wind.
(3.) Lower. The flare cloud reaches the Earth a few days after being ejected from the flare.
*Courtesy IPS Radio and Space Services.*

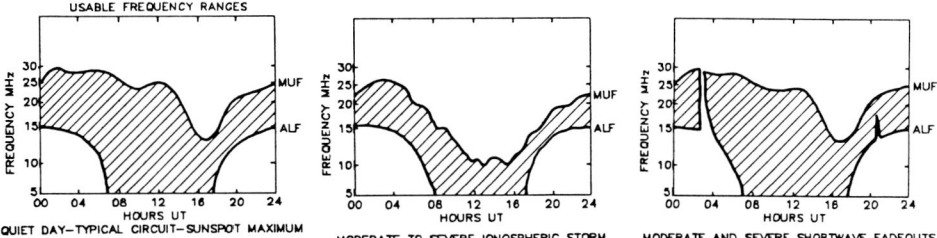

**Figure 8.12** The effects on a communications circuit of (a) two SWFs and (b) an ionospheric storm. In general, communications are possible at any frequency/time within the shaded areas. The severe SWF occurring at around 03 UT (local afternoon at the reflection point) caused a complete loss of communications at all frequencies for about 20 minutes. As expected, recovery started at the highest frequencies. The minor SWF at around 21 UT (local morning) affected only the lowest frequencies. The ionospheric storm shown in the middle panel decreased the nighttime MUF from over 20 MHz to around 10 MHz. Note that the lowest usable frequency was unaffected by the storm.
*Courtesy IPS Radio and Space Services.*

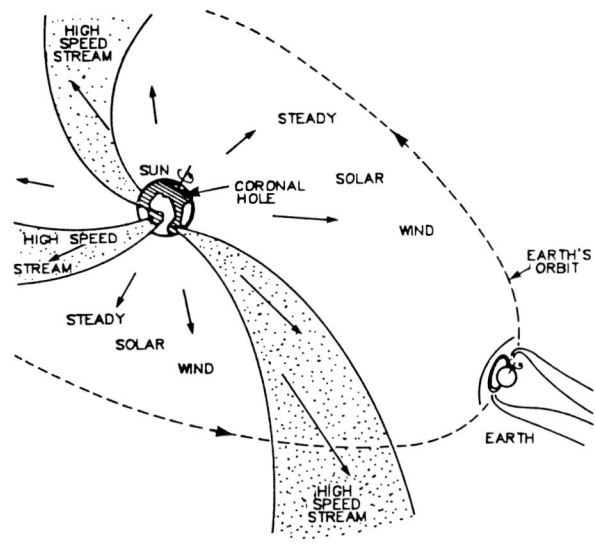

**Figure 8.13** View from above of high speed solar wind streams (HSSWS) emanating from coronal holes on the sun. Matter flows radially out from the sun, but the rotation of the sun gives rise to spiral streams. A HSSWS causes a geomagnetic and possibly an ionospheric storm as it sweeps over the Earth. Note the way the quiet solar wind has stretched out the lines of force of the Earth's magnetic field on the night-side of the Earth into a long tail. *Courtesy IPS Radio and Space Services.*

a rotating garden water sprinkler. Matter leaves the sun in a radial direction, straight out from the surface, but while it is travelling out into space the sun has rotated, so that the new material flowing into the HSSWS comes out from the sun in a direction different from that of the material which left earlier. Every HSSWS therefore ends up curved, the material which left the sun earlier being delayed with respect to later material. Slow streams are more curved than fast streams.

HSSWSs are important because they cause ionospheric storms as they sweep over the Earth, just as the plasma cloud from a solar flare causes storms. However HSSWSs are different in that their effects are not usually as marked or as devastating as those of a large solar flare, partly because the HSSWS does not overtake the Earth as fast as the cloud from a flare hits it. The effects also tend to last longer because they are felt for the whole time that it takes for the HSSWS to sweep over the Earth, which is typically a few days.

Coronal holes and their associated HSSWSs are a feature of declining solar activity, as illustrated in Figure 8.2. Particular hole/stream combinations have been observed to last almost a year, reappearing every 27 days or so in step with the rotation of the sun. The fact that HSSWSs are associated with long-lived features on the sun makes it relatively easy to predict their return, and in turn to predict their effects on the Earth. The only problems which arise are when a new hole appears, when a hole disappears, or when the HSSWS speeds up or slows down. If the HSSWS speeds up, it will straighten out to some extent and will overtake the Earth in its orbit around the sun a little earlier than expected. The reverse holds if the HSSWS slows down.

## 8.4 Sudden Disappearing Filaments

Disappearing filaments are the last of the three solar phenomena which can affect the ionosphere. We encountered filaments in Chapter 2, where we saw that they are relatively cool and large structures in the solar chromosphere which are seen as prominences when viewed on the edge of the sun. Filaments are often seen to disappear within a few hours, and it is surmised that all or part of the material of the filament has been blown out into space, in a fashion similar to solar flares. This is illustrated in Figure 8.14. These sudden disappearing filaments (SDFs) can affect the Earth's magnetic field although the effects are often small and hard to confirm [8]. SDFs also affect the ionosphere and HF communications to some extent, but no detailed studies have yet

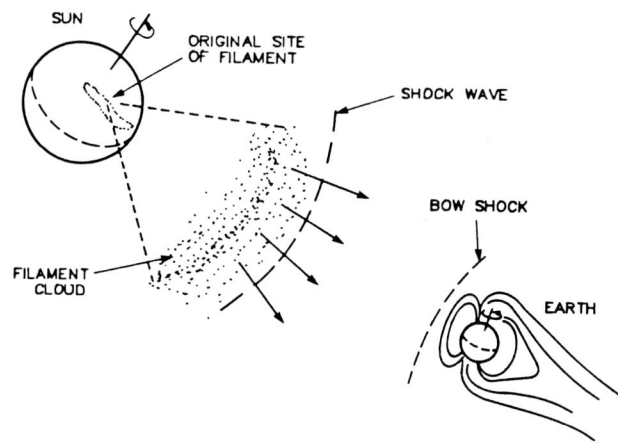

| PHENOMENA | SUN – EARTH DELAY | TERRESTRIAL EFFECT |
|---|---|---|
| SHOCK WAVE | 3–4 DAYS | Sudden commencement |
| FILAMENT CLOUD | 3–10 DAYS | • Magnetic storm<br>• Ionospheric storm<br>• Aurora |

**Figure 8.14** The matter from a large disappearing filament, ejected into interplanetary space and preceded by a shock wave. If the shock wave hits the Earth, it will compress the Earth's magnetic field and give rise to a sudden commencement magnetic storm. The following plasma cloud, like the cloud from a large flare, can cause a magnetic and possibly an ionospheric storm. The bow shock illustrated is a permanent feature of the Earth's magnetic field and marks its boundary with the magnetic field of the solar wind. *Courtesy IPS Radio and Space Services.*

been made of these effects. SDFs are a feature of high levels of solar activity.

## 8.5 Geomagnetic Effects

We have concentrated so far in this chapter on the effects of solar disturbances on the ionosphere and HF communications. The disturbances also affect the Earth's magnetic field, causing what are called **geomagnetic storms,** which are important to anyone, such as geophysical prospectors, concerned with measuring geographical variations of the Earth's magnetic field. Our main interest in the geomagnetic effects is that they are somewhat easier to talk about than ionospheric effects, many of the terms used in discussing the effects of solar disturbances arising from a consideration of what happens to the Earth's magnetic field.

Geomagnetic storms occur in conjunction with ionospheric storms and have the same causes—solar flares, HSSWSs, and SDFs. A geomagnetic storm usually consists of a small increase in the Earth's magnetic field, called the **initial phase,** followed by a large decrease, called the **main phase.** A geomagnetic storm is not really much of a storm—the field may change by only 100 nanoteslas out of a total of 30,000. Recall that a major ionospheric storm can drop $f_oF_2$ down by 50%.

The level of the disturbance of the Earth's magnetic field in a 3-hour period is defined by the **magnetic index, K.** K ranges from 0 (no disturbances) to 9 (very disturbed). $K_p$ is a planetary average (over the surface of the Earth) of the K index. A second index which is also used, $A_p$, is somewhat similar to $K_p$, except that it is a daily index.

A geomagnetic storm caused by a solar flare usually starts off with a sudden increase at the start of the initial phase. This is called a **sudden commencement,** or SC for short, and arises when the shock front from the flare hits the Earth's magnetic field and suddenly compresses it. A storm caused by a HSSWS, on the other hand, usually starts off gradually as the HSSWS overtakes the Earth. The onset of the storm is more insidious than for a flare-induced storm, and consequently it is described as a **gradual commencement** storm. Because storms caused by HSSWSs tend to recur every 27 days or so, in step with the sun's rotation, they are also called **recurrent storms.**

Figure 8.15 illustrates how the horizontal component of the Earth's magnetic field varies during typical geomagnetic storms due to (a) a large flare and (b) a HSSWS, while Figure 8.16 shows how the number of magnetic storm days (defined as days for which $A_p > 50$) varied over the last five solar cycles. The number of storm days does not vary dramatically over the cycle, since as one cause of geomagnetic activity starts to disappear, another takes over.

The rapidly changing magnetic field associated with a storm will induce electric currents in long pipelines and power lines [9, 10, 11]. The currents induced in the pipelines, such as the Alaskan oil pipeline, lead to increased erosion of the pipes. The currents induced in power lines can cause a whole power distribution system to fail, as in August 1972 and March 1989. Both of these effects are more important at higher latitudes [12].

## 8.6 Bartels Charts

Figures 8.17 and 8.18 give the Bartels magnetic charts for 1974 (low solar activity) and 1989 (high solar activity). These charts are very useful because they contain such a large amount of information presented in a very clear fashion. They were invented by J. Bartels, a pioneer in the study of the Earth's magnetic field. They are also known as **musical charts**.

The horizontal axis in the charts is a 27-day period corresponding to one rotation of the sun, with about 14 full rotations each year. The vertical lines in each of the 14 rows of plots represent the values of the magnetic index $K_p$. Each chart represents very disturbed periods ($K_p > 5$)

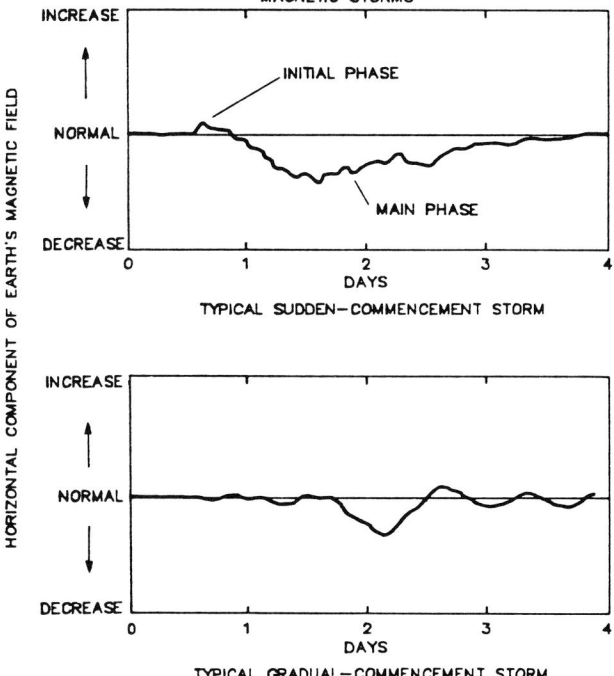

**Figure 8.15** Typical sudden commencement and gradual commencement magnetic storms. The former would be caused by flares or disappearing filaments, while the latter would be caused by HSSWSs. Note that even the largest decrease in the Earth's magnetic field during a storm is usually less than 1% of the undisturbed value. *Courtesy IPS Radio and Space Services.*

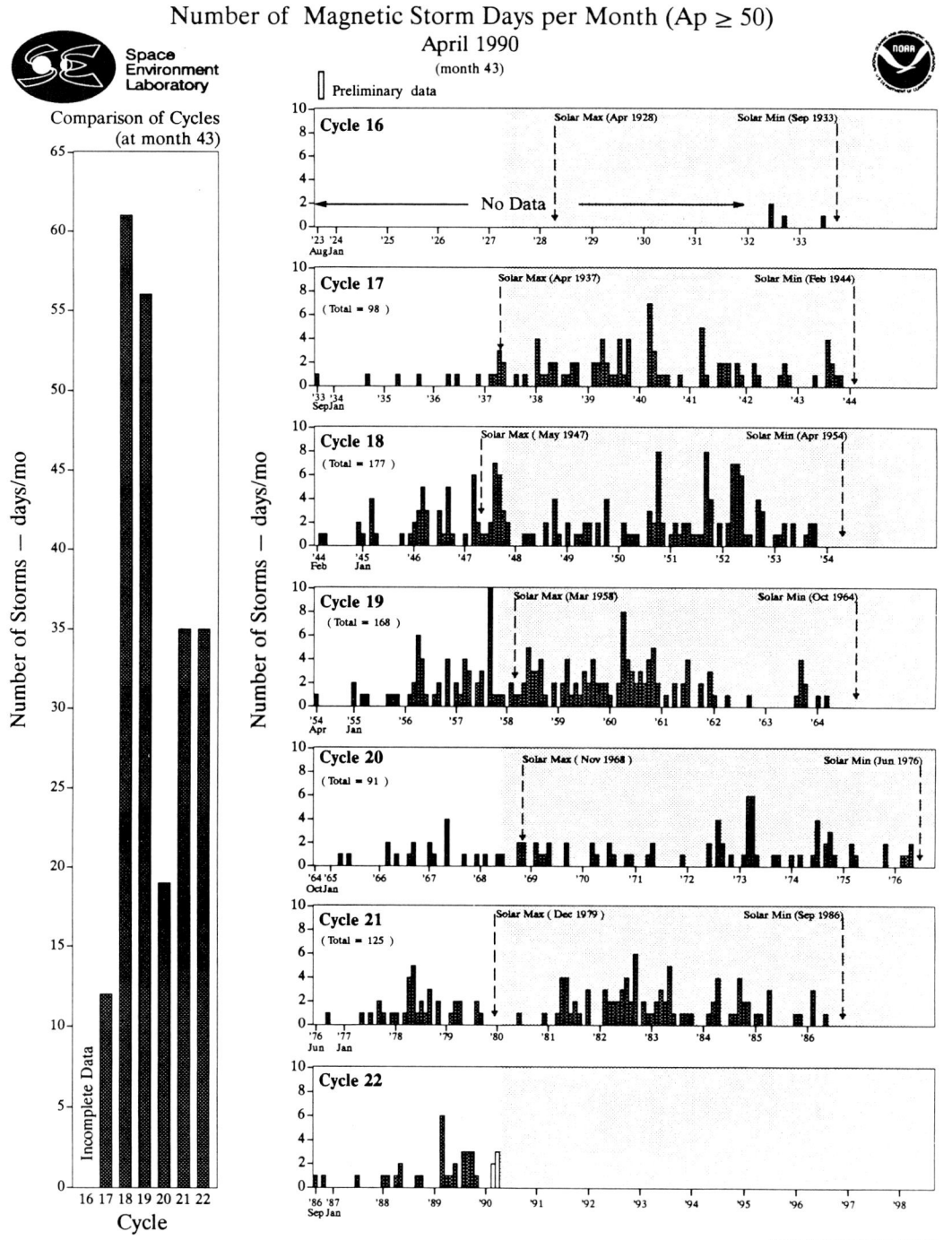

**Figure 8.16** The number of magnetic storm days per month for cycles 16 to 22. A storm day is defined as having $A_p \geq 50$. *Courtesy Space Environment Laboratory.*

as black vertical bars, so the more disturbed the magnetic field, the longer the black bar. Magnetically quiet conditions, when the field is not disturbed, are represented by short thin bars. Without worrying about the detail, we can summarize the charts by noting that solid black areas denote magnetically disturbed periods, while a lot of white space denotes magnetically quiet periods. The arrowheads denote storm sudden commencements which were described earlier. One advantage of the Bartels charts is that if magnetic storms are caused by the same

# Disturbances to Normal Communications

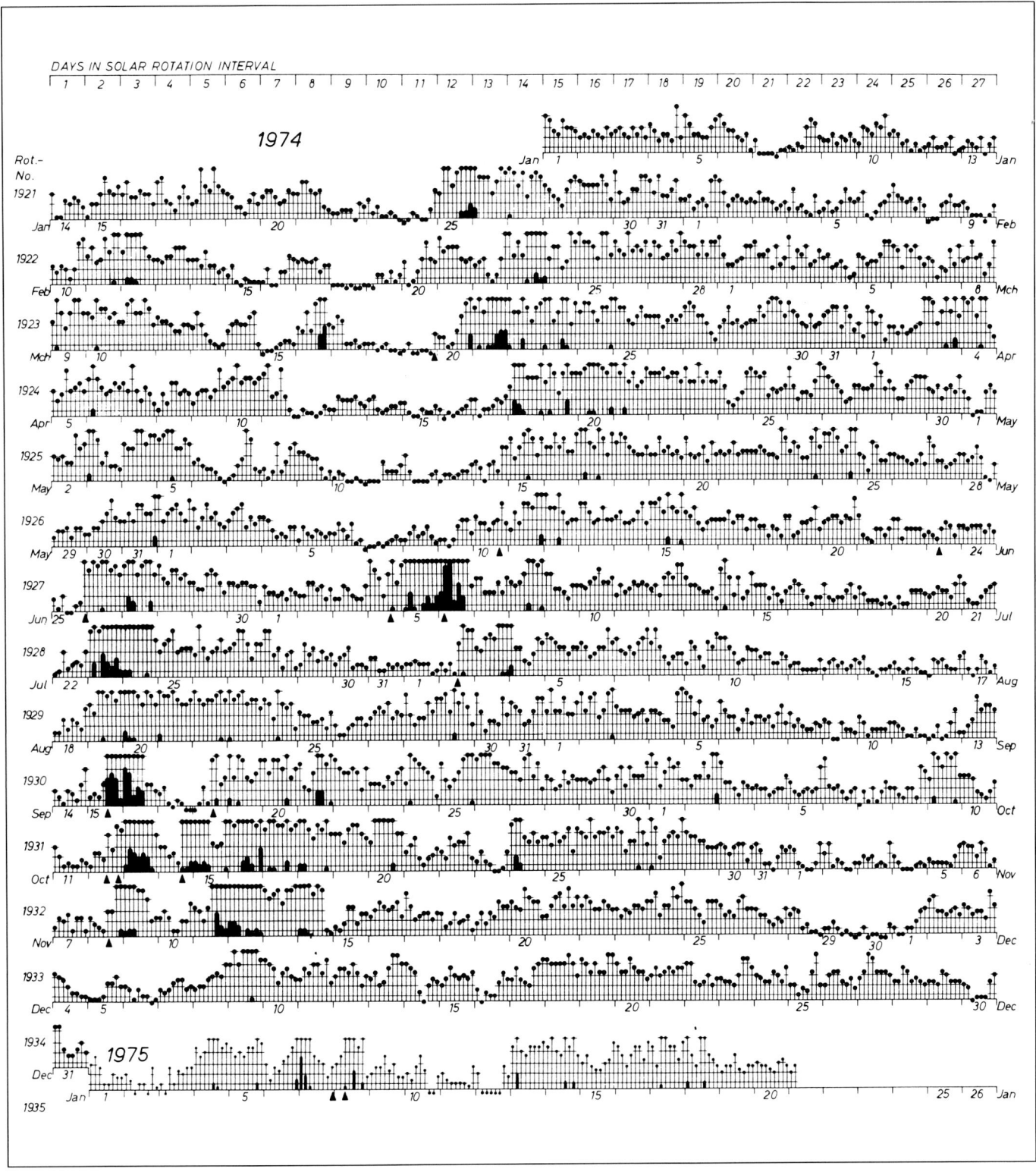

**Figure 8.17** Bartels charts of the planetary average 3-hourly values of the magnetic disturbance index, $K_p$, for a year of declining solar activity (1974). Each row of the plot is 27 days long, corresponding to the rotation period of the sun. Thus large values of $K_p$ which occur vertically below each other can be attributed to a feature of the sun which returns to face the Earth every 27 days. This feature is usually a coronal hole which is the source of a high speed stream in the solar wind. *Courtesy University of Göttingen.*

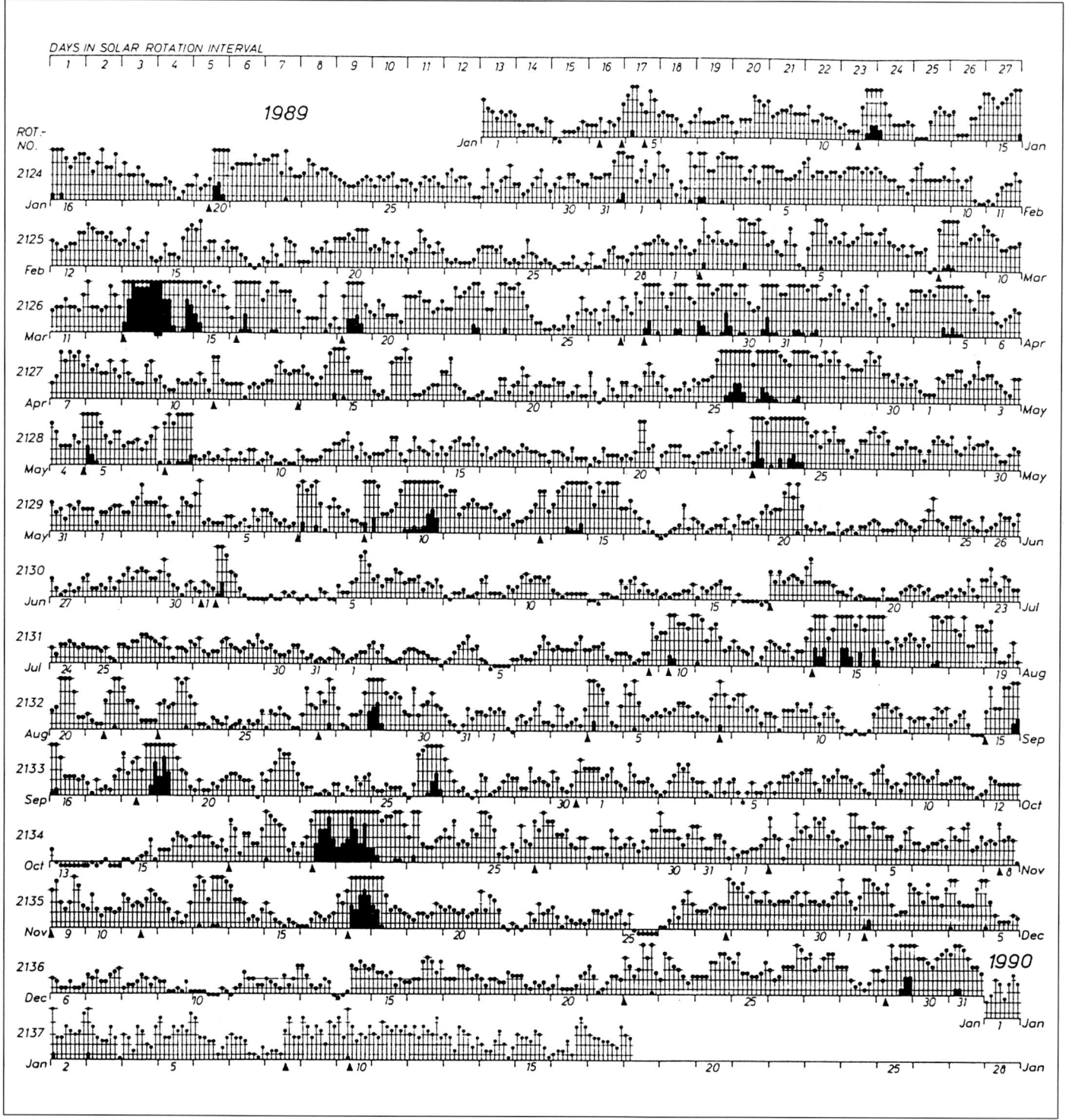

**Figure 8.18** Bartels chart of the planetary 3-hourly average values of the disturbance index, $K_p$, for a year of high solar activity (1989). The very disturbed periods, indicated by the thick black vertical bars, are attributable to either disappearing filaments or flares. There is no evidence of any recurrent activity in the field caused by high speed streams from coronal holes. *Courtesy University of Göttingen.*

feature on the sun on successive rotations, the disturbed (black) areas will lie one under the other.

The 1974 Bartels chart shows two long-lived series of storms which were due to HSSWSs from coronal holes. One series had disturbed periods starting on January 25, February 20, March 20, April 18, and May 15, with magnetically quiet periods preceding each of the disturbed periods. In fact the white space corresponding to the quiet periods probably stands out better than the recurrent disturbed periods. This particular series of disturbances was broken on July 4, 5, and 6 by a major storm which was caused by a large solar flare on July 3. A second series of recurrent storms due to a HSSWS started up on June 28, with recurrences on July 23, August 19, September 15, October 12, and November 8. The given starting times of the disturbances are only approximate, since the disturbances commence gradually. The starting times of the January to May series of storms are not aligned exactly one under the other, indicating that the speed of the HSSWS decreased from one rotation to the next, with the result that the stream took a little longer than 27 days to overtake the Earth. Other series of disturbed periods may suggest themselves. The magnetically quiet periods before the recurrent storms, as exemplified by the values of $K_p$ for 18 and 19 March, support the contention of many HF operators that a very good period for communications is often a portent of a disturbed period.

The 1989 Bartels chart does not show any of the recurrent features evident in the 1974 chart. During periods of high solar activity, most storms are caused by flares which are short-lived phenomena not lasting for a full solar rotation. Consequently the very disturbed periods are distributed more or less at random throughout the chart.

## 8.7 UT and Seasonal Control of Geomagnetic Disturbances

The probability that a given solar event (flare, HSSWS, or disappearing filament) will cause a geomagnetic effect depends on the season of the year and universal time, as well as on the solar event itself [13]. Figure 8.19 gives the number of disturbed days between 1932 and 1983 for which the magnetic index $A_p$ exceeded 36. The number of disturbed days can be seen to be twice as large during equinoctial months (March, April, September, October) as during solstitial months, indicating that the Earth's magnetic field is twice as susceptible to being disturbed by events on the sun during the equinoxes. This increased susceptibility is attributed to the fact that during the equinoxes the direction of the axis of the Earth's magnetic field is more nearly at right angles to the direction of the flow of the solar wind.

The rotation of the Earth (and therefore its magnetic field) about its axis also leads to situations in which the axis of the field is more nearly at right angles to the direction of the solar wind. This leads to a variation of susceptibility throughout a 24-hour period [14], as illustrated in Figure 8.20. Note that the time is universal time, the effect having nothing to do with whether it is day or night. The susceptibility is high during the two shaded S-shaped areas of Figure 8.20 and low during the areas within the closed contours, especially in the centers of these areas, which are marked LOW in the diagram.

For example, if a flare cloud hits the Earth between 15 and 18 UT in June, there is only a small chance of the magnetic field being disturbed. On the other hand, if the same cloud were to hit the Earth at the same UT, but in February or March, there would be a much higher probability that the Earth's magnetic field will be disturbed. In general then, the geomagnetic effects of a given flare, HSSWS, or disappearing filament will be greater during the equinoxes than during the solstices, with a similar story for effects on the ionosphere. A corollary to this statement is that it is quite possible for a small flare (HSSWS, filament) to have a larger effect on the Earth's magnetic field than a larger flare, if the small one occurs during the equinoxes and the large one occurs during the solstices.

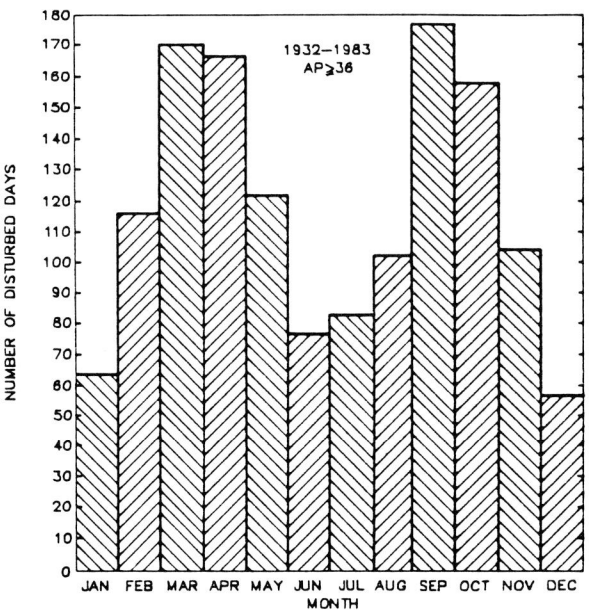

**Figure 8.19** The number of magnetically disturbed days in each month of the year, averaged over the years 1932 to 1983, for which the magnetic disturbance index, $A_p$, exceeded 36. The number of disturbed days is twice as great in the equinoxes as it is in the solstices. In other words, during the equinoxes the geomagnetic field is twice as susceptible to being disturbed by events on the sun. *Courtesy IPS Radio and Space Services.*

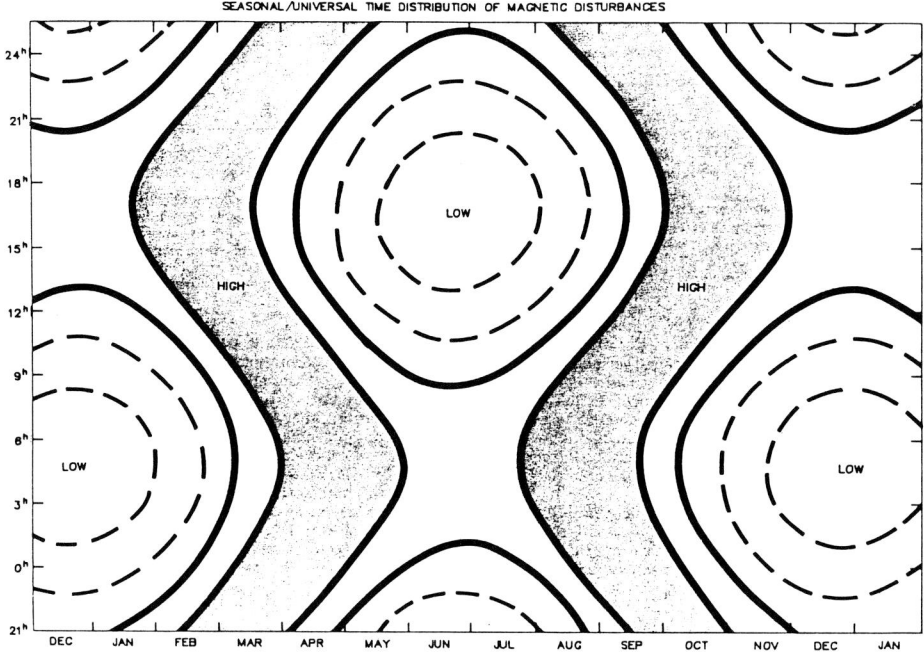

**Figure 8.20** The susceptibility of the geomagnetic field to being disturbed by events on the sun, as a function of universal time and month of the year. The values plotted in Figure 8.19 represent the values for a particular month, integrated across the 24 hours. *Courtesy IPS Radio and Space Services.*

The two recurrent storm sequences apparent in the Bartels $K_p$ chart of Figure 8.17 support these contentions, the storm effects being stronger during the equinoctial months. The largest storms in 1989 occurred in March and October.

## 8.8 References

1. Rodger, A. S., J. Hruska, J. A. Joselyn, and K. Marabushi. *Report of the Geomagnetic and Space Environment Working Group.* Solar-Terrestrial Predictions Workshop, Leura, Australia, October 1989.

2. Webb, D. F. *The Role of the Coronal Mass Ejection as a Geoeffective Phenomenon.* Solar-Terrestrial Predictions Workshop, Leura, October 1989.

3. Wright, C. S. *A Comparison of the Signatures of Disappearing Filaments and Flares in the Interplanetary Medium at 1 AU, and their Relationships to Geomagnetic Disturbances.* Solar-Terrestrial Predictions Workshop, Leura, October 1989.

4. Hewish, A. *IPS-Imaging and Short-term Solar-Terrestrial Predictions.* Solar Terrestrial Predictions Workshop, Leura, October 1989.

5. Sawyer, C., J. W. Warwick, and J. T. Dennett. *Solar Flare Prediction.* Colorado Associated University Press, Boulder, 1986.

6. Kahler, S. W. *The Role of the Big Flare Syndrome in Correlations of Solar Energetic Proton Fluxes and Associated Microwave Burst Parameters.* J. Geophys. Res. 87, pp. 3439–3448, 1982.

7. Cliver, E. W., L. F. McNamara, and L. C. Gentile. *Peak Flux Density Spectra of Large Solar Radio Bursts and Proton Emission from Flares.* J. Geophys. Res. 90(A7), pp. 6251–6266, 1985.

8. Wright, C. S., and L. F. McNamara. *The Relationships between Disappearing Solar Filaments, Coronal Mass Ejections, and Geomagnetic Activity.* Solar Physics 87, pp. 401–417, 1983.

9. Hruska, J., R. L. Coles, H. L. Lam, and G. Jansen van Beek. *The Major Magnetic Storm of 13–14 March, 1989: Its Character and Impact in Canada.* Solar-Terrestrial Physics Workshop, Leura, October 1989.

10. Lundby, S., B. E. Chapel, T. Watanabe, D. H. Boteler, and R. E. Horita. *Prediction of Geomagnetically Induced Current Levels in the British Columbia Hydro 500 kV System.* Solar Terrestrial Predictions Workshop, Leura, October 1989.

11. Jones, C., and C. S. Wright. *The Effects of Geomagnetic Disturbances on Long-Conductor Systems—A Review.* IPS Radio and Space Services Report IPS-TR-86-06, Sydney, December 1986.

12. Heckman, G. *Terrestrial Impacts of Active Region 5395 and Implications for the Remainder of Solar Cycle 22.* Solar Terrestrial Predictions Workshop, Leura, October 1989.

13. Thompson, R. J. *The Morphology of Geomagnetic Disturbances in the Australian Region*. IPS Radio and Space Services Technical Report IPS-TR-87-01, Sydney, January 1987.
14. Thompson, R. J. *The Sun and Solar-Induced Terrestrial Disturbances: A Review*. IPS Radio and Space Services Technical Report IPS-TR-85-12, Sydney, 1985.

## 8.9 Problems

1. What are the three features seen on the sun's surface which must be watched carefully because of resulting effects on the Earth's ionosphere and thus on HF communications?
2. List the three main effects of a solar flare and explain in a few words just what happens.
3. How long after the flare would the three effects start to be felt?
4. Explain the connection between solar coronal holes and disturbances to HF communications.
5. How long after a large coronal hole has passed through CMP (central meridian passage) would you expect to see the effects on the Earth of the HSSWS?
6. At what parts of the solar cycle would you need to worry about (a) solar flares and (b) coronal holes?
7. Under which of the following would an ionospheric storm be more likely to affect HF communications?
    a. During periods of low or high solar activity
    b. At low or high latitudes
    c. During summer or winter
8. What effects does an ionospheric storm have on the D, E, and $F_1$ layers?
9. How would you recognize a shortwave fadeout? What would you do about it?
10. You find that you can get through on only the lowest frequencies of your allocated set, in conflict with predictions supplied by your local frequency prediction organization. What is going on, and what can you do about it?
11. Your transpolar HF link has just dropped out. What can you do about it?
12. Describe the two types of geomagnetic storms and briefly explain what causes each type.
13. What is it that causes a series of recurrent storms, and why?
14. What change of frequency would you follow during (a) a SWF and (b) an ionospheric storm?
15. During what seasons of the year is the Earth's magnetic field most susceptible to the effects of disturbances on the sun?
16. Does the seasonal control of the severity of a geomagnetic storm show up in Figure 8.18?
17. In his book *Spy Catcher*, Peter Wright states that a successful raid during the Second World War was carried out under cover of a magnetic storm. Given that the (correct) year was 1943, what was the probable solar origin of the storm? Is this consistent with Wright's claim that he "had a good chance of predicting a storm of sufficient size"?

# Chapter 9

# Unusual HF Propagation Modes

## 9.1 Introduction

The propagation modes considered in this chapter exist on a fairly regular basis, but are not taken into account in normal HF propagation predictions because our knowledge of the ionosphere is still too limited to allow us to make reliable predictions of them. Many of them are associated with the equatorial and polar ionosphere, and knowledge of them may help to solve perplexing HF communications or interference problems. The subject matter of this chapter is also covered to some extent by CCIR Report 259, which gives special emphasis to sporadic E modes [1].

## 9.2 Mid-Latitude $E_s$ Modes

We have encountered sporadic E several times already in this book, in Chapters 3 and 4, where we found that it is a very thin and dense layer of electrons situated at about 100 km altitude and that it comes and goes in a sporadic fashion. The critical frequency, $f_oE_s$, of the $E_s$ layer at mid-latitudes often exceeds 5 MHz during the day and during a summer day can even exceed 10 MHz on a regular basis. For a 1,000 km circuit, which has an obliquity factor of 4.4 (see Figure 4.6), the MUF on a one-hop $E_s$ mode will therefore often exceed 22 MHz and perhaps even exceed 44 MHz.

The $E_s$ layer can thus support frequencies above, and sometimes well above, the normal F-layer MUFs on short circuits. On a 1,000 km circuit, a typical 1F MUF is 15–20 MHz during the day. Propagation via the $E_s$ layer will not of course take place if the antenna pattern has effectively zero gain at the low elevation angles involved (9°—see Figure 4.6). For a 2,000 km circuit, which has an obliquity factor of about 6, $E_s$ MUFs of up to 60 or 70 MHz could be anticipated, especially during a summer day. However the corresponding elevation angle is then only about 1°, and only very high power transmitters will provide enough energy at this sort of elevation angle for common antennas to be effective.

$E_s$ propagation modes have several advantages over normal E- and F-layer propagation modes. The most important of these is that they are far less susceptible to fading, because to the radio wave the $E_s$ acts like a good quality mirror. The signals are also stronger because they do not travel as far and thus suffer less distance attenuation, and also because the $E_s$ layers often lie below the altitude at which most HF absorption takes place. The third advantage is that the $E_s$ layers do not move up and down the way the F layer does, and therefore the signals suffer virtually no Doppler shift or spreading. This is an important consideration for very high speed digital data links.

## 9.3 Low-Latitude $E_s$ Modes

As we saw in Chapter 3, low-latitude $E_s$ is a regular feature of the daytime equatorial ionosphere in a band about 5° on either side of the magnetic equator. At vertical incidence, this type of $E_s$ is very transparent and has no great effect on signals passing through it. At high angles of incidence, corresponding to long circuits, the $E_s$ layer apparently acts as a good reflector and it is possible to get higher MUFs on transequatorial circuits than would be expected for normal propagation.

A study of propagation conditions for two circuits, Guam to N. W. Cape (Australia) and Manila to N. W. Cape showed that the MUF regularly exceeded 32 MHz, the upper limit of the observing equipment [2]. It therefore regularly exceeded the 2F MUF, which was typically 25–27 MHz during the day for the data analyzed. This suggests the presence of another propagation mode with an MUF higher than the normal MUF (for the expected 2F mode) during most of the day.

Because of the length of the circuits involved (about 5,000 km), it could be expected that propagation would involve at least two reflections from the ionosphere. If one reflection is from the normal F layer, the only way to increase the MUF on this hop is to increase the length of the hop and thus increase the obliquity factor. However this would mean that the length of the second hop must then be *shorter* than before, lowering the MUF on this hop. The MUF for the circuit would also be lowered, since it is set by the *lower* of the MUFs on the separate hops. Thus reflection on the second hop must take place from a lower layer which has a higher obliquity factor. For the two N. W. Cape circuits, it is found that the observed rate of occurrence of MUFs above 32 MHz can be explained by assuming that propagation takes place by a mode which involves a reflection from the equatorial $E_s$ layer, followed by a longer-than-normal F-layer second hop. This propagation mode is described as an $E_s$-F mode and is illustrated in Figure 9.1. There is also evidence for a mode in which the signals are reflected twice by the $E_s$ layer and then by the F layer. This is called a $2E_s$-F propagation mode, or in general, an $nE_s$-F mode.

There do not seem to have been any studies made of the signal characteristics of $nE_s$-F modes, but the observations made at N. W. Cape indicated that the strongest signals on the two circuits corresponded to the 2F mode. However the antenna characteristics did not favor the low angle $nE_s$-F modes.

## 9.4 Transequatorial Propagation

Transequatorial propagation (TEP for short) is the name given to propagation on circuits which cross the equator, more or less at right angles, and which have higher MUFs than the normal multihop modes. TEP was discovered by radio amateurs thirty or forty years ago, but it was much later before the propagation modes were deduced. There are two types of TEP which rely on different features of the equatorial ionosphere for their support and which have different characteristics [3, 4, 5]. These are called **afternoon-type** and **evening-type TEP**, according to the time of day when they occur. TEP has been observed regularly between the Mediterranean countries and southern Africa, between Japan/Korea and northern Australia, and between Central and South America. It has been found that it is the magnetic equator and not the geographic equator which is important in determining what circuits will support TEP modes.

Both types of TEP have been observed simultaneously on some circuits, at around 2000 local time. Experiments in Australia by radio amateurs (see Section 9.7) have shown that for locations in northern Australia, northern Japan transmitters (Hokkaido) can be heard on 50 MHz during the day and southern transmitters (Okinawa) can be heard during the evening. Transmitters in central Japan can be heard most of the afternoon and evening.

### 9.4.1 Afternoon-type TEP

Afternoon-type TEP has the following characteristics:

1. An MUF greater than the normal 2F MUF, i.e., greater than 40–50 MHz

2. A peak occurrence at around 1700 to 1900 local time, near the equinoxes and around solar cycle maximum

3. Path lengths of 6,000 km and sometimes longer

4. Normally strong steady signals with a low fading rate and a small Doppler spread (which means that the reflecting layer does not move much)

It has been established by raytracing simulations [6] that the propagation mode for these signals is a **super mode**, or FF mode, in which the signals are reflected twice by the F layer, on opposite sides of the equator, without an intervening ground reflection. This is illustrated in Figure 9.2. The super mode relies for its support on the particular electron density distribution which exists in the vicinity of the magnetic equator. As we saw in Chapter 3, the electron density distribution in the equatorial regions has maximum values (called "crests") at about 15° dip angle north and south of the magnetic equator, while the height of the maximum electron density,

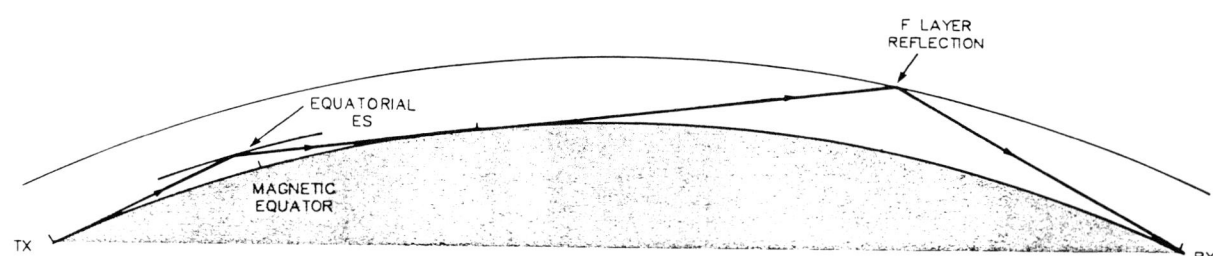

**Figure 9.1** Sketch of an $E_s$-F propagation mode supported on the first hop by equatorial $E_s$ and on the second by the F layer. The F-layer hop is longer than the normal hop, which would have a length equal to half the ground range, and can thus support frequencies higher than the normal 2F MUF. These higher frequencies are also supported by the $E_s$ layer because of its higher obliquity factor and high critical frequency.

# Unusual HF Propagation Modes

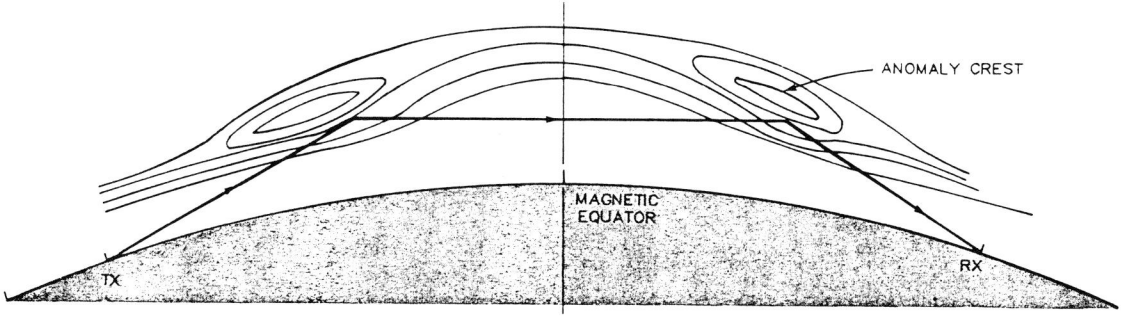

**Figure 9.2** The "super" mode or FF mode of propagation believed to be responsible for afternoon-type TEP. Because of the upward tilt of the crest of the equatorial anomaly at the first reflection point, the radio waves travel directly across to the opposite crest without an intervening ground reflection. At this crest, the waves are reflected back down to Earth at the receiver. The FF mode has a higher obliquity factor at each of its reflection points because of the tilts and can thus support higher frequencies than the normal 2F mode.

$h_mF_2$, has minimum values at these points and a maximum over the magnetic equator. It is thus possible for a ray leaving a suitably placed transmitter to be reflected from the first crest in such a direction as to miss the Earth and strike the ionosphere at the opposite crest and thence be reflected back to Earth. Such a propagation mode would have an MUF higher than the normal 2F MUF because the ionosphere at the crests is tilted upwards towards the magnetic equator and gives rise to larger angles of incidence, and consequently larger obliquity factors and MUFs. The MUFs are also high because the critical frequencies $f_oF_2$ are so high in the crests.

Maximum observed frequencies for the super mode regularly exceed 50 MHz, where they have been observed in the 50 MHz or 6m amateur band, and have been known to exceed 60 MHz. The high signal strengths associated with super modes are readily explained in terms of focussing effects clearly exhibited in the raytracing studies. Signals from a large range of elevation angles at the transmitter arrive at the receiver, rather than from just one particular elevation angle, giving higher than average field strengths. The signals also pass through the absorbing D region only twice, as against four times for a 2F mode, and are thus less attenuated. The super mode does not exist every day, even during the times of its maximum occurrence—the equinoxes at solar maximum. Whether or not it appears on a given day depends on how the crests of the equatorial ionosphere have developed on that day. The crests must be high and well separated before they will support the super mode of propagation. This is also the reason the super mode exists for only part of the day, mainly the late afternoon and early evening, in spite of the fact that the crests themselves exist for a much larger part of the day. Figures 9.3 and 9.4 illustrate the variation of $f_oF_2$ and $h_mF_2$ along the Yamagawa to Townsville circuit for selected days in August 1970, when TEP was observed or not observed.

The super mode is a very useful mode when it exists, offering strong, slowly fading, signals at frequencies well clear of the congested HF spectrum. However it has not yet been possible to evolve a reliable prediction scheme which would allow an HF communicator to switch to the higher frequencies with confidence of finding a super mode present.

### 9.4.2 Evening-type TEP

Like the afternoon-type TEP, the evening-type TEP relies on a feature of the equatorial ionosphere for its

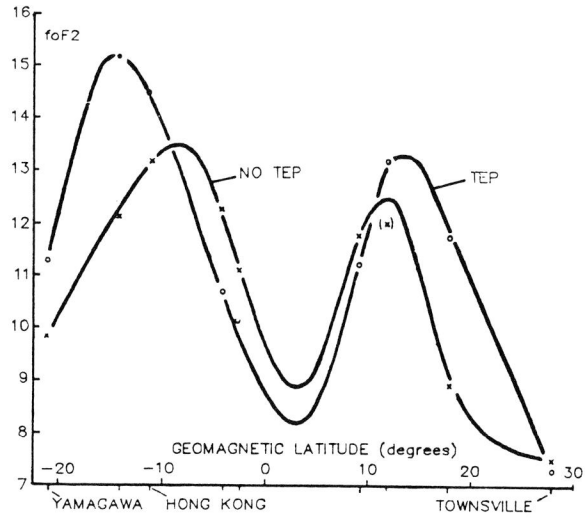

**Figure 9.3** The variation of $f_oF_2$ along the Yamagawa to Townsville circuit for selected days in August 1970 for which the FF mode existed (TEP) or did not exist (NO TEP). The maximum values of $f_oF_2$, which correspond to the crests of the equatorial anomaly, are seen to be higher in frequency and further from the geomagnetic equator on those days for which TEP was observed. This is especially true for the northern peak for this month. The distribution is more symmetric with respect to the magnetic equator during the equinoxes. The ionospheric stations marked are Chung Li, Hong Kong, Manila, Bangkok, Singapore, Vanimo, and Port Moresby.

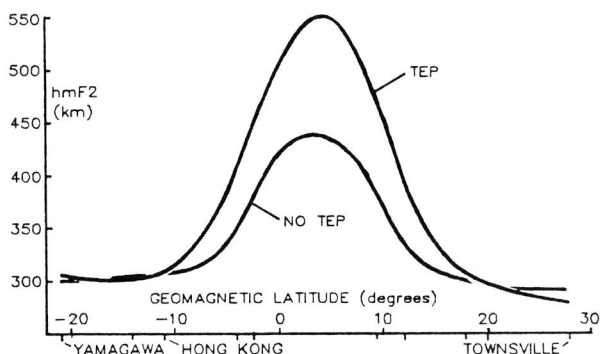

**Figure 9.4** The variations with geomagnetic latitude of the height of the peak of the $F_2$ layer, $h_mF_2$, for the conditions described in Figure 9.3. The main difference between the two curves is that the peak of the $F_2$ layer over the equator is significantly higher on those days for which TEP was observed.

support. (This is why TEP is confined to transequatorial circuits; the appropriate conditions do not exist on mid-latitude and high-latitude circuits.) However evening-type TEP generally supports higher frequencies than the afternoon type and has very different characteristics:

1. A peak occurrence at around 2000 to 2300 local time, near the equinoxes and around solar maximum

2. High signal strengths but with deep and rapid fading, with rates up to about 15 Hz, and a large Doppler spread which sometimes exceeds 40 Hz

3. Path lengths usually shorter than for the afternoon-type mode, being about 3,000 to 6,000 km

The occurrence of evening-type TEP is strongly correlated with the occurrence of range spreading on evening ionograms at equatorial stations along the circuit [7]. Figure 9.5 illustrates the fact that range spreading is a necessary, but not sufficient, condition for the TEP to occur. The occurrence rates peak in the equinoxes and drop off as the frequency increases.

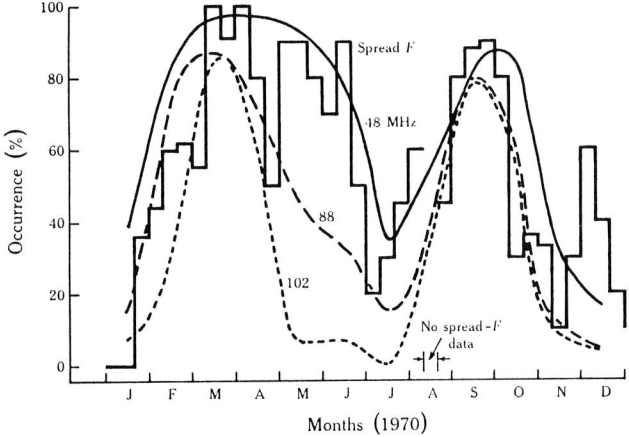

**Figure 9.5** Occurrence rates for 1970 of spread F before 0000 LT at Vanimo, New Guinea, and of reception at Yamagawa, Japan, of 48, 88, and 102 MHz transmissions from Darwin.

The propagation mode for evening-type TEP is probably a whispering gallery, or "field-guided" mode. Experiments have shown that at times when range spreading and evening-type TEP are observed, the equatorial ionosphere is threaded by "empty" tubes which are aligned along the field lines of the Earth's magnetic field and in which the electron density is much lower than that of the surrounding ionosphere. This is illustrated in Figure 9.6. Propagation is assumed to take place by the rays skidding around the walls of the tube, being reflected several times off the walls, and then emerging at the far end of the tube. Platt and Dyson have extended the ideas of earlier workers and provide many useful references [8]. The occurrence rates for both range spreading and evening-type TEP were found to be much higher for the 1970 solar maximum than for the much higher 1979 maximum, even though the occurrence rates correlate directly with the level of solar activity in any one cycle. This may be due to the higher E-layer electron densities and conductivities in 1979, which would tend to stabilize any fluctuations in the base of the $F_2$ region.

Maximum observed frequencies (MOFs) in excess of 100 MHz were regularly observed between Darwin (Australia) and Yamagawa (Japan) during the 1970 peak in solar activity. The very high MOFs arise from the very high angles of incidence, the rays just grazing the walls of the empty tubes. The large Doppler shifts are caused by the general upward movement of the tubes, which rise rapidly through to the top of the ionosphere after their creation near its base. The best circuits for this type of TEP are those for which rays can enter the tubes in a direction parallel to their axes, or in other words, the rays from the transmitter should be tangential to the Earth's magnetic field at the altitude where the rays enter the tube. To achieve the highest MUFs, the receiver should be similarly placed with respect to the Earth's magnetic field, which means that the circuit should be symmetric with respect to the magnetic equator. Darwin and Yamagawa are ideally placed for evening-type TEP, but are too close for afternoon-type TEP. Evening-type TEP also remains essentially unpredictable on a night-to-night basis.

## 9.5 Round-the-World Propagation

In a long series of experiments, it has been found that under certain conditions signals can travel right around the world, reappearing at the back end of the transmitting antenna and in the azimuth opposite to that of transmission. The elevation angles of these signals have been measured to be about 22–24°, while time delays have been found to be around 138 milliseconds(ms). If we consider only the elevation angles, we would deduce that propagation was via a very high order mode, something like a 35-hop mode. However with losses at each of 34

# Unusual HF Propagation Modes

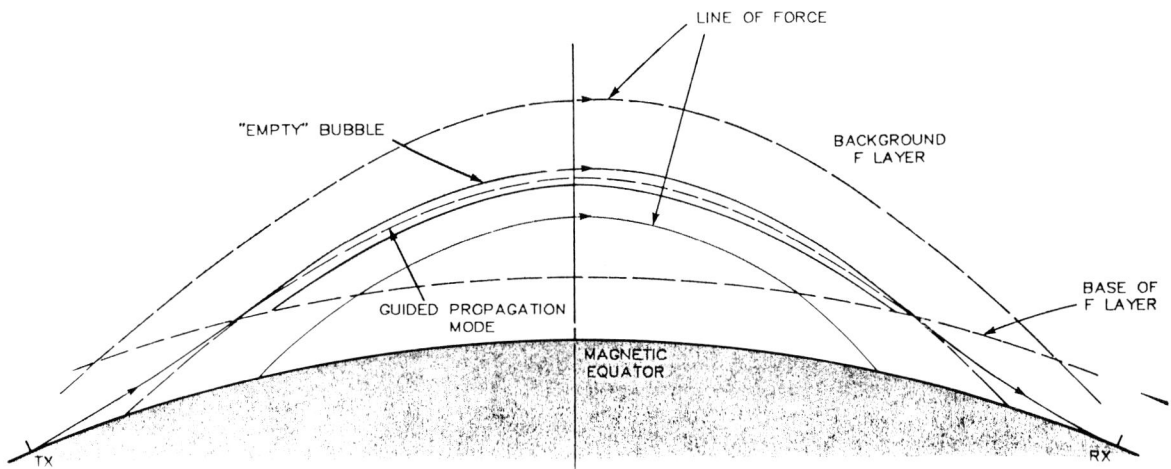

**Figure 9.6** Sketch of a field-guided propagation mode which is believed to be responsible for evening-type TEP on many circuits. Hollow "bubbles" are formed just after sunset as the F layer rises rapidly, and diffuse quickly along the lines of force of the Earth's magnetic field, yielding a family of field-aligned "empty" tubes embedded in the background ionosphere. VHF signals entering these tubes from favorable transmitter locations are guided around the tubes in a type of whispering gallery effect. The high frequencies supported by this mode are made possible by the large angle of incidence which the signals make with the walls of the tube—the ray path is nearly tangential to the walls—with a correspondingly high obliquity factor. At any one time, the transmitter and receiver may be linked by any number of bubbles which, because of their rapid rise through the F region, impose a range of Doppler shifts on the received signals.

ground reflections and 35 ionospheric reflections, and with 70 passages through the absorbing D region, there is no chance that any of the transmitted signal would be left to be detected by a receiver. The propagation time for a 35-hop mode would be about 160 ms, telling us again that this round-the-world (RTW) propagation cannot be by normal multihop propagation. It seems very likely that propagation is via a chordal mode (Figure 9.7) in which the signal skips from one part of the ionosphere to the next without hitting the Earth between ionospheric reflections [9, 10]. The conditions must be just right at the beginning and end of a RTW mode, firstly for the radio-wave energy to get into the mode and secondly to get the energy out of the mode again for it to reach the receiver.

The tilts in the ionosphere associated with dawn and sunset offer ideal methods of getting energy into and out of chordal modes. It is also possible to create artificial situations for the successful coupling of HF energy into and out of chordal modes. This is done by heating the F region with a high power radar (a few megawatts), thus creating a region of field-aligned irregularities which scatter the signal into or out of the mode.

It is unlikely that the normal communicator will ever have to make allowances for this type of propagation, although amateurs find it attractive, and a prediction scheme has been proposed [11]. RTW echoes also appear in some over-the-horizon backscatter ionograms.

## 9.6 Non-Great-Circle Propagation

The normal HF propagation path follows approximately a great circle from the transmitter to the receiver. However, for signals propagating in middle- to high-latitude regions and across the equator, this is not always the case. Substantial evidence indicates that propagation

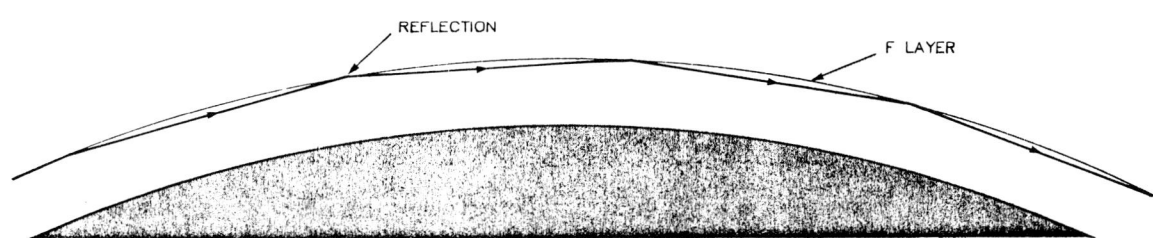

**Figure 9.7** The chordal mode of propagation involves multiple reflections from the F layer without any ground reflections. The high signal strengths associated with such modes are attributed to the fact that the signals do not pass repeatedly through the D region, only on the first and last hops, and do not suffer any losses at ground reflections.

often takes place by non-great-circle (NGC) propagation paths, which consequently must be taken into account. Possible causes of NGC propagation include ground side-scatter, deviation by strong horizontal gradients, and reflection from spread F irregularities in the auroral and equatorial zones. In each case, the propagation path makes a triangle with the great circle path at its base. This is illustrated in Figure 9.8.

On long middle- and high-latitude paths, NGC signals due to auroral side-scatter could be important to communications since there are occasions when the direct-path (great circle) signals are much lower in amplitude than the NGC signals. In certain cases, however, NGC propagation modes would be a liability rather than an asset. High-speed pulse transmissions might be impossible to receive correctly if the NGC signals were of comparable amplitude to the direct path signals, and the NGC travel time (propagation delay) exceeded the direct-path time by approximately the pulse width or more. This situation would lead to the multipath interference described in Chapter 7. NGC propagation on transequatorial circuits relies on side scattering by field-aligned irregularities in the nighttime equatorial F region.

## 9.7 6-Meter Observations of TEP

Extensive propagation experiments have been conducted between Australia and Japan, mainly in attempts to explain the phenomena of TEP which we met in Section 9.3. Stepped-frequency soundings (4 to 64 MHz) have been made between Yamagawa and Okinawa in Japan and Townsville and Adelaide in Australia, while CW transmissions from Darwin (on 32, 48, 72, 88, and 102 MHz) have been monitored at Yamagawa. The experiments provided considerable details of the conditions existing on the various circuits, but the nature of the conclusions that could be drawn was limited by the design of the experiments. Of particular interest here is the fact that little can be said concerning the effect of the latitudes of the terminals, because these were fixed.

When TEP is present, the maximum usable frequency on Japan-Australia circuits usually exceeds 50 MHz, which means that the 6m amateur band is open. Consequently, amateur radio operators work these circuits, collecting data which might be useful to supplement the data obtained on the fixed circuits. In fact, amateur radio operators made most of the initial observations on TEP. This data is available from all parts of Japan and Australia and might therefore be expected to contain information concerning latitudinal properties of TEP.

In this section, we describe results obtained by analysis of entries in the logbooks of some Australian operators. The section is openly didactic and is given as an example of the analysis of experimental data involving HF propagation. The data is far from ideal, not being the results of a planned experiment, and great care must be taken to distinguish between operator effects and effects arising from the properties of TEP. We commence by formulating an ideal mathematical model which facilitates the understanding of the data and the discussion of results.

### 9.7.1 Theory

For purposes of radio, Japan is divided into 10 areas, denoted by JA0, JA1 ... JA9. The approximate locations of these areas are shown in Figure 9.9. The data available for analysis consisted mainly of the times of day and the call sign of the operator contacted in Japan. These call signs were of the form JAabcd where "a" denoted the area (0, 1 ... 9) and "bcd" was a combination of letters unique to the particular operator. Only "a" has been included in the analyses since this contains the latitudinal information sought. The times recorded were usually the start

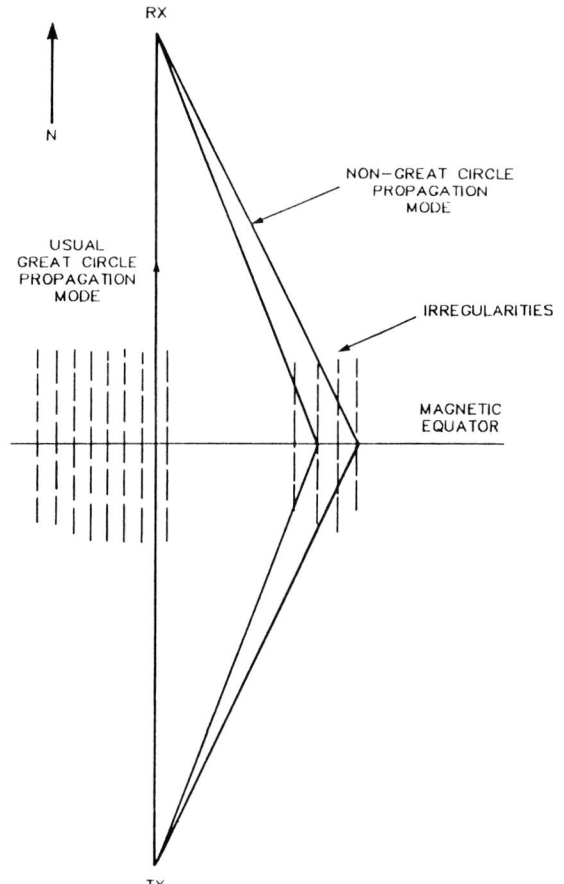

**Figure 9.8** Non-great-circle propagation due to scattering by field-aligned irregularities to one side of the great-circle path. This particular sketch is for a south-north path across the equator at night. A similar situation holds for a circuit of any orientation which passes near the auroral oval. It is also possible for the irregularities to scatter energy back towards the transmitter, in a process called backscatter.

# Unusual HF Propagation Modes

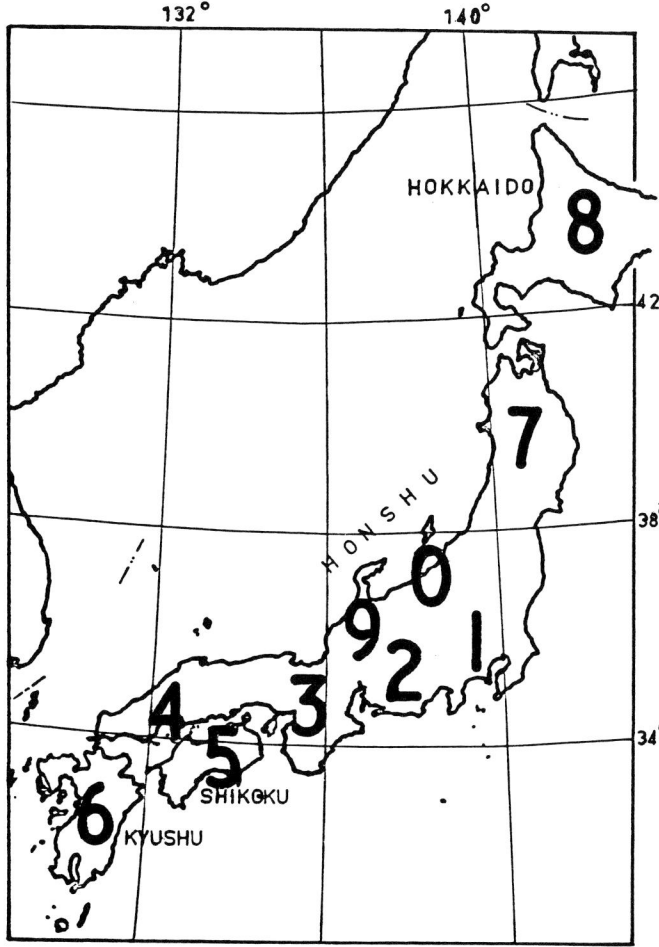

**Figure 9.9** The locations in Japan of the areas identified by their radio call signs JA0, JA1, . . . JA9.

time of the contact, noted to the nearest minute. The finish times were usually not recorded, nor were those times when attempts to make contacts were unsuccessful. The interval between successive contacts was of the order of 10 minutes, this being the time necessary to exchange the relevant information and pleasantries.

To derive a theoretical expression for the number of contacts made by an operator in Australia with a particular area i of Japan in a time t, $C_i(t)$, there are two factors which must be taken into account:

1. During each observing period, one evening for example, the same operator is not contacted more than once.
2. The ten areas of Japan have different populations, $p_i$.

The former of these factors arises because it is the aim of the operators to make as many different contacts as possible. Its effect is to decrease the number of operators available in a particular area at a given time, so that the number of contacts made must saturate to a finite number as time progresses. We shall call this effect the saturation effect. Provided the areas 0, 1 . . . 9 are chosen at random, the effect of the second factor is just a weighting effect taking into account the fact that when all areas are available for contacts, it will be the most heavily populated areas which are contacted most often.

Arguing by analogy with the theory of radioactive decay, we shall assume that the rate at which contacts are made with a particular area at time t depends on the number of uncontacted operators in that area, that is,

$$dC_i/dt = k_i \{ N_i - C_i(t) \} \quad (9.1)$$

where $N_i$ is the number of operators available in area i and $k_i$ is the proportionality factor for this effect for area i.

Equation 9.1 is not yet complete because although an operator may be available in area i, the chance of that operator being selected by the operator in Australia is only $N_i/N$, where N is the number of operators available in all of Japan. (This assumes that areas are selected at random.) Assuming a fixed proportion of amateur radio operators per head of population in each of the ten areas, that is,

$$N_i = gp_i \quad (9.2)$$

and denoting the total population of Japan by P, we have

$$dC_i/dt = k_i \{gp_i - C_i(t)\}. \quad (9.3)$$

Integrating with respect to t and using the initial condition

$$C_i(0) = 0 \quad (9.4)$$

yields

$$C_i(t) = gp_i \{1 - \exp(-k_i(p_i/P)t\}. \quad (9.5)$$

It should thus be possible, given the validity of the preceding assumptions, to determine the unknowns g and $k_i$. The parameter $k_i$ would then be a measure of the "activity" or "quality" of the circuit and would be a suitable basis for comparison of any two circuits, including circuits from different areas of Australia.

There are two limiting forms of equation 9.5, according as the exponent $k_i(p_i/P)t$ is very large or very small compared with unity:

(A)  $k_i(p_i/P)t \ll 1$   $C_i(t) = k_i g P^{-1} p_i^2 t$   (9.6)

(B)  $k_i(p_i/P)t \gg 1$   $C_i(t) = gp_i$.   (9.7)

Condition A corresponds to one or more of (a) a poor circuit ($k \approx 0$), (b) a short time t, or (c) a low population $p_i$, while condition B corresponds to one or more of (a) a very good circuit ($k \gg 0$), (b) a long time t, or (c) a high population $p_i$. The dependence of $C_i$ on $p_i$ will be quadratic when condition A is valid, and linear when condition B is valid. When neither condition is valid, the dependence will be given by equation 9.5, with possibly a rough power law dependency between $p^1$ and $p^2$.

## 9.7.2 Results

The data available for analysis unfortunately had some flaws (described elsewhere [12]) which made it difficult to draw too many conclusions from them. However there are two types of results which can still be obtained. Firstly there are those deduced from a study of the relationship between $C_i$ and $p_i$, and secondly there are histograms of the occurrence rate as a function of time.

One approach to the problem of comparing the qualities of circuits to different parts of Japan would be to look for departures from the best fit straight line on a log $C_i$ versus log $p_i$ graph. Any point lying significantly below the line would represent an area that is undercontacted relative to its population, and vice versa. The slope of the line would be 2 if limit A applied, and 1 if limit B applied. The occurrence histograms are readily constructed, but should be regarded with some caution since there is no way to remove operator effects from this set of data.

### 9.7.2.1 The Relationship Log C versus Log p

The observations considered here came from three sets of logbooks from observers stationed in Townsville, VK4ZBE (W. S. O'Donnell), VK4ZRG (R. L. Grommet), and VK4ZTK (R. C. Tulloch). There is great advantage in having the three sets of data since any conclusions drawn must hold for all three sets. This considerably reduces the possibility of statistical fluctuations or operator effects being mistaken for properties of TEP. The times are all 150° E standard times.

The data has been divided into two groups depending on whether the contact was made before 1900 h or after 2000 h, in an attempt to bring out any differences between afternoon- and evening-type TEP. All logarithms are to base 10 and populations are measured in units of $10^6$.

The log $C_i$ versus log $p_i$ results for the three sets of data are given in Figure 9.10. The solid lines are considered to be the straight lines best fitting the data and are constrained to pass through the data point for area 1 (which should have the least uncertainty associated with it since

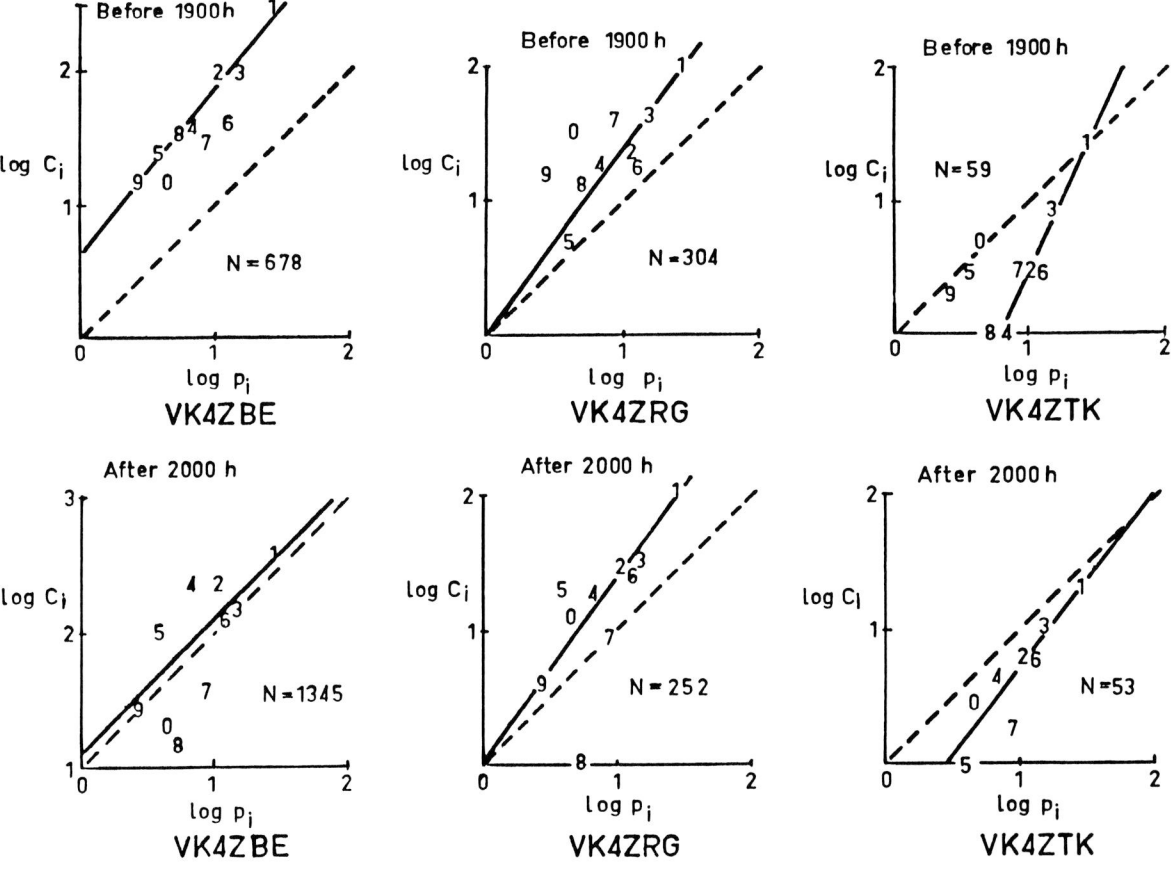

**Figure 9.10** Relationship between the number of contacts made by three Townsville radio amateurs and the population of the corresponding areas of Japan, before 1900 LT (135° E), and after 2000 LT. The dashed lines have a slope of unity.

it corresponds to the greatest number of contacts). The dashed lines have a slope of unity ($C_i \propto p_i$) and are shown for comparison.

It can be seen from Figure 9.10 that during the afternoon area 6 was undercontacted relative to its population by all three observers. No additional conclusions can be drawn which are consistent with all three sets of data. The position of a data point above the best-fit line is taken as indicating either a conscious effort on the part of the observer to contact that area in particular, or a statistical fluctuation. The alternative hypothesis, that the circuit to that area is significantly better than all other circuits, is rejected because it is not consistent with the data of all observers. The areas which were overcontacted are almost exclusively those with the lowest populations, i.e., areas 0, 5, and 9. The results for VK4ZTK show very clearly that attempts were made to increase the numbers of contacts with these areas. They should not have been contacted at all on the basis of the numbers of contacts with the other areas.

The steep slope of the VK4ZTK line in Figure 9.10 arises because most of the data considered corresponded to just one or two observations per day. The observing time is thus very small and the limit (A) is valid, yielding a slope of 2.

Figure 9.10 also shows that during the evening the areas 7 and 8 (in northern Japan) were significantly undercontacted by all three observers. Again, no additional conclusions can be drawn which are consistent with all three sets of data. Note that absence of a data point for any area indicates that it was not contacted at all.

### 9.7.2.2 The Occurrence Histograms

The diurnal variations of the occurrence rates of Japan-Townsville contacts for the three observers are shown in Figure 9.11. The occurrence rate peaks in the afternoon at or before about 1600 h and again at 1900–2000 h. The 2100–2200 h peak in the VK4ZBE data is probably a real effect, corresponding to the evening enhancement of the MOF associated with the opening up of the evening-type TEP mode. The 1900–2000 h peak is attributed to the reappearance of the afternoon-type TEP mode, although it may have been due to the reappearance of the operator in his shack after dinner.

### 9.7.3 Summary of Results

There are two definite conclusions which can be drawn concerning latitudinal effects on the Japan-Townsville circuits:

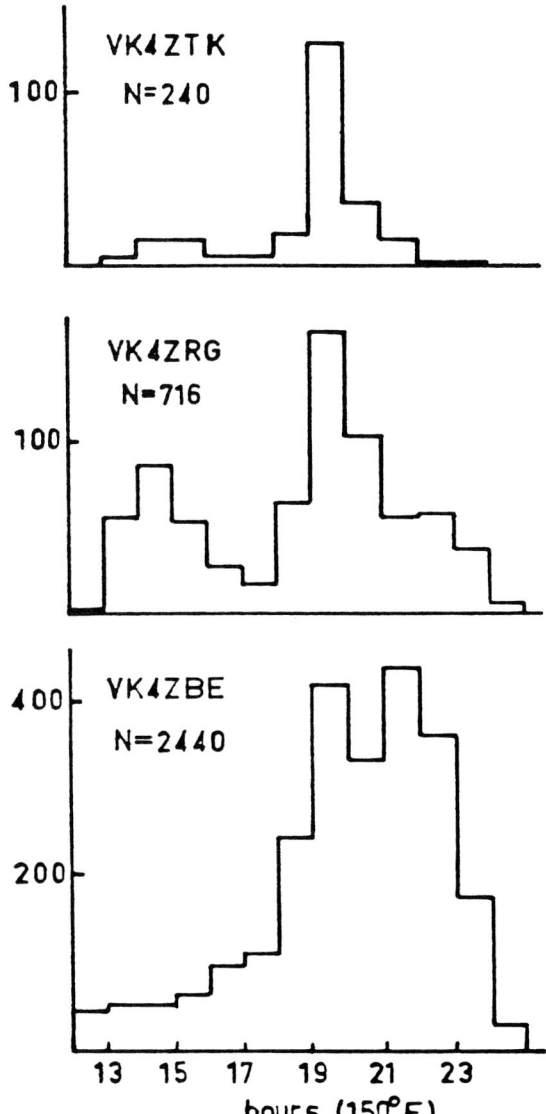

Figure 9.11 The diurnal variation of the occurrence rate of Japan-Townsville 50 MHz contacts for three Townsville observers.

1. During the afternoon (before 1900 h) the circuit to area 6 in southern Japan is significantly poorer than those to all other areas.

2. During the evening (after 2000 h) the circuits to areas 7 and 8 in northern Japan are significantly poorer than those to all other areas.

These conclusions are consistent with the properties of the two types of TEP described earlier. The same conclusions can be drawn for locations throughout Australia, but it is not possible to compare the qualities of the circuits to two different Australian locations because of the fairly large scatter in the data. The slopes of the lines in figures such as Figure 9.10 are determined by the length of the average observing period and usually cannot serve for comparisons of two sets of data. In spite of the satisfaction that can be derived from deducing *anything* from a set of observations, it is quite clear in retrospect that the 6m data observations are of limited use. The design of a better experiment is left to the reader (see problem 8 of Chapter 16).

## 9.8 References

1. CCIR Report 259-6. *VHF Propagation by Regular Layers, Sporadic E or Other Anomalous Ionization.* Dubrovnik, 1986.

2. McNamara, L. F. *Ionospheric Predictions on Transequatorial Circuits.* Proc. IRE (Aust.), pp. 117–126, May 1974.

3. McCue, C. G., and D. F. Fyfe. *Transequatorial Propagation: Task Bridger Introductory Review.* Proc. IRE (Aust.), pp. 1–10, January 1965.

4. Nielson, D. L. *Long-range VHF Propagation across the Geomagnetic Equator.* Stanford Research Institute Report, 1969.

5. Harrison, R. L. *V.H.F. Transequatorial Propagation.* Amateur Radio (Aust.) 40(5), p. 3; 40(6), p. 6, 1972.

6. Gibson-Wilde, B. C. *Relation between the Equatorial Anomaly and Transequatorial VHF Radio Propagation.* Radio Science 4, p. 797, 1969.

7. McNamara, L. F. *Evening-Type Transequatorial Propagation on Japan-Australia Circuits.* Aust. J. Physics, 26, pp. 521–543, 1973.

8. Platt, I. G., and P. L. Dyson. *VHF Transequatorial Propagation via Three Dimensional Waveguides.* J. Atmos. Terr. Physics 51(11/12), pp. 911–928, 1989.

9. Harrison, R. L. (Ed.). *Try the "Twilight Zone" for Consistent DX.* Radio Experimenters Handbook, pp. 15–17. Federal Publishing Company, Sydney, 1984.

10. Hortenbach, K. J., and F. Rogler. *On the Propagation of Short Waves over Very Long Distances: Predictions and Observations.* Telecomm. J. 46(VI), pp. 320–327, 1979.

11. Moo, C. A. *HF-RTW Propagation Study: Prediction Scheme and Synoptic Behaviour.* In Ionospheric Forecasting, AGARD Conference Proceedings No. 49, V. Agy (Ed.), France, 1970.

12. McNamara, L. F., and R. L. Harrison. *6-Meter Amateur Contacts between Australia and Japan.* Ionospheric Prediction Service Report IPS-R22, Sydney, April 1973.

## 9.9 Problems

1. Why do we need to concern ourselves with unusual propagation modes?

2. What feature of the mid-latitude ionosphere sometimes supports propagation at both HF and VHF?

3. At what time of day, season, and level of solar activity would you expect unusual propagation to occur at mid-latitudes?

4. What features of the equatorial, or low-latitude, ionosphere lead to the support of unusual propagation modes?

5. Describe and sketch the propagation modes associated with each of these phenomena.

6. At what time of day, season, and level of solar activity would you expect these propagation modes to occur?

7. Which of the unusual equatorial F-layer propagation modes supports the higher frequency, and which would be the most useful for communications?

8. If you had a north-south transequatorial circuit about 6,000 km long which was bisected by the magnetic equator, what unusual propagation modes would you expect to find on it during the equinoxes at high levels of solar activity?

9. What evidence is there that round-the-world propagation takes place via a chordal mode of propagation?

10. Under what conditions would you need to consider the possibility of non-great-circle propagation?

11. With regard to the analysis of the Japan-Australia TEP logbooks, show that equation 9.6 may alternatively be derived from equation 9.3 by neglecting the saturation effect, i.e., by assuming that the number of contacts made is small compared with the number of operators available.

# Chapter 10
# Propagation Prediction Programs

## 10.1 Introduction

The general methods used for predicting what the HF propagation conditions will be on a given circuit at a given time have already been outlined in Chapter 6. Historically, propagation predictions have been made by a central agency using a mainframe computer, and then distributed by mail to the frequency managers who share their allocated frequencies among their users according to their own internal priorities. Probably the best known mainframe program in the United States is IONCAP [1], but other governments have similar programs. Ideally, all such programs are based on the reports and recommendations of CCIR. Radio waves know no national boundaries, so a united effort is required to rationalize the frequency allocation procedures.

With the increasing amount of computer power available, it is becoming increasingly possible for communicators to make their own predictions and thus work to far more flexible and probably more correct frequency schedules than have traditionally been available. A number of propagation programs suitable for PCs now exist, and new ones are continually being added. Before the advent of powerful PCs, many valiant attempts were made to downsize the problems to suit the available computers. This led to some rather clever approximations and parameterizations, but the value of these efforts must now be questioned in the light of modern technologies.

The field of PC programs covers a wide range of levels of complexity. At one end, there is the "mainframe" version of IONCAP, which has been successfully ported to a (rather powerful in today's terms) PC without significant loss of function. At the other end, there are programs such as MINIMUF [2] which provided the best possible physical models for use in the very constrained computer environment of its day. The program chosen by each communicator is determined by many considerations, such as availability of the code, expected reliability, how closely it meets particular needs, user-friendliness, available computer assets, and, of course, the cost. The author uses IONCAP and ASAPS [3], which is the PC version of the program used by the Australian IPS Radio and Space Services, for all of the above reasons.

Sections 10.2 and 10.3 use output from ASAPS to illustrate some of the topics covered in earlier chapters which are best discussed in terms of particular cases. Other programs could have been used, but not all provide the wide range of useful outputs given by ASAPS. A brief review of some of the available computer programs is presented in Section 10.5. Note that it is not intended that this review be complete, which would be a daunting task.

## 10.2 Predictions of Usable Frequencies

Propagation prediction programs provide monthly median predictions of both the frequencies which the ionosphere is expected to support on the given circuit at the given time and the expected S/N levels for these frequencies. The predictions of frequency support tend to be quite reliable, although they have limited utility without the accompanying S/N ratios. Provided too many approximations have not been made with the model of the ionosphere, the major cause of errors will be a poor forecast of what the sunspot number or ionospheric index will be at the time.

### 10.2.1 The Boulder to St. Louis Circuit

This circuit was chosen to illustrate the various ionospheric aspects of frequency support mainly because it is the circuit which was used for the IONCAP manual. The T index used was 100, which corresponds very closely to a sunspot number of 100 (IONCAP is driven by the sunspot number), so the ASAPS and IONCAP results given in the IONCAP users guide can be directly compared if desired.

Figure 10.1 gives the GRAFEX output provided by ASAPS. The format is slightly different from that given in Figure 6.7, in order to better fit a PC screen, but the key definitions remain unaltered. Two ASAPS predictions were given in Section 6.12 for use with Problems 11 and 12.

Figure 10.2 gives a graphical presentation of the same information (similar to Figures 6.5 and 6.6). The graphs

```
=================================================================================
ASAPS GRAFEX FREQUENCY PREDICTIONS -------------------------------- 15 Aug 1992
=================================================================================
Circuit: boulder-st louis        Distance: 1299 KM    Date: January,1970
Tx: boulder          40.02 254.72  Bear: 92  281      T-index: 100
Rx: st louis         38.67 269.75  Path: Short Path   Circuit#: 1
=================================================================================
     First Mode
                                                                    Second Mode
1F 18-20 1E   5    |--------F r e q u e n c y  (MHz)---------|      2F 36-39 2E 14
UT  OWF  EMUF  ALF 1...5...10 ...15...20 ...25...30 ...35...40     OWF  EMUF  ALF UT
00  13.6  2.6  0.0 XMMMMMMMMM FFF%%..                              8.7  1.6  0.0 00
01  10.6  2.6  0.0 XMMMMMMMFF %%...                                7.1  1.6  0.0 01
02   8.6  2.6  0.0 XMMMMMMF%%  ..                                  5.8  1.6  0.0 02
03   6.8  2.6  0.0 XMMMMF%%..                                      5.0  1.6  0.0 03
04   6.0  2.6  0.0 XMMMM%%..                                       4.6  1.6  0.0 04
05   5.6  2.6  0.0 XMMMM%..                                        4.2  1.6  0.0 05
06   5.4  2.6  0.0 XMMMF%..                                        4.2  1.6  0.0 06
07   5.3  2.6  0.0 XMMMF%.                                         4.1  1.6  0.0 07
08   5.2  2.6  0.0 XMMMF%.                                         4.0  1.6  0.0 08
09   5.0  2.6  0.0 XMMM%%.                                         3.9  1.6  0.0 09
10   5.0  2.6  0.0 XMMM%%.                                         3.9  1.6  0.0 10
11   4.9  2.6  0.0 XMMM%%.                                         3.8  1.6  0.0 11
12   4.9  2.6  0.0 XMMM%%.                                         3.8  1.6  0.0 12
13   5.4  2.6  0.0 XMMMF%.                                         3.9  1.6  0.0 13
14   9.4  6.6  3.9    AMMMFFF% ..                                  5.6  0.0  2.7 14
15  13.8 10.6  5.7    SSMMMMM FFF%%%..                             8.7  6.0  3.4 15
16  15.5 12.3  6.5    ASSXMMM MBFFF%%%..                           9.8  7.2  3.8 16
17  17.1 13.2  6.9    ASSXMMM MMBFFFF%%. ..                       10.9  7.9  4.0 17
18  17.9 13.7  7.0    SSXXMM  MMMFFFF%%% ..                       11.4  8.2  4.1 18
19  18.3 13.7  7.1    ASSSXMM MMMFFFFF%% ...                      11.8  8.3  4.1 19
20  17.4 13.3  6.9    ASSXMMM MMMFFFF%%. ..                       11.4  7.9  4.0 20
21  17.1 12.3  6.5    ASSXMMM MMFFFFF%%. ..                       11.1  7.3  3.8 21
22  16.7 10.7  5.7    SSXMMMM MMFFFF%%%.  .                       10.9  6.1  3.4 22
23  15.4  6.8  4.0    AMMMMMM MFFFF%%%..                          10.1  0.0  2.6 23
UT  OWF  EMUF  ALF 1...5...10 ...15...20 ...25...30 ...35...40     OWF  EMUF  ALF UT
=================================================================================
```

**Figure 10.1** ASAPS frequency predictions in GRAFEX form for the Boulder to St. Louis circuit in January 1970. The body of the figure gives a coded description of the expected propagation modes and conditions, following the code given on Page 83, while the numbers give the Optimum Working Frequency (OWF; 90% chance of support by the F layer), the E-layer Maximum Usable Frequency (EMUF; 50% chance of support by the E layer), and the Absorption Limiting Frequency (ALF). Propagation conditions are given for both 1-hop (First Mode) and 2-hop (Second Mode) modes. The best propagation conditions occur for those frequencies and universal times corresponding to the letters F (First mode only) and S (Second mode only). The letter M indicates Mixed mode propagation, with both the first and second modes being supported. The dots and % signs indicate probabilities of support of between 10% and 50%, and between 50% and 90%.

show essentially the same information as in Figure 10.1, but the different style of presentation gives the complete propagation conditions at a glance. There is a graph of frequency versus time for both the first and second propagation modes, without the overlap necessarily encountered in Figure 10.1 when both modes are supported at a particular frequency. The rectangle on the right-hand side of the figure gives the key to the shading (the screen output is in color). The six shadings correspond to the six possible sets of propagation conditions. Reading from the bottom of the key, we have:

1. Below the ALF—Absorption will prevent the use of these frequencies.

2. Above the ALF, but below the EMUF—Propagation will be via reflection from the ionospheric E layer. The upper frequency limit (the EMUF) corresponds to a probability of 50% that propagation on this frequency will be supported (i.e., on 15 days of the month) by the E layer.

3. Above the EMUF, but below the OWF (or lower decile MUF for the F layer)—Propagation would be supported by the F layer on at least 90% of the days of the month (27 days).

4. Above the OWF, but below the FMUF—Propagation will be supported by the F layer on 50% to 90% of the days of the month.

5. Above the FMUF, but below the upper decile MUF for the F layer (UD FMUF)— propagation will be sup-

# Propagation Prediction Programs

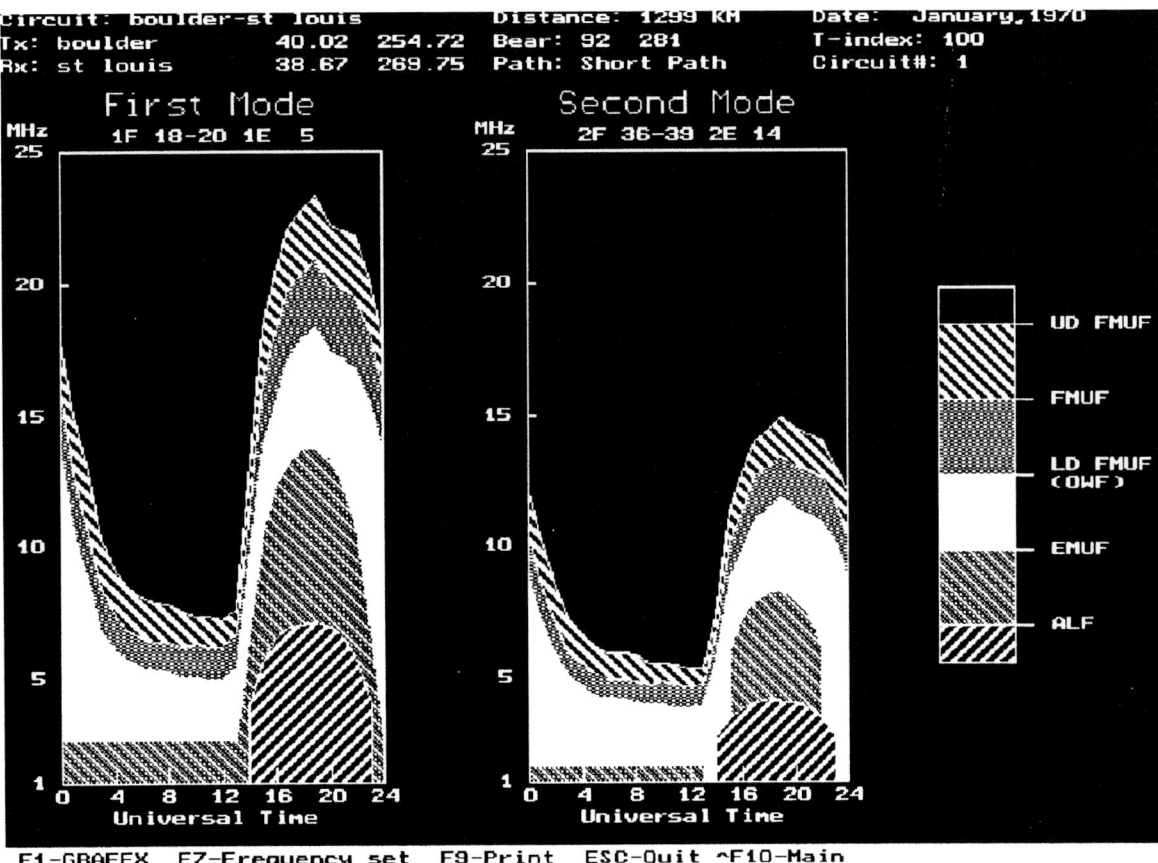

**Figure 10.2** ASAPS graphical display of the propagation conditions for the Boulder to St. Louis circuit in January 1970. There is a graph for each of the two propagation modes considered. The key to the shading (ASAPS actually works in color) is given by the rectangle on the far right. This display has the advantage that there is no requirement for the 1-hop and 2-hop results to be overlaid. The abbreviations in the key are ALF (Absorption Limiting Frequency), EMUF (E-layer Maximum Usable Frequency, which has a 50% chance of support by the E layer), LD FMUF or OWF (Lower Decile MUF for the F layer, or Optimum Working Frequency, which has a 90% chance of support), FMUF (F-layer MUF, which has a 50% chance of support), and the UD FMUF (Upper Decile FMUF, which has a 10% chance of support). The display will also indicate the allocated frequencies and the time for which each is recommended, if a set of frequencies is supplied.

ported by the F layer on 10% to 50% of the days of the month.

6. Above the UD FMUF—Propagation will be supported on fewer than 10% of the days of the month (3 days). This area of the screen is left blank.

It is an unfortunate fact of life that the HF band is so heavily congested that we are not able to use the frequency which is best at the time, but only the best *allocated* frequency. ASAPS therefore provides a **frequency plan** which recommends which of the allocated frequencies to use at a given time. Figure 10.3 gives the plan for the Boulder to St. Louis circuit using the frequencies allocated for amateur use. The plan is mostly self-evident. Note, however, that it is really indicating which of the allocated frequencies would be supported, not whether or not successful communications would be possible. We return to this point in Section 10.3.

It is of some interest to see how the predictions for the 2F mode are derived. Figures 10.4 and 10.5 show the GRAFEX predictions for each half of the 2-hop propagation mode from Boulder to St. Louis. Figure 10.4 is for a 650 km circuit from Boulder on a bearing equal to that of the bearing to St. Louis (92°). Figure 10.5 is for the circuit from that point to St. Louis, another 649 km. When these are put together, one would expect to end up with the 2F predictions which were given in Figure 10.1.

Figures 10.4 and 10.5 are combined to form the 2F part of Figure 10.1 using a few simple rules:

1. The ALF for the full 2-hop mode is equal to the *higher* of the ALFs for the two separate hops.

2. The EMUF for the full 2-hop mode is equal to the *lower* of the EMUFs for the two separate hops.

A similar rule applies to the OWF, with the OWF being set

```
==================================================================
ASAPS FREQUENCY PLAN PREDICTIONS ------------------------------ 15 Aug 1992
==================================================================
Circuit: boulder-st louis         Distance: 1299 KM     Date:   January,1970
Tx: boulder          40.02  254.72  Bear: 92   281       T-index: 100
Rx: st louis         38.67  269.75  Path: Short Path    Circuit#: 1
Selected frequency set: amateur
   3.000   7.000  10.000  14.000  21.000  28.000
==================================================================
        Mode: 1F      TakeOff Angle:18-20    | Mode: 1E  TakeOff Angle:5
     Probability > 90%    |  Probability 50-90%  |
==================================================================
Time(UT) Frequency(MHz) | Time(UT) Frequency(MHz) | Time(UT) Frequency(MHz)
 00-01      10.000      |  00-02      None        |  00-15      None
 01-02       7.000      |  02-04      7.000       |  15-22     10.000
 02-14       3.000      |  04-14      None        |  22-24      None
 14-15       7.000      |  14-15     10.000       |
 15-16      10.000      |  15-24      None        |
 16-23      14.000      |                         |
 23-24      10.000      |                         |
==================================================================
        Mode: 2F      TakeOff Angle:36-39    | Mode: 2E  TakeOff Angle:14
     Probability > 90%    |  Probability 50-90%  |
==================================================================
Time(UT) Frequency(MHz) | Time(UT) Frequency(MHz) | Time(UT) Frequency(MHz)
 00-01       7.000      |  00-15      None        |  00-16      None
 01-15       3.000      |  15-16     10.000       |  16-21      7.000
 15-17       7.000      |  16-24      None        |  21-24      None
 17-23      10.000      |                         |
 23-24       7.000      |                         |
==================================================================
```

**Figure 10.3** The ASAPS frequency plan for the Boulder to St. Louis circuit for January 1970 for the "amateur" frequency set. The recommended frequencies are given for both 1-hop and 2-hop modes, and for probabilities of 50% and 90% support by the ionosphere. Note that the "None" entries do not indicate that propagation will not be possible, just that it will not be possible on any of the allocated frequencies. The 90% results would normally be considered for routine communications, but it would be possible to use the higher frequencies at each Universal Time in the 50–90% column for more than 50% of the days of the month.

by the lower of the OWFs for the two separate hops. However it has been found experimentally [4] that the OWF on a 2-hop circuit is somewhat higher than predicted on the basis of two single hops, so ASAPS uses a different formula to calculate the 2F OWF from the one used for the 1F OWF. Thus while the OWF for the whole circuit is set by the lower of the OWFs for the two separate hops, the OWFs for the separate hops are effectively a little greater than those given in Figures 10.4 and 10.5.

### 10.2.2 Long-Path Propagation

The term "long path" means different things to different people. In the radio amateur literature, long-path propagation and gray-line openings seem to be synonymous. In the present context, the term is taken to mean that the signals go the long way around the Earth, and not via the more obvious "short path." Figures 10.6 and 10.7 give GRAFEX predictions for the circuit from Dallas to Sydney for November 1991. Figure 10.6 corresponds to short-path predictions, with a circuit length over the Pacific of 13,819 km. Figure 10.7, on the other hand, corresponds to long-path predictions, for a circuit length of 26,205 km, the long way around the Earth. Note that ASAPS does not define the propagation mode for such long circuits.

The long-path predictions in Figure 10.7 show that between 08 and 11 UT, propagation is unlikely on any frequency. The letter A indicates the likelihood of severe absorption, while the dots indicate a less than 50% chance of the frequency being supported. The general lack of support of frequencies below about 15 MHz is an absorption effect. On all multihop circuits, the circuit ALF is taken to be the highest of the ALFs for each of the separate hops of the complete propagation mode. On circuit longer than 12,000 km, ASAPS uses a hypothetical 12-hop propagation mode. For the OWF on long circuits, ASAPS does not work in terms of propagation modes of different order, but instead it makes predictions for the first and last hops, and ignores what happens in between. The logic here is that what happens on very long circuits is largely controlled by what happens at the first and last reflection points, which are called the control points. The OWF is taken as the lower of the OWFs for the two hops.

The high values for the ALF on the Dallas to Sydney long path throughout the whole 24 hours arise because one of

```
================================================================================
ASAPS GRAFEX FREQUENCY PREDICTIONS ----------------------------------- 15 Aug 1992
================================================================================
Circuit:                        Distance: 650 KM      Date:    January,1970
Tx: boulder      40.02  254.72  Bear: 92   277        T-index: 100
Rx:              39.57  262.31  Path: Short Path      Circuit#: 1
================================================================================
First Mode                                                    Second Mode
1F 36-39  1E 14      |--------F r e q u e n c y   (MHz)--------|   2F 56-59  2E 28
UT  OWF  EMUF  ALF   1...5...10 ...15...20 ...25...30 ...35...40   OWF  EMUF  ALF  UT
00  9.3  1.6   0.0   MMMMMMMMM% ..                                  8.0  1.0   0.0  00
01  7.3  1.6   0.0   MMMMMM%..                                      6.3  1.0   0.0  01
02  6.0  1.6   0.0   MMMMMM%..                                      5.3  1.0   0.0  02
03  4.8  1.6   0.0   MMMMM..                                        4.4  1.0   0.0  03
04  4.3  1.6   0.0   MMMM%.                                         4.0  1.0   0.0  04
05  4.0  1.6   0.0   MMMM..                                         3.7  1.0   0.0  05
06  3.9  1.6   0.0   MMMM.                                          3.6  1.0   0.0  06
07  3.8  1.6   0.0   MMMM.                                          3.6  1.0   0.0  07
08  3.8  1.6   0.0   MMMM.                                          3.6  1.0   0.0  08
09  3.6  1.6   0.0   MMMM.                                          3.4  1.0   0.0  09
10  3.6  1.6   0.0   MMMM.                                          3.4  1.0   0.0  10
11  3.6  1.6   0.0   MMMM.                                          3.3  1.0   0.0  11
12  3.5  1.6   0.0   MMMM.                                          3.3  1.0   0.0  12
13  3.6  1.6   0.0   MMMM.                                          3.3  1.0   0.0  13
14  5.4  1.6   0.0   MMMMM%.                                        4.6  0.0   1.5  14
15  8.6  6.0   3.1   AMMMMM%. .                                     7.1  3.5   2.6  15
16  9.7  7.2   3.6   AXMMMMM% %.                                    8.1  4.4   2.9  16
17 11.0  7.9   3.9   XMMMMMM %%..                                   9.2  4.8   3.1  17
18 11.5  8.2   4.0   XMMMMMM M%%.                                   9.7  5.0   3.2  18
19 12.0  8.3   4.0   XXMMMMM M%%..                                 10.2  5.1   3.2  19
20 11.6  8.1   4.0   XMMMMMM M%%.                                  10.0  5.0   3.2  20
21 11.2  7.6   3.8   XMMMMMM M%..                                   9.6  4.6   3.1  21
22 10.9  6.7   3.4   AXMMMMM %%..                                   9.3  4.1   2.8  22
23 10.2  5.0   2.6   MMMMMMM %..                                    8.8  2.8   2.2  23
UT  OWF  EMUF  ALF   1...5...10 ...15...20 ...25...30 ...35...40   OWF  EMUF  ALF  UT
================================================================================
```

**Figure 10.4** ASAPS GRAFEX predictions for the first hop of a 2-hop mode between Boulder and St. Louis. The circuit is from Boulder to the midpoint. These predictions are combined with those of Figure 10.5 to produce the 2-hop predictions of Figure 10.1. On a 2-hop circuit, the OWF for the whole circuit is nominally equal to the lower of the OWFs for each of the separate hops, while the ALF for the whole circuit is equal to the higher of the ALFs for the two separate hops.

the two control points is always in daylight. These are clearly illustrated by the large area of crosshatching in the lower part of Figure 10.8. The dominant feature in Figure 10.9 for the short path, on the other hand, is the white area corresponding to propagation between the EMUF and the lower decile FMUF.

The philosophy of the control point method is to concentrate on what is known to be true, and to exclude what is not known to be true. What we *do* know to be true is that the signals hit the ionosphere at least somewhere around the two control points, even though we cannot be sure exactly where these are. What we do *not* know is how many hops the signals make along the circuit. This means that we do not know what ionospheric conditions will exist at each of the reflection points, and so we are unable to make reliable predictions for the circuit as a whole. However the control point method does have its limitations, and does not make any use of what we *do* know about the ionosphere along the circuit [5].

The usual approach to deciding whether or not a procedure is a good one is to consider what happens in the limit. When the circuit is taken to have its maximum length of 40,000 km (the full circumference of the Earth), and it goes over both poles, the only ionosphere taken into account is that 1,500 km to the north and south of the transmitter/receiver. Clearly this is not a sensible procedure, because we know that the signals have to be supported by the high-latitude ionosphere over the poles, one of which could be in complete darkness. The signals must also be strong enough to overcome the absorption as they cross through the equatorial ionosphere. The logic behind the control point method is that the two points should represent the worst that can happen on the circuit—the lowest optimum working frequency (OWF) for one hop (at night), and the highest ab-

```
===========================================================================
ASAPS  GRAFEX  FREQUENCY  PREDICTIONS ----------------------------- 15 Aug 1992
===========================================================================
Circuit:                        Distance: 649 KM        Date:  January,1970
Tx:                  39.57  262.31  Bear: 97    281     T-index: 100
Rx: st louis         38.67  269.75  Path: Short Path    Circuit#: 1
===========================================================================
First Mode                                                    Second Mode
1F 36-39  1E 14       |--------F r e q u e n c y  (MHz)---------|   2F 56-59  2E 28
UT  OWF  EMUF  ALF   1...5...10  ...15...20 ...25...30  ...35...40  OWF EMUF  ALF UT
00  8.6  1.6   0.0   MMMMMMMM%%  ..                                 7.5  1.0  0.0 00
01  6.9  1.6   0.0   MMMMMMM%..                                     6.1  1.0  0.0 01
02  5.6  1.6   0.0   MMMMMM..                                       5.0  1.0  0.0 02
03  4.7  1.6   0.0   MMMMM..                                        4.3  1.0  0.0 03
04  4.4  1.6   0.0   MMMM%.                                         4.0  1.0  0.0 04
05  4.2  1.6   0.0   MMMM%.                                         3.9  1.0  0.0 05
06  4.1  1.6   0.0   MMMM..                                         3.8  1.0  0.0 06
07  4.0  1.6   0.0   MMMM.                                          3.7  1.0  0.0 07
08  3.8  1.6   0.0   MMMM.                                          3.5  1.0  0.0 08
09  3.7  1.6   0.0   MMMM.                                          3.5  1.0  0.0 09
10  3.7  1.6   0.0   MMMM.                                          3.4  1.0  0.0 10
11  3.6  1.6   0.0   MMMM.                                          3.4  1.0  0.0 11
12  3.6  1.6   0.0   MMMM.                                          3.4  1.0  0.0 12
13  4.3  1.6   0.0   MMMM%                                          3.8  1.0  0.0 13
14  7.2  5.0   2.7      MMMMF%.                                     5.9  2.8  2.3 14
15  9.3  6.8   3.4      AXMMMMM% ..                                 7.8  4.1  2.8 15
16 10.5  7.6   3.8       XMMMMMM %%.                                8.8  4.6  3.1 16
17 11.3  8.1   4.0       XMMMMMM F%..                               9.6  5.0  3.2 17
18 11.8  8.3   4.1       SXMMMMM M%%..                              9.9  5.1  3.3 18
19 11.9  8.3   4.0       XXMMMMM M%%..                             10.1  5.1  3.2 19
20 11.4  7.9   3.9       XMMMMMM M%%.                               9.9  4.9  3.1 20
21 11.2  7.3   3.6      AXMMMMMM M%..                               9.6  4.4  2.9 21
22 11.0  6.1   3.1      AMMMMMMM %%..                               9.3  3.6  2.6 22
23 10.1  2.4   1.4     MMMMMMMMM %..                                8.7  0.0  1.6 23
UT  OWF  EMUF  ALF   1...5...10  ...15...20 ...25...30  ...35...40  OWF EMUF  ALF UT
===========================================================================
```

**Figure 10.5** ASAPS GRAFEX predictions for the second hop of a 2-hop mode between Boulder and St. Louis. The circuit is from the midpoint to St. Louis. These predictions are combined with those of Figure 10.4 to produce the 2-hop predictions of Figure 10.1. On a 2-hop circuit, the OWF for the whole circuit is nominally equal to the lower of the OWFs for each of the separate hops, while the ALF for the whole circuit is equal to the higher of the ALFs for the two separate hops.

sorption limiting frequency (ALF) for the other (day). If this is not true, predictions based on the control point method should be viewed with some skepticism [6].

### 10.2.3 General Effects of Increasing the Circuit Length

The most obvious effect of increasing the circuit length is the appearance and disappearance of various propagation modes, mainly because of elevation angle considerations and changes in the number of hops assumed to exist on the circuit. Some changes in the ALF, MUF, and OWF also occur as the length of a hop increases.

1. The ALF increases as the hop length increases. This is because the signals travel at lower angles, and therefore spend more time traversing the D and lower E regions, where the HF absorption takes place. As we shall see in Section 10.3, HF absorption is significantly less on a 2-hop propagation mode than on a 1-hop mode, which is why the 2F ALF is lower than the 1F ALF for the Boulder to St. Louis circuit (see Figure 10.1).

2. Provided the ionosphere at the different reflection points has about the same critical frequencies, the MUF and OWF will both increase with increasing circuit length, simply because of increases in the obliquity factor M(D) which was introduced in Section 4.3.

In summary, then, the effects of increasing the circuit length are:

1. The dominant propagation modes change.
2. The ALF increases.
3. The MUF and OWF increase.

Note that the lower and upper frequency limits, i.e., the ALF and the OWF, both increase as the hop length increases. It would have been rather inconvenient if the OWF were to

```
=================================================================================
ASAPS GRAFEX FREQUENCY PREDICTIONS -------------------------------- 15 Aug 1992
=================================================================================
Circuit: dallas-sydney              Distance: 13819 KM    Date: November,1991
Tx: dallas            32.78  263.20 Bear: 249  71         T-index: 139
Rx: sydney           -33.87  151.18 Path: Short Path      Circuit#: 1
=================================================================================
First Mode                                                      Second Mode
F  0-5             |--------F r e q u e n c y  (MHz)---------|  NONE
UT  OWF  MUF  ALF  1...5...10 ...15...20 ...25...30 ...35...40  OWF  MUF  ALF UT
00  27.8 34.0 14.8                AFFFFF FFFFFFF%%% %%%.......  0.0  0.0  0.0 00
01  28.4 33.2 14.7                AFFFFF FFFFFFFF%% %%%..       0.0  0.0  0.0 01
02  25.3 28.5 14.6                AFFFFF FFFFF%%%..             0.0  0.0  0.0 02
03  21.5 24.3 14.3                FFFFFF F%%%.                  0.0  0.0  0.0 03
04  19.6 22.1 13.7               AFFFFF% %%.                    0.0  0.0  0.0 04
05  17.9 20.6 12.8               AFFFF%%% ...                   0.0  0.0  0.0 05
06  16.0 18.8 11.4               FFFF%%%.. ..                   0.0  0.0  0.0 06
07  14.9 17.5  9.0            AF FFFF%%%...                     0.0  0.0  0.0 07
08  13.7 16.0  0.0 FFFFFFFFFF FFF%%...                          0.0  0.0  0.0 08
09  12.3 15.1  0.0 FFFFFFFFFF FF%%%..                           0.0  0.0  0.0 09
10  11.3 14.4  0.0 FFFFFFFFFF F%%%...                           0.0  0.0  0.0 10
11  10.7 13.6  0.0 FFFFFFFFFF %%%...                            0.0  0.0  0.0 11
12  10.3 13.1  0.0 FFFFFFFFFF %%%..                             0.0  0.0  0.0 12
13  12.4 14.6  0.0 FFFFFFFFFF FF%%..                            0.0  0.0  0.0 13
14  19.2 21.1  0.0 FFFFFFFFFF FFFFFFFFF% %..                    0.0  0.0  0.0 14
15  19.7 24.9  9.9          A FFFFFFFFF% %%%%...                0.0  0.0  0.0 15
16  17.6 23.3 11.7               AFFFFF%%% %%%..                0.0  0.0  0.0 16
17  17.3 22.9 12.7               AFFFF%%% %%...                 0.0  0.0  0.0 17
18  18.6 24.7 13.3               FFFFF%% %%%%...                0.0  0.0  0.0 18
19  22.0 27.8 13.7               AFFFFFF F%%%%%%... ..          0.0  0.0  0.0 19
20  24.1 28.7 14.1                FFFFFF FFFF%%%%.. .....       0.0  0.0  0.0 20
21  25.4 30.3 14.5                AFFFFF FFFFF%%%%% .......     0.0  0.0  0.0 21
22  25.8 30.8 14.7                AFFFFF FFFFF%%%%% ........    0.0  0.0  0.0 22
23  26.7 32.3 14.8                AFFFFF FFFFFF%%%% %%.......   0.0  0.0  0.0 23
UT  OWF  MUF  ALF  1...5...10 ...15...20 ...25...30 ...35...40  OWF  MUF  ALF UT
=================================================================================
```

**Figure 10.6** ASAPS short-path GRAFEX predictions for the circuit from Dallas to Sydney for November 1991. The short path is the one which goes across the Pacific Ocean. These predictions can be compared with the long-path predictions given in Figure 10.7. On such long circuits, ASAPS uses the control point method, working in terms of a 2-hop mode with one hop at the transmitter end and the other at the receiver end of the circuit. The letter F at a particular time and frequency indicates a 90% chance of that frequency being supported, while the dots and % signs indicate probabilities of between 10 and 50%, and between 50 and 90%. The two blank areas below about 15 MHz correspond to high absorption at all frequencies below 15 MHz.

*decrease* with circuit length, while the ALF continued to increase.

## 10.3 Field Strength Predictions

The better propagation prediction programs also provide what are generically called field strength predictions. The most important of these is the S/N ratio, since it is this ratio which determines whether or not communications at any of the allocated frequencies currently being supported by the ionosphere will be successful. The factors contributing to the field strength of the received signal have already been reviewed in Section 6.8. The derivation of the formulas for field strength and S/N ratio is given in Appendix A.

In general, the accuracy of the calculated field strengths leaves a little to be desired. One of the problems is the difficulty associated with obtaining reliable observations of field strength, since it is not possible to reduce the actual observations to usable form without making some sort of assumption concerning the propagation modes which are present, and their elevation angles. The measurement of the elevation angles by what are now routine techniques [7] would go a long way towards improving the usefulness of such observations. The "excess system loss" term included by IONCAP possibly has its origin in vagaries of the field strength data bases, although scattering of energy out of the great-circle path by field-aligned irregularities probably also contributes to the discrepancies [8].

### 10.3.1 Factors Affecting the Received S/N Ratios

The key factors in determining the received signal-to-noise (S/N) ratios are the transmitter power, the antenna

```
================================================================================
ASAPS GRAFEX FREQUENCY PREDICTIONS ------------------------------- 15 Aug 1992
================================================================================
Circuit: dallas-sydney             Distance:  26205 KM      Date:  November,1991
Tx: dallas           32.78  263.20 Bear:  69   251          T-index: 139
Rx: sydney          -33.87  151.18 Path:  Long Path         Circuit#: 1
================================================================================
First Mode                                                       Second Mode
F   0-5              |--------F r e q u e n c y  (MHz)--------|     NONE
UT  OWF  MUF   ALF  1...5...10 ...15...20 ...25...30 ...35...40 OWF  MUF  ALF UT
00  20.4 22.9  13.7            AFFFFFF %%.......                0.0  0.0  0.0 00
01  21.3 22.8  14.2            FFFFFF F%.....                   0.0  0.0  0.0 01
02  19.1 21.3  14.5            AFFFF% %..                       0.0  0.0  0.0 02
03  16.7 19.0  14.6            AF%%..                           0.0  0.0  0.0 03
04  15.4 17.7  14.5            A%%..                            0.0  0.0  0.0 04
05  14.6 16.8  14.6            A%..                             0.0  0.0  0.0 05
06  14.4 16.5  14.7            A%..                             0.0  0.0  0.0 06
07  13.9 16.5  14.7            A%..                             0.0  0.0  0.0 07
08  13.2 16.0  14.8            A..                              0.0  0.0  0.0 08
09  13.2 15.9  14.8            A..                              0.0  0.0  0.0 09
10  12.9 15.6  14.7            A..                              0.0  0.0  0.0 10
11  12.4 14.5  14.5            A.                               0.0  0.0  0.0 11
12  17.5 19.8  14.2            FFF%%. ..                        0.0  0.0  0.0 12
13  17.6 21.1  13.8            AFFF%%% %..                      0.0  0.0  0.0 13
14  16.4 20.1  13.3            FFF%%%% ..                       0.0  0.0  0.0 14
15  15.9 19.5  12.9            AFF%%%%. .                       0.0  0.0  0.0 15
16  15.1 18.5  12.7            AFF%%%%..                        0.0  0.0  0.0 16
17  13.7 17.5  12.9            A%%%%..                          0.0  0.0  0.0 17
18  12.3 16.0  12.8            A%%..                            0.0  0.0  0.0 18
19  11.8 15.3  12.5            A%%.                             0.0  0.0  0.0 19
20  14.2 18.5  11.8            AFF%%%%..                        0.0  0.0  0.0 20
21  15.4 20.1  10.6            AFFFF%%%%% ....                  0.0  0.0  0.0 21
22  16.9 21.4  11.7            AFFFF%%%% %......                0.0  0.0  0.0 22
23  18.9 22.2  12.9            AFFFFF%% %%......                0.0  0.0  0.0 23
UT  OWF  MUF   ALF  1...5...10 ...15...20 ...25...30 ...35...40 OWF  MUF  ALF UT
================================================================================
```

**Figure 10.7** ASAPS long-path GRAFEX predictions for the circuit from Dallas to Sydney for November 1991. The long path is the one which goes up over the pole on a bearing of 69°. These predictions can be compared with the short-path predictions given in Figure 10.6. According to the ASAPS model, propagation via the long path is severely limited by absorption at the low frequencies and a nighttime ionosphere at the high frequencies. On such long circuits, ASAPS uses the control point method, working in terms of a 2-hop mode with one hop at the transmitter end and the other at the receiver end of the circuit. The letter F at a particular time and frequency indicates a 90% chance of that frequency being supported, while the dots and % signs indicate probabilities of between 10% and 50%, and between 50% and 90%. The blank area below about 15 MHz corresponds to high absorption at all frequencies below 15 MHz, the letter A indicating high absorption.

gains, the loss of energy along the circuit, and the local radio noise level at the receiving site.

For a given circuit, epoch and equipment, the S/N ratio differs with the frequency and the propagation mode. Figure 10.10 gives the ASAPS S/N predictions for the 1 F mode on the Boulder to St. Louis circuit, for frequencies of 3, 6, 9, 12, 15, 18, 21, and 24 MHz, and a transmitter power of 1 kW. During the day, the S/N ratios reach 41 dB (in a 3 kHz bandwidth) at the highest supported allocated frequency. These ratios are not quite so high at the MUF, and are a few dB lower at the OWF, in keeping with the general rule that the S/N increases as the frequency increases (and with the standard operational procedure of using the highest frequency possible). The S/N ratios at the OWF are significantly lower at night than during the day, in spite of the lower path losses, because the noise levels at the lower nighttime OWFs are about 20 dB higher than at the daytime OWFs, as can be seen from Figure 10.11.

The huge size of the pathloss (typically 120 dB, or a factor of $10^{12}$) is not important, provided that the signal level is significantly greater than the radio noise level at the receiving site. As explained in Chapter 7, this noise includes galactic noise, atmospheric noise, man-made noise, and interference, as well as noise internal to the receiver. Galactic noise becomes important at the higher end of the HF band, and at very quiet receiving sites. The last sources of noise are very specific to the equipment, the receiving site, operating frequency, and antenna gains and orientation, so they

# Propagation Prediction Programs

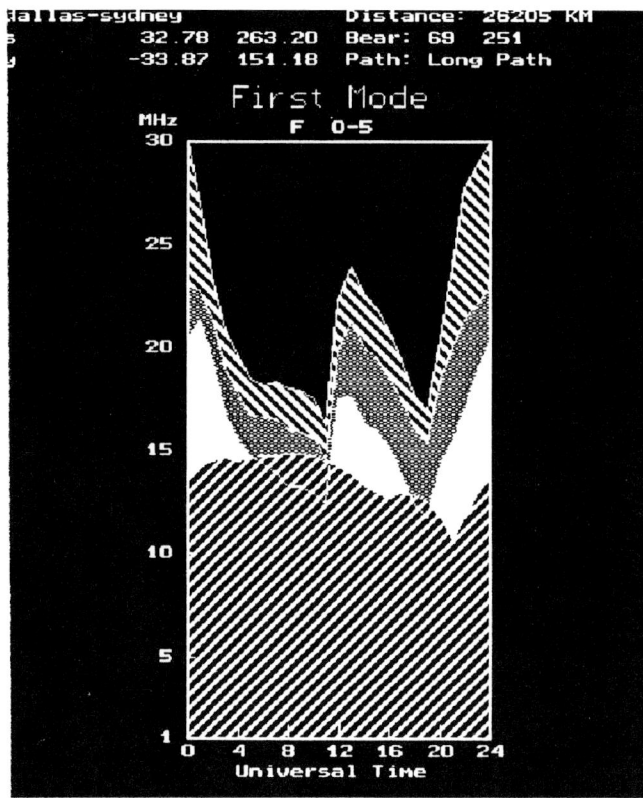

**Figure 10.8** Graphical output from ASAPS for the long-path Dallas to Sydney circuit of Figure 10.7. On such long circuits, ASAPS uses the control point method, working in terms of a 2-hop mode with one hop at the transmitter end and the other at the receiver end of the circuit. The key to the shading has already been given in Figure 10.2. The ALF remains above 10 MHz for all 24 hours, and reaches 15 MHz between 04 and 12 UT. The best propagation conditions correspond to the white areas, and these do not exist for all 24 hours.

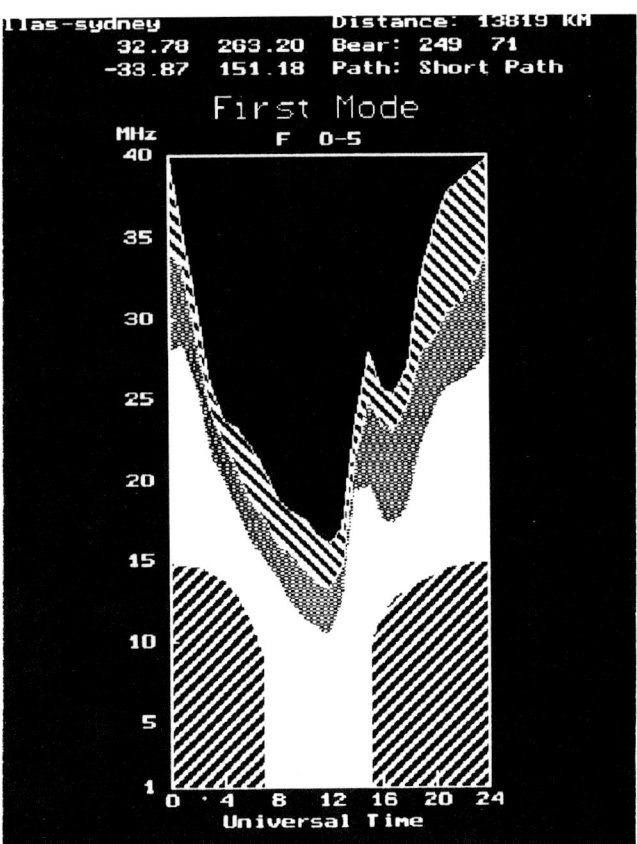

**Figure 10.9** Graphical output from ASAPS for the short-path Dallas to Sydney circuit of Figure 10.6. On such long circuits, ASAPS uses the control point method, working in terms of a 2-hop mode with one hop at the transmitter end and the other at the receiver end of the circuit. The key to the shading has already been given in Figure 10.2. The best propagation conditions correspond to the white areas, and these have a typical diurnal variation.

are not normally taken into detailed account by a propagation prediction program. Instead, the programs include general man-made noise levels at a frequency of 3 MHz which are found to be representative:

| | |
|---|---|
| Industrial areas | $-125$ dBW |
| Residential | $-136$ dBW |
| Rural | $-148$ dBW |
| Remote | $-164$ dBW |

The noise level at any other frequencies can be derived from these values by using calibration curves based on an extensive series of worldwide noise observations. ASAPS defaults to a noise level of $-148$ dBW, since most receiving sites will be in relatively quiet rural areas. However any required noise level, such as the high ones encountered on board a ship, may usually be specified.

Figure 10.11 tabulates the noise levels ("All Noise", in dBW) as a function of frequency and UT for the receiver end (St. Louis) of the Boulder to St. Louis circuit. The noise level decreases (gets more negative) as the frequency increases, which is one reason for using the highest possible frequency. Man-made (or anthropogenic) noise decreases with frequency, f, approximately as

$$N(f) = -148 - 29 \log_{10}(f/3) .$$

Therefore, applying the above equation to a 3MHz signal, the noise level at a rural site, is $-148$ dB; and for 30 MHz, the noise level is $-148 - 29 = -177$ dB.

Absorption has very little effect at higher frequencies. At 24 MHz, for example, there is no diurnal variation of the noise level. At the lowest frequency of 3 MHz, however, the noise levels are about 10 dB higher at night than during the day. This is because there is no HF absorption at night, so distant transmitters (including thunderstorms) provide a finite and unwanted signal at the receiving site, thus raising the noise level.

The total pathloss (made up of the components given in

```
================================================================================
Circuit: boulder-st louis        Distance: 1299 KM      Date: January,1970
Tx: boulder         40.02  254.72 Bear: 91   281        T-index: 100
Rx: st louis        38.66  269.75 Path: Short Path      Circuit#: 1
FSet: long                        Antenna: ISOTROPIC    RxNoise: -148 dBW/Hz
Mode: 1F                SIGNAL/NOISE
================================================================================
UT   MUF   OWF    MUF  OWF  3.0  6.0  9.0 12.0 15.0 18.0 21.0 24.0
00  16.0  13.6    40   38   29   33   35   37   40   ..   ..   ..
01  12.8  10.6    37   36   29   33   34   37   39   ..   ..   ..
02  10.5   8.6    35   34   29   33   34   37   ..   ..   ..   ..
03   8.4   6.8    32   31   27   31   33   ..   ..   ..   ..   ..
04   7.3   6.0    30   29   24   29   32   ..   ..   ..   ..   ..
05   6.8   5.6    29   28   24   28   ..   ..   ..   ..   ..   ..
06   6.6   5.4    29   28   24   28   ..   ..   ..   ..   ..   ..
07   6.4   5.3    29   28   23   29   ..   ..   ..   ..   ..   ..
08   6.4   5.2    30   28   22   29   ..   ..   ..   ..   ..   ..
09   6.2   5.0    30   27   22   29   ..   ..   ..   ..   ..   ..
10   6.3   5.0    30   27   22   29   ..   ..   ..   ..   ..   ..
11   6.2   4.9    31   29   25   30   ..   ..   ..   ..   ..   ..
12   6.2   4.9    32   30   27   32   ..   ..   ..   ..   ..   ..
13   6.7   5.4    32   31   27   32   ..   ..   ..   ..   ..   ..
14  11.0   9.4    37   34   ..   25   33   39   ..   ..   ..   ..
15  16.3  13.8    40   38   ..   ..   30   36   39   41   ..   ..
16  18.1  15.5    39   36   ..   ..   22   33   36   39   ..   ..
17  19.8  17.1    40   37   ..   ..   ..   31   35   38   41   ..
18  20.4  17.9    40   38   ..   ..   ..   30   34   38   41   ..
19  20.9  18.3    40   38   ..   ..   ..   29   32   37   41   ..
20  19.9  17.4    39   35   ..   ..   ..   29   31   36   40   ..
21  19.8  17.1    39   35   ..   ..   21   30   32   36   41   ..
22  19.5  16.7    40   35   ..   ..   31   33   33   37   41   ..
23  18.1  15.4    40   37   ..   29   33   35   37   40   ..   ..
================================================================================
```

**Figure 10.10** ASAPS S/N ratios for a 3 kHz bandwidth for the 1F mode on the Boulder to St. Louis circuit. The S/N ratios are given for the MUF (50%), the OWF (90%), and for the frequencies in the frequency set "long"; the transmitter power is 1 kW; the 3 MHz noise level at the receive site is −148 dbW/Hz; and the transmitting antenna is an isotropic antenna with zero gain.

Table 6.1) at the OWF is about 10 dB lower at night than during the day (Figure 10.12), mainly because of the lack of HF absorption. Every HF signal reflected by the ionosphere suffers from absorption as it passes through the lower regions of the ionosphere, both on the way up and on the way down. Absorption has been mentioned at various places throughout this book, and it seems a useful idea to pull the ideas together in the one place, before proceeding further. In summary, the important properties of absorption are as follows:

1. Decreases as frequency increases, approximately as $1/f^2$. For example, if the frequency doubles from 4 to 8 MHz, the absorption goes down by a factor of $2^2 = 4$, or 6 dB.

2. Greatest in the middle of the day, and virtually nonexistent at night.

3. Greater in summer than in winter, except on occasional "winter anomaly" days, when the absorption reaches summertime values.

4. Greater at equatorial latitudes than at mid-latitudes.

5. Sometimes very high at high latitudes, but essentially unpredictable at these latitudes.

6. Less for steeper paths through the lower ionosphere (i.e., on shorter hops) for a given frequency.

Points 2, 3, and 4 are all aspects of the fact that absorption is higher when the sun is higher in the sky (closer to the zenith, hence at smaller zenith angles). The effect of point 6 on the Boulder to St. Louis circuit, for example, is to make the 2F ALF lower than the 1F ALF.

Figures 10.13 and 10.14 give the S/N ratios for the 1E mode and the 2F mode, respectively, for the Boulder to St. Louis circuit. Figure 10.13 shows, as expected, that the frequencies supported by the E layer are substantially lower than those supported by the F layer (Figure 10.10). At the same frequency (12 MHz during the day, for example), the F mode offers a 4 or 5 dB S/N advantage over the E mode. The S/N for the same 12 MHz frequency on a 2F mode is

```
================================================================================
Circuit: boulder-st louis          Distance: 1299 KM       Date:    January,1970
Tx: boulder           40.02 254.72  Bear: 91   281          T-index: 100
Rx: st louis          38.66 269.75  Path: Short Path        Circuit#: 1
FSet: long                          Antenna: ISOTROPIC      RxNoise: -148 dBW/Hz
Mode: 1F                 ALL NOISE
================================================================================
UT   MUF  OWF   MUF   OWF   3.0   6.0   9.0  12.0  15.0  18.0  21.0  24.0
00  16.0 13.6  -166  -162  -144  -151  -155  -160  -164  -168  -171  -173
01  12.8 10.6  -161  -158  -144  -151  -155  -160  -164  -168  -171  -173
02  10.5  8.6  -158  -154  -144  -151  -155  -160  -164  -168  -171  -173
03   8.4  6.8  -153  -150  -142  -149  -154  -160  -165  -169  -171  -173
04   7.3  6.0  -150  -147  -139  -147  -153  -161  -166  -169  -171  -173
05   6.8  5.6  -149  -146  -139  -147  -153  -161  -166  -169  -171  -173
06   6.6  5.4  -148  -146  -139  -147  -153  -161  -166  -169  -171  -173
07   6.4  5.3  -149  -145  -138  -147  -156  -163  -167  -169  -171  -173
08   6.4  5.2  -149  -145  -137  -148  -158  -164  -167  -169  -171  -173
09   6.2  5.0  -149  -144  -137  -148  -158  -164  -167  -169  -171  -173
10   6.3  5.0  -149  -144  -137  -148  -158  -164  -167  -169  -171  -173
11   6.2  4.9  -150  -146  -140  -149  -158  -163  -167  -169  -171  -173
12   6.2  4.9  -150  -147  -142  -150  -157  -163  -167  -169  -171  -173
13   6.7  5.4  -152  -149  -142  -150  -157  -163  -167  -169  -171  -173
14  11.0  9.4  -162  -158  -142  -150  -157  -163  -167  -169  -171  -173
15  16.3 13.8  -168  -165  -148  -155  -159  -163  -166  -169  -171  -173
16  18.1 15.5  -168  -165  -148  -156  -160  -162  -164  -168  -171  -173
17  19.8 17.1  -170  -167  -148  -156  -160  -162  -164  -168  -171  -173
18  20.4 17.9  -171  -168  -148  -156  -160  -162  -164  -168  -171  -173
19  20.9 18.3  -171  -167  -148  -156  -160  -161  -163  -167  -171  -173
20  19.9 17.4  -169  -164  -148  -156  -159  -160  -161  -165  -170  -173
21  19.8 17.1  -169  -164  -148  -156  -159  -160  -161  -165  -170  -173
22  19.5 16.7  -168  -163  -148  -156  -159  -160  -161  -165  -170  -173
23  18.1 15.4  -167  -163  -148  -155  -158  -160  -163  -167  -170  -173
================================================================================
```

**Figure 10.11** The noise levels at the receive site (St. Louis) as a function of frequency and universal time. The 3 MHz noise level is −148 dbW/Hz during the day, but increases to −137 dbW/Hz at night, when absorption vanishes. The noise level decreases as the frequency increases.

even lower than for the E mode, and the mode would suffer about a 12 dB disadvantage relative to the 1F mode.

### 10.3.2 Multi-Hop Circuits

In Section 10.2.1, we looked at the details of the two separate hops on the 2-hop Boulder to St. Louis circuit, showing how the two separate predictions are combined to produce the 2F predictions. If we now do the same thing for the field strength predictions, we find, for example, that the pathloss on 12 MHz during the day is 119 dB for each hop. Now the 1F pathloss (for the whole path) is given by ASAPS as 127 dB, so we clearly do not just add two losses of 119 dB each and end up with a total loss of 238 dB! This is because the energy going into the second hop is not from a point source, but from an area. Therefore, the correct way to calculate the total loss is to work out the inverse-square distance loss for the whole circuit length, which will be 3 dB more (twice as much) than for one hop, and then add in the other losses which occur at the ionospheric and ground reflections.

ASAPS is not set up to give the details of the losses, but IPS has provided them for the following two illustrative cases at a frequency of 10 MHz on the Boulder to St. Louis circuit:

### Example I

At 00 UT (near sunset), the loss on the 2F mode is approximately 12 dB higher than on the 1F mode. This difference is made up as follows:

Extra ground reflection loss = 8.1 dB

Extra spatial attenuation = 1.9 dB

Extra polarization coupling loss = 0.9 dB

Less horizon focus gain = 1.0 dB (higher angle means lower focus gain)

Less absorption loss = 0.0 dB (near sunset, the loss is 0.7 dB for both modes)

Less sporadic-E obscuration loss = 0.0 dB

The total extra loss is 11.9 dB, which comes from adding up all the contributions.

```
===================================================================================
Circuit: boulder-st louis        Distance: 1299 KM      Date:    January,1970
Tx: boulder        40.02  254.72  Bear: 91   281        T-index: 100
Rx: st louis       38.66  269.75  Path: Short Path      Circuit#: 1
FSet: long                        Antenna: ISOTROPIC    RxNoise: -148 dBW/Hz
Mode: 1F                  LOSSES
===================================================================================
UT   MUF   OWF    MUF   OWF   3.0   6.0   9.0  12.0  15.0  18.0  21.0  24.0
00   16.0  13.6   121   119   110   113   116   118   120    ..    ..    ..
01   12.8  10.6   119   117   110   113   116   118   120    ..    ..    ..
02   10.5   8.6   117   116   110   113   116   118    ..    ..    ..    ..
03    8.4   6.8   116   114   110   113   116    ..    ..    ..    ..    ..
04    7.3   6.0   115   113   110   114   117    ..    ..    ..    ..    ..
05    6.8   5.6   115   113   110   114    ..    ..    ..    ..    ..    ..
06    6.6   5.4   115   113   110   114    ..    ..    ..    ..    ..    ..
07    6.4   5.3   114   113   110   114    ..    ..    ..    ..    ..    ..
08    6.4   5.2   114   113   110   114    ..    ..    ..    ..    ..    ..
09    6.2   5.0   114   113   110   114    ..    ..    ..    ..    ..    ..
10    6.3   5.0   114   112   110   114    ..    ..    ..    ..    ..    ..
11    6.2   4.9   114   112   110   114    ..    ..    ..    ..    ..    ..
12    6.2   4.9   114   112   110   114    ..    ..    ..    ..    ..    ..
13    6.7   5.4   114   113   110   114    ..    ..    ..    ..    ..    ..
14   11.0   9.4   120   119    ..   121   119   120    ..    ..    ..    ..
15   16.3  13.8   123   122    ..    ..   124   122   123   123    ..    ..
16   18.1  15.5   124   124    ..    ..   133   124   124   124    ..    ..
17   19.8  17.1   125   125    ..    ..    ..   126   125   125   125    ..
18   20.4  17.9   125   125    ..    ..    ..   127   125   125   126    ..
19   20.9  18.3   126   125    ..    ..    ..   127   125   125   126    ..
20   19.9  17.4   125   125    ..    ..    ..   126   125   125   125    ..
21   19.8  17.1   125   124    ..    ..   133   124   124   124   125    ..
22   19.5  16.7   124   123    ..    ..   124   122   123   123   124    ..
23   18.1  15.4   122   121    ..   121   119   120   121   122    ..    ..
===================================================================================
```

**Figure 10.12** The total path losses on the Boulder to St. Louis circuit for the 1F mode. During the day (14-01 UT), the losses decrease as the frequency increases, because of the lower effects of absorption. The losses are consistently lower at night because of the lack of HF absorption at these times.

### Example II

At 19 UT (near noon), when absorption is near its maximum value for the day, the difference is only 6 dB; this can be accounted for as follows:

Extra ground reflection loss = 7.9 dB

Extra spatial attenuation = 1.7 dB

Extra polarization coupling loss = 0.9 dB

Less horizon focus gain = 1.1 dB

Less absorption loss = 5.3 dB (19.8 dB for 1F, 14.5 dB for 2F)

Less sporadic-E obscuration loss = 0.5 dB

The total extra loss is 5.8 dB, which comes from (7.9 + 1.7 + 0.9 + 1.1) − (5.3 + 0.5) = 5.8 dB.

The interesting thing to note here is that the 2F mode actually suffers less absorption than the 1F mode, even though the signals travel through the lower part of the ionosphere twice as often. This is because absorption decreases so quickly as the path length gets shorter and the raypath gets steeper. Approximate ground reflection losses may be derived from Figure 10.15. There is, of course, no ground reflection for the 1F mode, which accounts for a large part of the S/N advantage of the 1F mode over the 2F mode.

### 10.3.3 Effects of Antenna Patterns

All of the field strength calculations that we have considered so far were made using isotropic transmitting and receiving antennas, with the gain set equal to 0 dB. Isotropic antennas radiate energy equally in all directions, and so have no counterpart in real life. We therefore look in this section at what effects the choice of a correct or incorrect antenna can have, considering especially a Near Vertical Incidence Skywave (NVIS) circuit.

For NVIS circuits, the radio waves propagate nearly straight up and down. NVIS HF communications are used for very short circuits for which the ground wave has been severely attenuated, such as in mountainous jungle, or for somewhat longer circuits (100 to 300 km, say) when there would be no ground wave anyway. NVIS communications tend to be regarded as fairly difficult. This is partly because

```
================================================================================
Circuit: boulder-st louis        Distance: 1299 KM      Date:    January,1970
Tx: boulder             40.02  254.72   Bear: 91   281       T-index: 100
Rx: st louis            38.66  269.75   Path: Short Path     Circuit#: 1
FSet: long                              Antenna: ISOTROPIC   RxNoise: -148 dBW/Hz
Mode: 1E                        SIGNAL/NOISE
================================================================================
UT  MUF   OWF    MUF  OWF   3.0   6.0   9.0  12.0  15.0  18.0  21.0  24.0
00  2.5   2.4    28   28    ..    ..    ..    ..    ..    ..    ..    ..
01  2.5   2.4    28   28    ..    ..    ..    ..    ..    ..    ..    ..
02  2.5   2.4    28   28    ..    ..    ..    ..    ..    ..    ..    ..
03  2.5   2.4    26   26    ..    ..    ..    ..    ..    ..    ..    ..
04  2.5   2.4    23   23    ..    ..    ..    ..    ..    ..    ..    ..
05  2.5   2.4    23   23    ..    ..    ..    ..    ..    ..    ..    ..
06  2.5   2.4    23   23    ..    ..    ..    ..    ..    ..    ..    ..
07  2.5   2.4    22   22    ..    ..    ..    ..    ..    ..    ..    ..
08  2.5   2.4    21   21    ..    ..    ..    ..    ..    ..    ..    ..
09  2.5   2.4    21   21    ..    ..    ..    ..    ..    ..    ..    ..
10  2.5   2.4    21   21    ..    ..    ..    ..    ..    ..    ..    ..
11  2.5   2.4    24   24    ..    ..    ..    ..    ..    ..    ..    ..
12  2.5   2.4    26   26    ..    ..    ..    ..    ..    ..    ..    ..
13  2.5   2.4    26   26    ..    ..    ..    ..    ..    ..    ..    ..
14  6.6   6.3    20   20    ..    21    ..    ..    ..    ..    ..    ..
15 10.5  10.0    27   28    ..    24    28    ..    ..    ..    ..    ..
16 12.3  11.7    26   27    ..    ..    26    25    ..    ..    ..    ..
17 13.2  12.5    25   26    ..    ..    24    26    ..    ..    ..    ..
18 13.6  12.9    25   26    ..    ..    23    26    ..    ..    ..    ..
19 13.6  12.9    23   24    ..    ..    23    25    ..    ..    ..    ..
20 13.2  12.5    22   23    ..    ..    24    24    ..    ..    ..    ..
21 12.3  11.7    23   25    ..    ..    26    23    ..    ..    ..    ..
22 10.6  10.1    25   27    ..    25    28    ..    ..    ..    ..    ..
23  6.8   6.5    24   26    ..    26    ..    ..    ..    ..    ..    ..
================================================================================
```

**Figure 10.13** ASAPS S/N ratios for the 1E mode on the Boulder to St. Louis circuit. Comparison with Figure 10.10 shows that during the day, working at the E-layer OWF rather than the F-layer OWF leads to a S/N penalty of about 12 dB. This is mainly because the E-layer OWF is so much lower than the F-layer OWF.

the available frequency range is not as wide as on longer circuits, so the choice of frequency must be made more carefully (and preferably in real time, as described in Section 6.9). It is also partly because the ubiquitous whip antenna transmits virtually no energy straight up. At 10 MHz, for example, the antenna pattern for a 35 foot (vertical) whip antenna has a maximum gain of −4 dB, and this occurs for elevation angles of 20° to 40°. The gain at 90° is −30 dB, which means that virtually no energy would be radiated straight up from a whip. A vertical whip antenna would thus be useless for an NVIS circuit.

Figures 10.16 and 10.17 dramatically illustrate the signal level advantage offered by a horizontal dipole over a vertical whip for NVIS circuits in the Dallas area, with S/N differences greater than 30 dB. The HHWD antenna used in the calculations is a half-wave horizontal dipole mounted a quarter of a wavelength above the ground. If it is not practical to use a dipole, such as with vehicle operations, it is still possible to use a whip quite successfully for NVIS communications. All that is necessary is to tilt it forward over the vehicle using a "whip-tilt adapter" [9], so that it is almost horizontal. Simply bending a whip tends to break it over time.

Note that ASAPS uses the antenna which is specified only for transmission. The receiving antenna is always an isotropic antenna with zero gain. In practice, it is found that the choice of receiving antenna is not particularly important, because it also picks up the noise unless some sort of beam steering allows noise or interference sources localized in azimuth to be avoided.

### 10.3.4 The Best Usable Frequency

The working rules for selecting the best frequency of operation can be summarized as follows: *Always use as high a frequency as possible*. If we consider only ionospheric support, and not the available S/N ratios, this rule simply says to work at the OWF (assuming that a 90% probability of ionospheric support is required). When S/N ratios are also considered, the rule actually stays the same, because the S/N increases as the frequency increases. However the available S/N ratio depends very strongly on the antennas

```
=================================================================================
Circuit: boulder-st louis          Distance: 1299 KM      Date:    January,1970
Tx: boulder           40.02 254.72 Bear: 91    281        T-index: 100
Rx: st louis          38.66 269.75 Path: Short Path       Circuit#: 1
FSet: long                         Antenna: ISOTROPIC     RxNoise: -148 dBW/Hz
Mode: 2F                  SIGNAL/NOISE
=================================================================================
UT   MUF  OWF    MUF  OWF   3.0   6.0   9.0  12.0  15.0  18.0  21.0  24.0
00  10.3  8.7    24   23    21    23    23    ..    ..    ..    ..    ..
01   8.5  7.1    22   22    21    22    23    ..    ..    ..    ..    ..
02   7.0  5.8    22   22    21    22    ..    ..    ..    ..    ..    ..
03   5.9  5.0    20   20    19    20    ..    ..    ..    ..    ..    ..
04   5.5  4.6    17   17    16    17    ..    ..    ..    ..    ..    ..
05   5.1  4.3    16   16    15    17    ..    ..    ..    ..    ..    ..
06   4.9  4.2    16   16    15    ..    ..    ..    ..    ..    ..    ..
07   4.9  4.1    16   16    14    ..    ..    ..    ..    ..    ..    ..
08   4.8  4.0    16   15    14    ..    ..    ..    ..    ..    ..    ..
09   4.7  3.9    16   15    14    ..    ..    ..    ..    ..    ..    ..
10   4.8  3.9    16   15    14    ..    ..    ..    ..    ..    ..    ..
11   4.7  3.8    17   17    16    ..    ..    ..    ..    ..    ..    ..
12   4.6  3.8    19   19    18    ..    ..    ..    ..    ..    ..    ..
13   4.7  3.9    19   19    19    ..    ..    ..    ..    ..    ..    ..
14   6.5  5.6    18   15    -7    16    ..    ..    ..    ..    ..    ..
15  10.1  8.7    21   18    ..     9    19    ..    ..    ..    ..    ..
16  11.3  9.8    20   17    ..    -3    16    21    ..    ..    ..    ..
17  12.6 10.9    20   17    ..   -23    13    19    ..    ..    ..    ..
18  13.0 11.4    19   17    ..   -31    11    18    ..    ..    ..    ..
19  13.4 11.8    19   17    ..   -32    11    17    ..    ..    ..    ..
20  13.0 11.4    17   16    ..   -23    12    17    ..    ..    ..    ..
21  12.8 11.1    18   18    ..    -4    15    18    ..    ..    ..    ..
22  12.6 10.9    20   20    ..    10    19    20    ..    ..    ..    ..
23  11.7 10.1    24   24    -3    22    24    24    ..    ..    ..    ..
=================================================================================
```

**Figure 10.14** ASAPS S/N ratios for the 2F mode on the Boulder to St. Louis circuit. Comparison with Figure 10.10 shows that during the day, working via the 2F mode at a frequency of 12 MHz rather than via the 1F mode at the same frequency leads to a S/N penalty of about 15 dB. This is mainly because of the losses at the ground reflection. It would also have been possible to choose a higher frequency for the 1F mode (such as 18 MHz) with a higher S/N, so the actual penalty is higher.

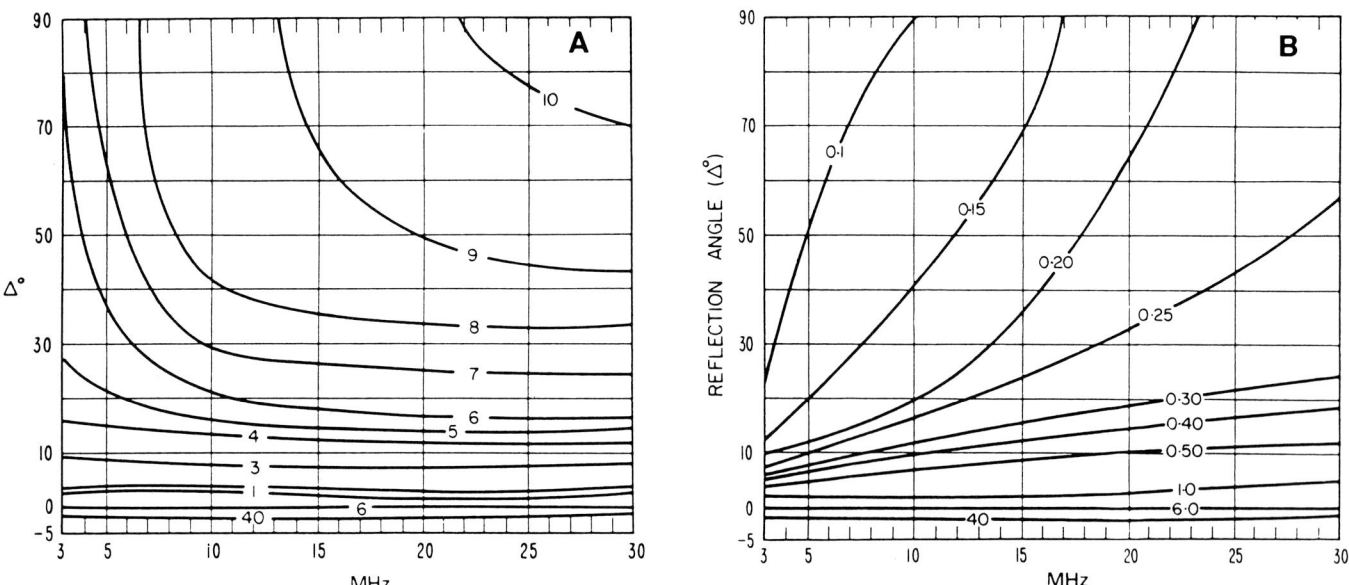

**Figure 10.15** The reflection losses in dB for random polarization, as a function of elevation angle for (A) poorly conducting ground and (B) the sea. At 10 MHz and 10°, for example, the losses are about 3.5 dB and 0.4 dB. For the 2-hop mode on the Boulder to St. Louis circuit, with a frequency of 10 MHz and an elevation of 40°, the loss would be about 8 dB.

# Propagation Prediction Programs

```
===============================================================================
Circuit: dallas                         Distance: 0-300 KM    Date:    March,1991
Tx: dallas            32.77  263.20     Bear: 0  180          T-index: 149
Rx: dallas            32.77  263.20     Path: District        Circuit#: 1
FSet: nvis                              Antenna: WH35         RxNoise: -148 dBW/Hz
Mode: 1F                     SIGNAL/NOISE
===============================================================================
UT  MUF  OWF    MUF  OWF   2.0  3.0  4.0  5.0  6.0  7.0  8.0  9.0 10.0 12.0
00 12.2 11.1   -15  -16   -32  -24  -20  -18  -18  -18  -17  -17  -16  -15
01 11.0 10.1   -16  -16   -19  -17  -16  -16  -16  -16  -16  -16  -16  -15
02  9.4  8.6   -17  -17   -19  -18  -16  -16  -16  -16  -16  -17  -17   ..
03  8.3  7.6   -17  -18   -23  -21  -19  -19  -18  -18  -17  -17   ..   ..
04  7.6  6.6   -19  -20   -26  -24  -23  -21  -20  -20  -18   ..   ..   ..
05  7.3  6.0   -20  -20   -26  -24  -23  -22  -20  -20  -18  -17   ..   ..
06  6.8  5.6   -21  -21   -26  -24  -23  -22  -21  -20  -18   ..   ..   ..
07  6.7  5.5   -21  -22   -28  -26  -24  -23  -21  -20  -18   ..   ..   ..
08  6.6  5.5   -21  -22   -29  -27  -26  -23  -22  -20  -18   ..   ..   ..
09  6.3  5.3   -22  -23   -29  -27  -26  -23  -22  -20   ..   ..   ..   ..
10  6.2  5.2   -22  -23   -29  -27  -26  -23  -22  -20   ..   ..   ..   ..
11  5.7  4.8   -21  -21   -24  -22  -22  -21  -21   ..   ..   ..   ..   ..
12  5.9  5.1   -18  -18   -19  -18  -18  -18  -18   ..   ..   ..   ..   ..
13  7.8  6.9   -17  -18    ..  -24  -21  -20  -18  -17  -16   ..   ..   ..
14 10.2  9.1   -13  -15    ..  -35  -26  -23  -20  -18  -16  -15  -14   ..
15 11.6 10.3   -10  -11    ..   ..  -24  -18  -15  -13  -12  -11  -11   -9
16 12.4 10.9   -9   -9     ..   ..  -30  -20  -15  -12  -11  -10   -9   -9
17 13.0 11.5   -10  -10    ..   ..  -36  -23  -18  -14  -12  -11  -11  -10
18 13.5 11.9   -10  -10    ..   ..  -38  -24  -19  -15  -13  -12  -11  -10
19 13.7 12.1   -11  -12    ..   ..  -38  -25  -19  -15  -14  -13  -12  -12
20 13.8 12.3   -13  -13    ..   ..  -36  -23  -18  -15  -14  -13  -13  -13
21 13.4 12.2   -13  -13    ..   ..  -30  -20  -16  -13  -12  -12  -12  -13
22 13.1 11.9   -12  -12    ..   ..  -22  -16  -13  -11  -11  -11  -11  -12
23 12.7 11.6   -13  -13    ..  -27  -18  -15  -14  -13  -13  -13  -14  -13
===============================================================================
```

**Figure 10.16** ASAPS S/N ratios for a 35-foot whip transmitting antenna on a Near Vertical Incidence Skywave (NVIS) circuit centered on Dallas. These ratios should be compared with those in Figure 10.17, which are much higher because of the more appropriate antenna gain pattern. The transmitter power is 10 watts.

being used. This is why ASAPS provides as one of its options the **Best Usable Frequency** or **BUF**.

Figure 10.18 shows the BUF output for the Boulder to St. Louis circuit, for the same frequency set used earlier as in Figure 10.10, a 1 kW transmitter, and a minimum elevation angle of 3° (the default). It can be seen that the propagation mode corresponding to the BUF is the 1F mode (column 2) for all 24 hours. This is what we found when we considered the modes separately. Columns 3 (Probability), 4 (Angle), 6 (S/N), 7 (S/N at the MUF), 8 (S/N at the OWF), 10 (MUF), and 11 (OWF) are all self-explanatory and offer no surprises. Column 5 gives the noise field strength (which is related to the noise power by equation A1.10 of Appendix A) at the BUF. Column 9 gives the BUF, which is the frequency which ASAPS recommends at each UT. A brief look shows that the BUF is the highest of the allocated frequencies (in the frequency set *long*) which still lies below the OWF. If the %Days is changed to some figure other than 90, the BUF will change so that it corresponds to the highest allocated frequency for which propagation is predicted for that percentage of the days of the month.

If the transmitting antenna were not really suitable for the Boulder to St. Louis circuit, and of no use below and elevation angle of 20°, so that the minimum angle is set to 20°, the BUF will no longer correspond to the 1F mode at all hours. This is because the elevation angles for the 1F mode lie between 18 and 20°, and lie below the new cutoff elevation angle of 20°.

## 10.4 Ionospheric Indices

Ionospheric indices were introduced in Section 3.4, but with no discussion of just what they are. We are now at the point where such a discussion should prove most useful. Consider the Washington monthly median noon values of $f_oF_2$ for December plotted against sunspot number in Figure 10.19 (which is an edited version of Figure 3.13). There is a data point for each of the 35 years 1934 through 1968. The straight line with an intercept of 5.9 MHz is the least-squares-fit regression line. It is clear that it is rare for a point to lie on the line.

Now we know that the observed values of $f_oF_2$ which have

```
=================================================================================
Circuit: dallas                      Distance: 0-300 KM      Date:    March,1991
Tx: dallas            32.77  263.20  Bear: 0   180           T-index: 149
Rx: dallas            32.77  263.20  Path: District          Circuit#: 1
FSet: nvis                           Antenna: HHWD           RxNoise: -148 dBW/Hz
Mode: 1F                SIGNAL/NOISE
=================================================================================
UT   MUF   OWF    MUF  OWF   2.0  3.0  4.0  5.0  6.0  7.0  8.0  9.0  10.0 12.0
00   12.2  11.1   20   19    4    12   15   17   17   18   18   18   19   20
01   11.0  10.1   19   19    17   19   19   19   19   19   19   19   19   20
02   9.4   8.6    18   18    17   18   19   19   19   19   19   18   18   ..
03   8.3   7.6    18   17    13   15   16   16   17   17   18   18   ..   ..
04   7.6   6.6    16   15    10   12   12   14   15   16   17   ..   ..   ..
05   7.3   6.0    15   15    10   12   12   13   15   15   17   18   ..   ..
06   6.8   5.6    14   14    10   12   12   13   14   15   17   ..   ..   ..
07   6.7   5.5    14   13    8    10   11   12   14   15   17   ..   ..   ..
08   6.6   5.5    14   13    7    9    9    12   13   15   17   ..   ..   ..
09   6.3   5.3    13   12    7    9    9    12   13   15   ..   ..   ..   ..
10   6.2   5.2    13   12    7    9    9    12   13   15   ..   ..   ..   ..
11   5.7   4.8    14   14    12   14   13   14   14   ..   ..   ..   ..   ..
12   5.9   5.1    17   17    17   18   17   17   17   ..   ..   ..   ..   ..
13   7.8   6.9    18   17    ..   12   14   15   17   18   19   ..   ..   ..
14   10.2  9.1    22   20    ..   1    9    12   15   17   19   20   21   ..
15   11.6  10.3   25   24    ..   ..   11   17   20   22   23   24   24   26
16   12.4  10.9   26   26    ..   ..   5    15   20   23   24   25   26   26
17   13.0  11.5   25   25    ..   ..   -1   12   17   21   23   24   24   25
18   13.5  11.9   25   25    ..   ..   -3   11   16   20   22   23   24   25
19   13.7  12.1   24   23    ..   ..   -3   10   16   20   21   22   23   23
20   13.8  12.3   22   22    ..   ..   -1   12   17   20   21   22   22   22
21   13.4  12.2   22   22    ..   ..   5    15   19   22   23   23   23   22
22   13.1  11.9   23   23    ..   ..   13   19   22   24   24   24   24   23
23   12.7  11.6   22   22    ..   9    17   20   21   22   22   22   21   22
=================================================================================
```

**Figure 10.17** ASAPS S/N ratios for a half-wave horizontal dipole transmitting antenna on a Near Vertical Incidence Skywave (NVIS) circuit centered on Dallas. The use of the HHWD in place of a 35-foot whip increases the S/N ratios by about 35 dB, illustrating the importance of choosing the correct type of antenna. The transmitter power is 10 watts.

been plotted are correct, and it seems a good assumption that the regression line represents the "true" relationship between $f_oF_2$ and R. We can therefore assume that for those points not lying on the line, the values of R did not represent accurately the solar EUV flux for that month, and search for a more representative value of R. Consider the point A, which has the coordinates (72,8.5). We can "put the point on the line" by moving it horizontally to the point (50,8.5). The new value of R corresponding to the point A is called the ionospheric index for that particular month and year, and we shall denote it by I. Similarly, the ionospheric indices derived for points B and C are 95 and 147.

Ionospheric indices can be derived from the regression line for each month with data available at each ionosonde site, once the site has been in operation over a solar cycle, and can then be continually revised as new data becomes available. Suppose the procedure has been implemented for N sites suitably distributed around the world, for a particular 11-year solar cycle. There will then be an index for each of N sites and 11*M months. However there is little value in having N values of the index for the same epoch (each month), so some sort of average value, denoted as $\langle I_N \rangle$, is calculated for each month of data. This procedure not only reduces the number of indices by N, but it also provides a worldwide index which can be used to characterize the ionosphere as a whole, eliminating regional differences.

When $f_oF_2$ is plotted against $\langle I_N \rangle$ rather than against a smoothed sunspot number, the scatter about the regression line will be reduced, and the points will no longer show any **saturation** effects at high levels of solar activity, or any **hysteresis** effects. The term "saturation" describes the failure of the $f_oF_2$ values to continue increasing as the sunspot number increases to very high values. Notice that the curve looks like the voltage versus time curve for a capacitor under charge. The term "hysteresis" comes from the field of magnetism and is used to indicate the difference between the plots for up-going and down-going parts of a solar cycle. The scatter will not disappear completely, because the indices obtained for each site will not be identical. Different ionospheric indices may be derived, depending on which of the possible 24 hourly values are used, the number and distribution of contributing ionosonde sites, and the averaging

```
===============================================================================
Circuit: boulder-st louis        Distance: 1299 KM        Date:   January,1970
Tx: boulder            40.02   254.72   Bear: 91   281    T-index: 100
Rx: st louis           38.66   269.75   Path: Short Path  Circuit#: 1
FSet: long                              Antenna: ISOTROPIC RxNoise: -148 dBW/Hz
Required S/N: 0 dB     TxPwr: 1000W       %Days: 90       Min. Angle: 3 deg.
Modes: 1F 1E 2F 2E                      BEST USABLE FREQ. BandWidth: 3.0kHz
===============================================================================
UT   MODE   PROB   ANGLE   NOISE   S/N   SN@MUF   SN@OWF    BUF      MUF    OWF
00   1F     99     18      -31     37    40       38       12.000   16.0   13.6
01   1F     99     18      -29     34    37       36        9.000   12.8   10.6
02   1F     99     18      -28     33    35       34        6.000   10.5    8.6
03   1F     99     19      -26     31    32       31        6.000    8.3    6.8
04   1F     90     21      -24     29    30       29        6.000    7.3    6.0
05   1F     99     19      -22     24    29       28        3.000    6.8    5.5
06   1F     99     20      -22     24    29       28        3.000    6.6    5.4
07   1F     99     20      -21     23    29       28        3.000    6.4    5.3
08   1F     99     20      -20     22    30       28        3.000    6.4    5.2
09   1F     99     20      -20     22    30       27        3.000    6.2    5.0
10   1F     99     19      -20     22    30       27        3.000    6.3    5.0
11   1F     99     19      -23     25    31       29        3.000    6.2    4.9
12   1F     99     18      -25     27    32       30        3.000    6.2    4.9
13   1F     99     18      -25     27    32       31        3.000    6.7    5.4
14   1F     99     19      -31     33    37       34        9.000   11.0    9.4
15   1F     99     18      -34     36    40       38       12.000   16.3   13.8
16   1F     99     19      -34     36    39       36       15.000   18.1   15.5
17   1F     99     19      -34     35    40       37       15.000   19.8   17.1
18   1F     99     19      -34     34    40       38       15.000   20.4   17.9
19   1F     99     19      -35     37    40       38       18.000   20.9   18.3
20   1F     99     19      -30     31    39       35       15.000   19.9   17.4
21   1F     99     19      -30     32    39       35       15.000   19.8   17.1
22   1F     99     18      -30     33    40       35       15.000   19.5   16.7
23   1F     99     18      -32     37    40       37       15.000   18.1   15.4
===============================================================================
```

**Figure 10.18** The Best Usable Frequency (BUF) predicted by ASAPS for the Boulder to St. Louis circuit. The BUF is the highest allocated frequency (in the frequency set "long") which satisfies the minimum angle (3°), the percentage of days for support (90%), and the required S/N threshold (0 dB). The transmitter power is 1,000 watts, and the transmitting antenna is an isotropic antenna. The S/N is given for the BUF, the MUF, and the OWF.

techniques. The index $IF_2$ used by CCIR is based on noon values of $f_oF_2$ at selected sites, whereas the T index uses all 24 hourly values.

By reducing the scatter in the $f_oF_2$ versus index plots relative to that produced by the direct use of the sunspot number, the use of an ionospheric index allows more reliable values of $f_oF_2$ at index values of 0 and 100 to be derived. This makes for more accurate worldwide mapping of $f_oF_2$. With the reverse situation, deriving an index from an observed value of $f_oF_2$ and the index-based world maps will lead to more accurate estimates of what is happening at some other location.

## 10.5 Programs Available

A comprehensive compilation of the available PC-based programs was prepared in 1988 by Bradley and Vernon [10] under the auspices of CCIR Interim Working Party 6/1. This report is 245 pages long and lists 22 programs, so obviously a similar review will not be attempted here. A comprehensive review of many of the available programs has just recently been published [11]. Broadly speaking, the programs can be grouped into several overlapping categories:

a. MINIMUF-based programs, which are designed for speed and minimum computer requirements

b. Other programs based on simplified models of the ionosphere and HF propagation, also designed for speed and minimum computer requirements

c. CCIR-based programs, which follow very closely the CCIR reports and recommendations

d. IONCAP-based programs, which aim to do all that IONCAP will do, but in a more user-friendly fashion

Ignoring such questions as availability, cost, and computer requirements, some of the attributes of an ideal PC-based prediction program would be:

1. User-friendly, simple (and *interesting*) to use

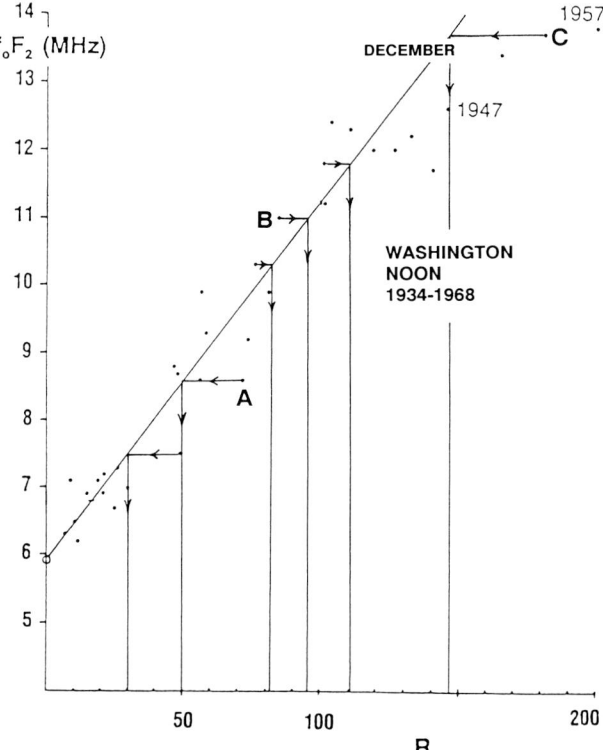

**Figure 10.19** Illustrating the derivation of an ionospheric index. The data points are for all Decembers between 1934 and 1968, as described in Chapter 3. The sunspot number which would have placed the data point on the line for each month is used in conjunction with similar values from other stations to derive an ionospheric index for that month.

2. Provides detailed information which is useful to the more experienced operators, but can be ignored to begin with by an inexperienced one

3. Provides an accuracy not limited by unnecessary approximations

4. Provides LUFS, MUFS, OWFs, and statistical variability of these parameters for at least two propagation modes

5. Allows the operator to anticipate awkward propagation conditions, such as multimode fading

6. Provides estimates of S/N ratios and Best Usable Frequency for specified parameters such as transmitter power and antenna patterns

7. Allows the operator to consider the support offered by different propagation modes and antennas

8. Provides results in an acceptable timeframe

9. Provides hard copy of all relevant information

10. Provides for the ready update of the ionospheric index

11. Is supported by an organization with extensive experience in HF communications

Having no real computer limitations, the author's preferences are for categories (c) and (d), with a view to making as few approximations as possible, on the grounds that there are already enough imponderables to worry about. The author uses ASAPS for PC work and reserves IONCAP for mainframe studies. ASAPS satisfies all of the above requirements (1 through 11) for an ideal PC-based program. Some of the available programs are definitely overkill for users with just two allocated frequencies, but the more complicated ones tell more of the story of what is going on.

Damboldt and Suessmann [12] adopt a pragmatic approach, arguing that their very simple program MINIFTZ4 reproduces the observed CCIR field strengths just as well as the more complicated methods. MINIPROP PLUS™ [13] is based on the work of Fricker [14], and falls under (b). In the author's limited experience of programs other than IONCAP and ASAPS, MINIPROP PLUS™ is exceptionally user-friendly and complete. The MINIMUF-based program DX [15] is another thorough and useful program, specifically designed for amateurs. Both MINIPROP PLUS™ and DX also provide details of the day-night terminator, which makes it very easy to go after the fascinating gray-line modes.

There are several IONCAP-based PC programs, provided by the Institute for Telecommunications Science (ITS) in Boulder [16] and by The Voice of America (VOA) in Washington. VOACAP is a mainframe version of IONCAP with some of the known "bugs" removed by VOA [any program as complex as IONCAP has a 100% chance of having a few bugs], while PC-VOACAP [17] is a menu-driven PC version of VOACAP. IONCAST, prepared by the Naval Research Laboratories (NRL) in Washington, replaces the card-image input of IONCAP by an interactive procedure. PC-VOACAP is a descendant of IONCAST and VOACAP, and reportedly pushes the PC capabilities to the limit.

The different programs require different indices to be supplied, but some are more flexible than others. Thus the IONCAP derivatives use the sunspot number; MINIPROP PLUS™ and DX use the 10.7 cm solar flux, but will accept a sunspot number; ASAPS works internally in terms of the ionospheric index T, but will also accept the sunspot number and flux.

## 10.6 Limitations of the Programs

As with virtually all the mainframe programs, nowhere is the accuracy as good as we would like it to be, and none of the programs really does a good job at high latitudes. In this regard, it will be interesting to see how well the Russian program SPARC [18] performs. SPARC is designed for high-latitude communications and accepts real time observations of $f_oF_2$ and the magnetic index $k_p$.

The antennas form an integral part of a communication system, and the different programs offer different solutions. For example, MINIPROP PLUS™ assumes a horizontal dipole at right angles to the circuit, but allows the gains to be changed; DX allows the choice of six antenna gains; and IONCAP and ASAPS offer a wide selection of antennas to choose from, with the possibility of the user providing his own.

It appears that all of the current PC programs, with the possible exception of the IONCAP derivatives, ignore $E_s$. This is not an unreasonable procedure, since the properties of $E_s$ are very difficult to predict, but the possibility of support by $E_s$ above the normal OWF should always be taken into account. This is especially true for summer days at mid-latitudes.

The programs are almost exclusively concerned with the quiet ionosphere, with the effects of storms being considered on only an ad hoc basis. It is worth bearing in mind that the next few years of the declining solar cycle will provide some rather large recurrent storms due to coronal holes and HSSWS (see Section 8.3). These storms will lead to decreases in the OWFs, especially during the summer and equinoxes. Coming at a time when the OWFs are decreasing because of the lower levels of solar flux, these storms will have severe effects on communications. The best way to keep on top of what is going on is to receive regular reports from the various Regional Warning Centers. See, for example, Rosenthal [19] and Rosenthal and Hirman [20].

## 10.7 Conclusions

The wide variety of PC-based propagation prediction programs currently available will ensure that the serious amateur has available a tool suitable for his own particular way of doing things. With its indefatigable worldwide base, and its penchant for practical observations, the amateur community can play a significant role in improving these programs. It is in the nature of the radio amateur to ensure that every program will be extensively and severely tested. We can therefore expect to see a natural attrition of those programs found wanting and continuing improvements to the programs of choice.

## 10.8 References

1. Teters, L. R., J. L. Lloyd, G. W. Haydon, and D. L. Lucas. *Estimating the Performance of Telecommunication Systems Using the Ionospheric Transmission Channel. Ionospheric Communications Analysis and Prediction Program User's Manual.* NTIA Report 83-127, U.S. Dept. of Commerce, July 1983.

2. Rose, R. [K6GKU]. MINIMUF-3: *A Simplified HF MUF Prediction Algorithm.* Technical Report 186, Naval Ocean Systems Center, San Diego, CA, 1978. See also QST December 1982, page 38.

3. ASAPS—*Advanced Stand-Alone Prediction System.* IPS Radio and Space Services, P. O. Box 5606, West Chatswood, N.S.W. 2057, Australia.

4. McNamara, L. F. *The Accuracy of MUF Predictions within Australia.* Ionospheric Prediction Service Report IPS-R28, Sydney, November 1975.

5. Brown, R. R. [NM7M]. *Long-Path Propagation. A Study of Long-Path Propagation in Solar Cycle 22.* March 1992.

6. Romberg, C. [N5CR]. Conrad is prepared to go on record as saying, "The long path is always open."

7. McNamara, L. F. *The Ionosphere: Communications, Surveillance and Direction Finding.* Chapters 11 through 14. Krieger Publishing Company, Malabar, Florida, 1991.

8. Tsedilina, E. E. *HF Radio Wave Field Strength on Long Mid-Latitude Paths.* Accepted for publication in *Radio Science*, 1993.

9. Christinsin, A. S. *High Frequency NVIS Vehicular Communications for Command and Control.* Presented at the Chesapeake Chapter of AFCEA symposium "Modern HF and Emerging Technologies," Annapolis, MD, November 1991.

10. Bradley, P. A., and A. Vernon. *Available Microcomputer-Based Propagation Prediction Programs.* CCIR IWP 6/1 Document 320 (Rev. 1), 20 April 1988.

11. d'Avignon, J. [VE3VIA]. *Propagation Programs.* Grove Enterprises, Brasstown, NC, Fall 1992.

12. Damboldt, T., and P. Suessman. FTZ *High-Frequency Sky-Wave Field Strength Prediction Method for Use on Home Computers.* Forschungsinstitut der DBP beim FTZ, Postfach 5000, D-6100 Darmstadt, Germany.

13. Shallon, S. C. *MINIPROP PLUS™.* W6EL Software, 11058 Queensland Street, Los Angeles, CA 90034–3029.

14. Fricker, R. *A Microcomputer Program for the Critical Frequencies and Height of the F layer of the Ionosphere.* Fourth International Conference on Antennas and Propagation, Institution of Electrical Engineers (U. K.), 1985, pp. 546–550.

15. Satterlee, B. [KD6SC]. *DX.BAS Propagation Predic-*

*tion Program, IBM-PC Version 4.3.* 8210 Blossom Hill Ct., Lemon Grove, CA 91945-4232, August 1990.

16. Stewart, F. G. *IONCAP Version PC 2.2.* DoC/NTIA/ITS, 325 South Broadway, Boulder, CO 80303-3328.

17. PC-VOACAP. Voice of America, USIA/VOA, 330 Independence Avenue SW, Washington, DC 20547.

18. SPARC—*System Prediction Analysis Radio Communication.* Institute of Applied Geophysics, Rostokinskaya 9, 129226 Moscow, Russia. [Professor A. D. Danilov].

19. Rosenthal, D. A. [N6TST]. *Behind the Solar Forecast.* World Radio TV Handbook, 1990, pp. 541–544.

20. Rosenthal, D. A., and J. W. Hirman. *A User's Guide to the Space Environment Services Center Geophysical Alert Broadcasts.* NOAA Technical Memorandum ERL SEL-80. NOAA, Boulder, CO, 1989.

# Appendix A

# Formulas for S/N and Field Strength

The intensity of a radio wave can be expressed either as the power available at the input terminals or as the field strength at the receiving antenna. The power, $p_r$, of an electromagnetic wave at the terminals of a receiving antenna is given by

$$p_r = p\, A\, g_r, \qquad (A1.1)$$

where

- $p_r$ is the received signal power
- $p$ is the power density of the signal at the antenna
- $A$ is the effective area of the antenna
- $g_r$ is the receiver gain

For transmission through free space, power is proportional to the square of the amplitude, $a$,

$$p = a^2/(119.9\,\pi) \qquad (A1.2)$$

where the term $1/(119.9\pi)$ is the impedance of free space.

For an isotropic antenna in free space,

$$A = c^2/(4\pi f^2) \qquad (A1.3)$$

where $c$ is the velocity of light in free space, and $f$ is the radiowave frequency.

Substituting A1.2 and A1.3 into A1.1 gives

$$p_r = \{a^2/(119.9\pi)\}\,\{c^2/(4\pi f^2)\}\, g_r. \qquad (A1.4)$$

Grouping the constants gives

$$p_r = \{c^2/(4\pi^2 119.9)\}(1/f^2)a^2 g_r. \qquad (A1.5)$$

Recall now that a power, $p$, is converted to dB by

$$P = 10\, \log_{10} p. \qquad (A1.6)$$

Taking logarithms to base 10 of equation A1.6, and writing all variables measured in dB as upper case letters, we get

$$P_r = 12.8 - 20\, \log f + 20\, \log a + G_r. \qquad (A1.7)$$

Note that the frequency is now measured in MHz (which is why the constant term is 12.8 and not 132.8).

The field strength, E, of an electromagnetic wave is defined as the power propagated by its E vector. It is usually expressed relative to a reference amplitude, $a_0$, of 1 $\mu$V/m (microvolt per meter)

$$E = 20\, \log (a/a_0) \qquad (A1.7)$$
$$= 20\, \log a + 120\text{ dB above }1\,\mu\text{V/m}. \qquad (A1.8)$$

(The 120 comes from $-20$ times $-6$.)

Equation A1.8 gives

$$20\, \log a = E - 120. \qquad (A1.9)$$

Substituting A1.9 into A1.6 gives the received power in dB relative to a watt as

$$P_r[\text{dBW}] = E[\text{dB}\mu] + G_r - 20\, \log f[\text{MHz}] - 107.2 \qquad (A1.10)$$

where dB$\mu$ is the shorthand for the field strength unit of dB above 1 $\mu$V/m, and the square brackets enclose the units for each term.

The received signal power is also equal to the signal power at the terminals of the transmitter, $P_t$, less the total loss of power in the system, $L_s$,

$$P_r = P_t - L_s. \qquad (A1.11)$$

The field strength at the receiver terminals is therefore given by

$$E = P_t - L_s - G_r + 20\, \log f + 107.2. \qquad (A1.12)$$

The system loss, $L_s$, is equal to the losses along the raypath, L, less the gains of the transmitting and receiving antennas

$$L_s = L - G_t - G_r. \qquad (A1.13)$$

$G_t$ and $G_r$ are the gains of the transmitting and receiving antennas over an isotropic antenna in the direction along the raypath.

---

This appendix is based on: Caruana, John. *HF Propagation Modes and Signal Intensities. Prediction and Observation.* IPS Radio and Space Services Series R Reports, IPS-R40, Sydney, Australia, 1982.

Combining A1.11 and A1.13 yields the power available at the input terminals of the receiver

$$P[\text{dBW}] = P_t[\text{dbW}] + G_t - L + G_r . \quad (A1.14)$$

Combining A1.12 and A1.13 yields the field strength at the receiving antenna

$$E = 107.2 + 20 \log f + P_t + G_t - L . \quad (A1.15)$$

The power, P, and the field strength, E, can therefore be calculated for a given transmitter power, if we know the loss along the raypath, L, and the antenna gains at the elevation angle of the raypath. Our main concern here is with the signal to noise ratio. Suppose that the noise power in 1 Hz is N dbW. Subtracting N from both sides of A1.14 (remember that we are now working in dB) gives

$$P - N = P_t + G_t + G_r - L - N . \quad (A1.16)$$

If we now define the term (S/N) to be P − N, and the term (N/P) to be −(L + N), we get

$$(S/N) = P_t + G_t + G_r + (N/P) . \quad (A1.17)$$

Now S/N ratios are normally given for something like a 3 kHz bandwidth, while the (N/P) values are given for a 1 Hz bandwidth. So, for general bandwidth of b Hz, we have

$$(S/N)_b = P_t + G_t + G_r + (N/P)_1 - 10 \log_{10} b \quad [\text{dBW}] . \quad (A1.18)$$

For a 3 kHz bandwidth, for example, the last term is 34.77. Increasing the bandwidth leads to the inclusion of more noise, so the S/N ratio goes down.

Equation A1.18 is the simplest one to use with the various parameters provided by ASAPS. If you would rather use the separate pathloss and noise terms, L and N, the equation to use is

$$(S/N)_b = P_t + G_t + G_r - L - N - 10 \log_{10} b \quad [\text{dBW}] . \quad (A1.19)$$

# Appendix B

# Answers to Problems

## Chapter 2

1. (a) The photosphere and sunspots.
   (b) The chromosphere, sunspots (only the big ones), plages, prominences, filaments and fibrils. Which two of these are really the same thing?
   (c) The corona and coronal holes.
   (d) Radio emission from the whole sun, or selected parts of it if we have a radio interferometer.
   Observing at all possible wavelengths helps to build up a more complete of what the sun is doing.

2. (a) Cloud cover prohibits optical observations.
   (b) Bad "seeing" conditions limits the detail that we can make out in an image of the sun.
   (c) The Earth's atmosphere absorbs short wavelength, high energy radiation from the corona.
   (d) Can you think of a few more?

3. The fibril structure around active regions and streamers in the corona both indicate that material on the sun is preferentially aligned along lines of force of the sun's magnetic field.

4. Count the number of sunspot groups, G, then the number of individual sunspots, N. The sunspot number is then equal to $10G + N$.

5. The sunspot number oscillates between zero and around 110 (on average), with a period of around 11 years.

6. The last maximum (this chapter was written in 1991) was in July 1989. By some measures, this was the second highest maximum ever observed.

7. The main problem would be making an incorrect forecast of the satellite drag.

## Chapter 3

1. Energetic ultraviolet light from the sun strips electrons off the neutral atoms of the Earth's atmosphere, leaving heavy positively charged ions embedded in a sea of light, negatively charged electrons. The process is called photoionization.

2. The free electrons. The ions are far too heavy to be able to respond to the oscillating electric field of the radio waves.

3. By recombination at higher altitudes, and attachment at lower altitudes.

4. Because the ultraviolet light from the sun covers a wide range of wavelengths and therefore energies, and there is more than one type of neutral atom or molecule which can be photoionized.

5. The D, E, and $F_1$ layers effectively disappear completely. The $F_2$ layer survives the night, but the electron density can get very low. The electrons are lost by recombination and attachment, and are not replenished by the sun. The rate at which electrons are lost is much lower at the higher altitude of the $F_2$ region.

6. $f_oE$ is the maximum frequency which will be reflected vertically from the E layer. Similar descriptions hold for $f_oF_1$ and $f_oF_2$.

7. The sun will be directly overhead, and the zenith angle will be zero.

8. Equation 3.4 tells us that $f_oE = 3.6\,\text{MHz}$. Note that the sun is directly overhead.

9. Diurnal, seasonal, location, solar activity, height. These variations are fairly regular and predictable. Significant variations also occur from day to day. See Figures 3.10 and 3.11.

10. Remembering that winter in Canberra occurs in June, we choose the top right-hand panel of Figure 3.10. We then find that $f_oE = 3.5\,\text{MHz}$ and $f_oF_2 = 12\,\text{MHz}$. There is no curve for $f_oF_1$ because the $F_1$ layer does not exist as a separate layer at mid-latitudes in winter except during disturbed conditions.

11. There are two good reasons. First, we do not know enough about the ionosphere to allow us to do much better than use a monthly average. Second, the amount of data we have to manipulate is decreased by a factor of 10.

12. Assume the 31 values are arranged in increasing order. Then the 16th value will be the median, with 15 values lying above it, and 15 values lying below it. The 28th value will be the upper decile value, and 28 values would lie below it.

13. Figure 3.13 indicates that $f_oF_2$ under these conditions would be 10.2 MHz. The scatter about the straight line of best fit indicates an uncertainty of about 0.5 MHz.

14. Sunspots are associated with plage areas, which are the source of the EUV light which photoionizes the Earth's neutral atmosphere and creates the ionosphere. Thus more sunspots means more plage areas, which means more EUV light, which means more photoionization and thus more electrons and higher critical frequencies.

15. $E_s$ layers occur at E-region heights of about 90 to 130 km. They are only a few km thick, and come and go more or less at random.

16. The equatorial anomaly. The electrons formed at low altitudes near the magnetic equator drift upwards to higher altitudes and then diffuse down along lines of force of the Earth's magnetic field. The processes of drift and diffusion, which are together called "the fountain effect," result in the presence of two clumps of electrons situated on either side of the magnetic equator.

17. The electron gyrofrequency is the frequency at which an electron spirals around a magnetic field line. Typical values for the ionosphere are 1 MHz near the equator, and 1.6 MHz near the poles.

18. (a) The crests of the equatorial anomaly appear on either side of the magnetic equator, not the geographic equator.
    (b) The belt of equatorial $E_s$ is centered on the dip equator.

19. Field-aligned irregularities occur in the F region at all latitudes. They are restricted to the hours of darkness at low and mid-latitudes, but can also occur during the day at high latitudes.

## Chapter 4

1. The ground wave can be used only for ranges up to about 100 km, whereas the range of the sky wave is virtually unlimited.

2. The ordinary and extraordinary rays travel via both the high raypath and low raypath. The high and low raypaths coincide at the MUF. Under many conditions, the high rays and the extraordinary low rays are heavily attenuated, leaving only one useful raypath.

3. The maximum usable frequency (MUF) depends on the critical frequency, $f_o$, of the reflecting layer, and on the geometry of the circuit, MUF = $f_c$ sec I, where I is the angle of incidence of the raypath at the reflecting layer.

4. For vertical incidence, the angle of incidence is 0°, cos 0° = 1, and sec 0° = 1. Thus MUF = $f_oF_2$. In fact $f_oF_2$ is defined to be the MUF for vertical incidence on the $F_2$ layer. (Actually the MUF for a vertical incidence circuit would be the critical frequency for the extraordinary mode of propagation, $f_xF_2$, which is about half the gyrofrequency greater than $f_oF_2$.)

5. The elevation angle is about 27° and the obliquity factor is about 2.0. Note that the obliquity factor includes the term 1.1 which takes the curvature of the Earth into account. If reflection occurs from an $E_s$ layer, the range would decrease to about 400 km.

6. The $F_2$ layer will reflect vertically incident signals at frequencies up to and including $f_oF_2$. It will reflect obliquely incident signals at frequencies even greater than $f_oF_2$. Thus there is no skip zone in this instance at 5 and 10 MHz. To determine the skip distance for 20 MHz, use the fact that 20 MHz must be the MUF. Using equation 4.2, we have 20 = 10 x obliquity factor. The obliquity factor is therefore 2. Referring to the right-hand panel of Figure 4.6, we see that for a 1-hop F layer circuit, an obliquity factor of 2 corresponds to a circuit length of 1,000 km. The radius of the skip zone is thus 1,000 km.

    An operating frequency of 100 MHz would require an obliquity factor of 10.0. Such a high obliquity factor is not possible for an F-layer reflection. Thus the 100 MHz could never be reflected in the F-layer and the skip zone would cover the whole Earth. Only with a $E_s$ layer is it possible to obtain sufficiently high critical frequencies and obliquity factors to support propagation at 100 MHz occasionally.

7. Because it was expected that the skip zone at 27 MHz would be so large as to preclude sky-wave communication between operators in different parts of the country. CB operators are supposed to use the ground wave.

8. We know that:
   (1) Each reflection at the ground leads to an extra loss of about 3 dB.
   (2) An $E_s$ layer is an excellent reflector at oblique incidence.
   (3) The E layer is a very lossy reflector.
   
   The required order is therefore probably $1E_s$, 1F, 2F, 1E, 2E.

9. Multipath interference occurs when signals from the transmitter travel to the receiver by two or more propagation modes. Different propagation modes almost in-

variably have different propagation times, causing the multipath signals to interfere with each other. Multipath interference can only be avoided by eliminating all but one propagation mode. This can be done by the astute choice of operating frequency or sometimes by choosing an antenna which favors one of the propagation modes.

10. (a) During the day we must consider whether the E layer will reflect our signals before they can get up to the F layer. Let us suppose $f_oE = 3$ MHz. Then for E-layer reflection, we would require an obliquity factor of at least 3.33. The left-hand panel of Figure 4.6 shows that a 1,000 km 1-hop E-layer mode would have an obliquity factor of 4.4. Thus the E layer would reflect the signals in this case. In fact it would reflect signals with frequencies up to $4.4 \times 3 = 13.2$ MHz. The corresponding elevation angle is about 9°. Table 4.1 tells us that we would need to use a vertical monopole with a ground screen to obtain these low angles. For a 1,000 km F-layer mode, which would have an elevation angle of about 28°, the obliquity factor is about 1.9. An operating frequency of 10 MHz would require an $f_oF_2$ value of $10/1.9 = 5.3$ MHz, which is exceeded at most mid-latitude locations at noon. Thus F-layer propagation is also possible for elevation angles of 28°. In fact, if we used a vertical monopole, our signals would travel by both E- and F-layer modes, depending on the elevation angles of the transmitting antenna, which would range from about 10° to 30°. To avoid having two propagation modes, we could use a horizontal dipole and feed most of the radiated power into the 2F propagation mode.

(b) At night, we have only the F layer to consider. We could use either a vertical monopole with ground screen, or a horizontal dipole. If we use the latter, we should attempt if possible to avoid putting too much energy into the 2F mode because of the ensuing problems of multipath interference.

11. (a) Absorption is virtually zero at midnight because the D and E regions have almost completely vanished due to attachment and recombination.

(b) Absorption is greater in summer when electron densities in the D region are higher. Remember that more electrons mean more loss of energy to the surrounding sea of neutral atoms or molecules.

(c) High solar activity when the electron densities in the D region are higher.

(d) Both of these frequencies would normally be reflected by the F layer, and not the E layer, at vertical incidence or for short circuits. The absorption at 4 MHz will be about $(8/4)**2 = 4$ times as great as that at 8 MHz.

(e) The 2 MHz signal would usually be reflected by the E layer while the 4 MHz signal would usually be reflected by the F layer, at vertical incidence. The 2 MHz signal would suffer the greater amount of absorption. It would also suffer more deviative absorption than the 4 MHz signal, especially if $f_oE$ happened to be just greater than 2 MHz.

(f) Absorption on winter anomaly days is about the same as on summer days. Recall, however, that we can use higher frequencies during winter at mid-latitudes because of the mid-latitude seasonal anomaly.

(g) The extraordinary mode is always more heavily absorbed than the ordinary mode, especially at low frequencies.

12. (a) Absorption, which increases as the operating frequency decreases.

(b) Antenna efficiency, which decreases as the operating frequency decreases, unless special efforts are made to ensure that the size of the antenna is as large as required. This is an expensive business at low frequencies and long wavelengths.

13. (a) 50% or 15 days. On the other half of the days, the actual MUF at the time will be lower than the predicted median MUF and you would have to use a lower frequency than the predicted MUF.

(b,c) The OWF and FOT are the same thing. You would expect successful communications on 90% or 27 days of the month.

## Chapter 5

No answers are given for this chapter.

## Chapter 6

1. (1) Predict the general level of solar activity.
   (2) Set up a model of the ionosphere for the level predicted in (1).
   (3) Calculate the geometry and propagation modes for the given circuit.
   (4) Calculate the propagation parameters MUF, LUF, field strength (pathloss), and elevation angles.

2. Assuming that maximum hop lengths for E and F layer reflections are 2,000 and 3,000 km respectively, the modes to consider would be
   (1) 5,000 km. 2F and 3F. E-layer propagation would involve three hops, and the attenuation would be too severe.
   (2) 7,000 km. 3F and 4F. The 2F mode would not be supported.
   (3) 12,000 km. The control point method would be used. The probable mode structure is ignored and reflection is assumed to occur 1,500 km in from

each end of the circuit. The lower of the MUFs for these two 3,000 km hops is taken as the MUF for the circuit. Why is it that the lower of the MUFs is chosen?

3. If you do not receive HF predictions and believe that you need them, contact your national organization (see CCIR Recommendation 313), or your local organization if you are a radio amateur. As regards the other parts to this question, one would hope that after reading this book you will give the right answers!

4. The $F_2$ region is also affected by neutral winds, electric fields, and the Earth's magnetic field. It also suffers from the effects of ionospheric storms, as discussed in Chapter 8.

5. Local time maps have less structure than universal time maps, and are thus easier to represent in computer models. They are not "real," since the local time is not the same over the surface of the Earth.

6. The lower limit is set by absorption and the increasing radio noise levels. The upper limit is set by the electron density at the reflection point. Both limits are also a function of the hop length, with both increasing as the hop length increases.

7. The minimum number you would require is probably 5, with the fifth one being necessary for dawn in winter at sunspot minimum.

8. Inverse-square law attenuation with distance (the free-space loss) and nondeviative absorption. However, at times the polarization coupling loss and the $E_s$ obscuration loss may dominate.

9. Basically, there is just one advantage—that of knowing what the ionosphere is doing at the time we are using it. This knowledge becomes particularly useful during disturbed conditions such as a SWF (shortwave fadeout) or during an ionospheric storm. These phenomena are described in Chapter 8.

10. The main advantage is that he would know what the ionosphere is doing, especially if he is working relatively short circuits.

11. You need only one frequency (5 MHz) if the predictions are exact. If you can use two frequencies, you should probably try 14 MHz during the day and 4 MHz during the night. During the day, a frequency of 21 MHz would be supported by only the 1F mode, and you would need an antenna which favors the elevation angles of 17°–19°. To find the nighttime value of $f_oE$, you can use Figure 4.6 (E modes) and Equation 4.2, for hop lengths of 650 (EMUF = 1.6 MHz) and 1,300 km (EMUF = 2.6 MHz). You will find that $f_oE$ is set to about 0.8 MHz.

12. (a) Yes, because the BUF is the highest allocated frequency which is supported by only one mode, and on 90% of the days of the month.
    (b) During the day, for example, the S/N at the OWF is 42 dB for a 3 kHz bandwidth, a 1 kW transmitter, and an isotropic antenna. To proceed further, you would need to specify the S/N ratio which is considered adequate before adding on the 10 dB fading margin, as well as accounting for the antenna gains.

13. Noise levels are higher during the night because HF absorption has virtually ceased to exist. The noise is greater at lower frequencies, and the OWFs are lower at night than during the day.

## Chapter 7

1. "Fading" is the term used to describe the rise and fall in signal strength encountered on most HF circuits. It is described by its depth relative to the average signal level (shallow or deep) and its periodicity. The latter is usually measured in Hz and ranges from fast (20 Hz, say) to slow (0.1 Hz, or a 10-second period, for example).

2. (1) Movement of the ionosphere and changes of the propagation path length.
   (2) Rotation of the plane of polarization of the radio wave.
   (3) Variation with time of the absorption due to the and lower E regions.
   (4) Skip fading at the MUF, as the edge of the skip zone moves to and fro across the receiver.

3. Multipath fading would be expected when the signals can travel by more than one propagation mode and when the travel times for the different modes are very similar. This is true for 1F and 2F modes, and for 1F and 2E modes. The fading can be minimized by ensuring that the signals can travel by only one mode (by choosing the appropriate antenna and/or frequency), or at least ensuring that the signals on one mode are much stronger than on the other.

4. Multipath (or multimode) fading and interference fading are the same thing. Flutter fading is just fast interference fading. Selective fading is also a particular type of interference fading, in which the fading rate changes rapidly with frequency.

5. Polarization fading results when the plane of polarization of the incident wave rotates with respect to the plane of polarization of the receiving antenna. When the two planes are lined up, the signal strength is greatest. When they are at right angles, the strength of the

# Appendix B

signal reaching the receiver will be zero. Crossed antennas are the solution.

6. The most likely cause of your problem is that the MUF for your circuit has dropped down to 15 MHz, and you are on the edge of the 15 MHz skip zone. Your first move should be to change to a lower frequency. Then ask yourself why you had not changed to a lower frequency before 15 MHz became unsuitable.

7. The five types of diversity techniques are: space diversity; frequency diversity; angle-of-arrival diversity; polarization diversity; time diversity. Diversity techniques are used to overcome the effects of severe fading. They are usually expensive, which is the main reason why they are not used more often.

8. "$E_s$ screening" is the term used to describe reflection of signals by an $E_s$ layer when we really would have liked them to go up to the F layer. Consequently it is a bad thing. However, if we organize things well, the same $E_s$ layer could be put to good use, giving us good strong signals on a circuit for which it is at the midpoint.

9. (1) Large horizontal gradients in $f_oF_2$ and $h_mF_2$. These cause problems in mapping the F region, and in calculating accurate values of the MUF.
   (2) Rapid increase of $f_oF_2$ at sunrise. Selecting an appropriate operating frequency is particularly hard at these times, especially when only two frequencies have been allocated.
   (3) Scattering by field-aligned irregularities. If the received signal arrives by an unwanted scatter mode by itself, or in conjunction with a normal propagation mode, the signals will suffer from flutter fading, Doppler shifting, and Doppler spreading.

10. The main feature is the auroral oval. HF signals are seriously degraded if reflected from the auroral E and F regions. Equatorwards of the oval (south of, in the northern hemisphere; and vice versa), lies the mid-latitude trough which is a region of low critical frequencies. MUFs for circuits with reflection points in this region will be significantly lower than those which avoid it. The auroral absorption zone lies near the auroral oval and can be responsible for severe absorption at HF.

11. (a) Horizontal; (b) Vertical; (c) Vertical. These choices of antenna polarization would reduce the possibility of polarization mis-match. The chances of encountering problems at mid-latitudes are rather small and a horizontally polarized antenna is recommended as a means of reducing the problem of noise.

12. (1) Atmospheric—distant thunderstorms.
    (2) Local—thunderstorms, industrial.
    (3) Galactic—emission from our Galaxy.
    (4) Interference—other operators on the same frequency.

13. (a) Put the receiver well away from centers of population.
    (b) Use a narrower bandwidth.
    (c) Use signal-coding techniques.
    (d) Use a horizontally polarized receiving antenna to eliminate local man-made noise. (Note, however, that you need a vertically polarized antenna for long circuits.)

14. Take the nighttime value of $f_oE$ to be 0.6 MHz, and the MF band frequency to be 1,200 kHz. Then the obliquity factor is 2.0 (Equation 4.2). Figure 4.6 (1-hop E modes) then shows that the skip distance is around 450 km. In other words, it is possible to hear via sky wave any MF transmitters which are no closer than 450 km.

15. Because it cannot be summer simultaneously in the two hemispheres.

## Chapter 8

1. Solar flares, coronal holes, sudden disappearing filaments.

2. (a) X rays. These lead to increased absorption and a shortwave fadeout.
   (b) Protons. These lead to increased absorption in the polar regions, causing a PCA.
   (c) Plasma cloud. If a plasma cloud from a flare hits the Earth, an ionospheric storm will result.

3. (a) X rays. X rays are emitted throughout the lifetime of the flare, and take about 8 minutes to reach the Earth.
   (b) Protons. Clouds of protons can take anything from 10 minutes to a few days to arrive at the Earth.
   (c) Plasma cloud. A plasma cloud can take from 1 to 4 days to reach the Earth. The average is about 2 days.

4. The connection between these two phenomena is the high speed solar wind stream or HSSWS which escapes from the corona through the coronal hole.

5. About 4 days.

6. (a) Solar maximum; (b) On the downward side of the solar cycle.

7. (a) Low; (b) High; (c) Summer.

8. None, as far as we are concerned here.

9. (a) The first thing to do is to ensure that it really is an

SWF, not an equipment problem. How would you do this?
(b) The second thing to do is to send priority messages by some other means, if possible.
(c) The third thing to do is to wait until the flare has died down. In the meantime, search the high end of the HF band for transmitters of opportunity, to get some idea of how the ionosphere is recovering. Recovery from an SWF starts with the highest frequencies.
(d) Fourth, reestablish your link at the highest of your allocated frequencies which is less than the OWF predicted for normal conditions.
(e) Finally, fill in your log book.

10. There is probably an ionospheric storm in progress. You will have to keep using the lower frequencies for a day or two.

11. Try avoiding the polar regions by using relay stations at lower latitudes. Even if you are also having SWF problems, they will probably go away well before the direct HF link opens up again.

12. (a) Sudden commencement storm—caused by flares; (b) Gradual commencement storm—caused by HSSWS.

13. Recurrent geomagnetic and ionospheric storms are due to HSSWS's from coronal holes which rotate with the sun and therefore pass over the Earth every 27 days or so.

14. (a) Try a higher frequency; (b) Try a lower frequency.

15. The equinoxes.

## Chapter 9

1. (a) Because they may help us establish communications on an otherwise unobtainable circuit, and on a higher frequency than can normally be used.
   (b) Because they can lead to interference from distant transmitters which are usually expected to have no possible effect.
   (c) Because they may make a supposedly secure communications link insecure.
   (d) Because they are there.

2. The $E_s$ layer.

3. Any unusual mode is probably an $E_s$ mode. Mid-latitude $E_s$ is usually a daytime summer phenomenon. The level of solar activity seems to be irrelevant.

4. (a) The equatorial $E_s$ belt, which follows the magnetic dip equator and is about 10 degrees in width.
   (b) The two crests, or peaks, of the equatorial anomaly.
   (c) Field-aligned irregularities in the F region.

5. (a) The equatorial $E_s$ layer will support long-range $E_s$ propagation, just like mid-latitude $E_s$. Combination modes such as $E_s$-F and $2E_s$-F have also been observed.
   (b) The high critical frequencies and tilts associated with the peaks of the equatorial anomaly support the afternoon-type TEP mode, described as a super mode or FF mode.
   (c) The field-aligned irregularities in the equatorial F region support the evening-type TEP mode. This mode may be either a field-guided mode in which ducted propagation occurs along empty field-aligned tubes, or by continuous forward scattering by the irregularities.

6. (a) Low-latitude $E_s$ modes are a daytime phenomenon. The seasons and level of solar activity are relatively unimportant.
   (b) Afternoon-type TEP occurs mainly between 1400 and 2000 local time, during the equinoctial months and around solar cycle maximum.
   (c) Evening-type TEP occurs between sunset and about midnight, during the equinoctial months and around solar cycle maximum.

7. The evening-type TEP mode has been observed to support frequencies greater than 100 MHz, whereas the afternoon-type mode rarely supports frequencies greater than 60 MHz. The afternoon-type mode is more useful for communications because its signal is characterized by slow fading and small Doppler shifts. The evening-type signal, on the other hand, suffers from deep and rapid fading, and large Doppler shifts.

8. A circuit such as this would support both afternoon and evening-type TEP. Around 1900 LT, it is possible that both propagation modes would be supported. A shorter circuit would favor the evening-type mode, and vice versa. TEP modes usually do not occur during the solstitial months or during low levels of solar activity.

9. The observed elevation angles suggest something like a 35-hop F-layer propagation mode. However, the signal for such a mode would be far too small to detect, and the observed time delays do not correspond to a 35F mode.

10. Non-great circle propagation modes are likely to occur whenever it is possible for signals to be scattered towards the receiver by field aligned irregularities which lie to one side of the normal great circle path. Thus they can occur on circuits crossing the magnetic equator (at night) and on circuits near the auroral oval.

11. Just ignore the term $C_i(t)$ in equation 9.3, and integrate once with respect to time. It would help if you insert the term $(p_i/P)$ which is missing from the right-hand side of the equation.

# Bibliography

## Texts and Reports Overlapping the Current Text

1. Abramowitz, M., and I. E. Stegun (Eds.). *Handbook of Mathematical Functions*. National Bureau of Standards, Applied Mathematics Series, No. 55, Washington, 1964.

2. Boithias, L. *Radio Wave Propagation*. McGraw-Hill Book Company, New York, 1987.

3. Bruzek, A., and C. J. Durrant. *Illustrated Glossary for Solar and Solar-Terrestrial Physics*. D. Reidel Publishing Company, Dordrecht, Holland, 1977.

4. David, P., and J. Voge. *Progagation of Waves*. Pergamon Press, Oxford, 1969.

5. Davies, K. *Ionospheric Radio Propagation*. U.S. Department of Commerce, National Bureau of Standards Monograph 80, Washington, 1965.

6. Davies, K. *Ionospheric Radio Waves*. Blaisdell Publishing Co., Waltham, Mass., 1969.

7. Davies, K. *Ionospheric Radio*. Peter Peregrinus Ltd., London, 1990.

8. Donnelly, R. F. (Ed.). *Solar Terrestrial Predictions Proceedings, Vol. 1 Prediction Group Reports*. No. 003-023-00041-9, U.S. Department of Commerce, U.S. Government Printing Office, Washington, D.C. 20402, 1979.

9. Donnelly, R. F. (Ed.). *Solar Terrestrial Predictions Proceedings, Vol. 2 Working Group Reports and Reviews*. No. 003-017-00471-6, U.S. Department of Commerce, U.S. Government Printing Office, Washington, D.C. 20402, 1979.

10. Donnelly, R. F. (Ed.). *Solar Terrestrial Predictions Proceedings, Vol. 3 Solar Activity Predictions*. No. 003-017-00473-2, U.S. Department of Commerce, U.S. Government Printing Office, Washington, D.C. 20402, 1979.

11. Donnelly, R. F. (Ed.). *Solar Terrestrial Predictions Proceedings, Vol. 4 Predictions of Terrestrial Effects of Solar Activity*. No. 003-017-00479-1, U.S. Department of Commerce, U.S. Government Printing Office, Washington, D.C. 20402, 1980.

12. Gething, P. J. D. *Radio Direction Finding*. Peter Peregrinus Ltd. on behalf of the Institution of Electrical Engineers, London, 1978.

13. Gibson, E. G. *The Quiet Sun*. NASA SP-303, NASA, Washington, 1973.

14. Goodman, J. M. *The Science and Technology of HF Communications*. Van Nostrand Reinhold, New York, 1991.

15. Goodman, J. M. (Ed.). *The Effect of the Ionosphere on Communication, Navigation, and Surveillance Systems*. Proceedings of the Ionospheric Effects Symposium IES-87. Naval Research Laboratory, Washington, 1987.

16. Goodman, J. M. (Ed.). *The Effect of the Ionosphere on Radiowave Signals and System Performance*. Proceedings of the Ionospheric Effects Symposium IES-90. Naval Research Laboratory, Washington, 1990.

17. Hall, M. P. M., and L. W. Barclay, *Radio Wave Propagation*. Peter Peregrinus Ltd., London, 1989.

18. Hargreaves, J. K. *The Upper Atmosphere and Solar-Terrestrial Relations*. Van Nostrand Reinhold, New York, 1979.

19. Hoover, K. E. *Ionospheric Modelling for HF Radio Source Location From a Single Site*. Ph.D. Thesis, University of Illinois. 1976.

20. Hunsucker, R. D. *Radio Techniques for Probing the Terrestrial Ionosphere*. Springer Verlag, 1991.

21. Jursa, A. S. (Ed.). *Handbook of Geophysics and the Space Environment*. Air Force Geophysics Laboratory, United States Air Force. NTIS Document Accession Number ADA 167000, 1985.

22. Kennedy, H. D., and W. Wharton. *Direction-Finding Antennas and Systems*. In Antenna Engineering Handbook, by R. C. Johnson and H. Jasik (Eds.). McGraw Hill, New York, 1984.

23. Picquenard, A. *Radio Wave Propagation*. MacMillan, New York, 1974.

24. Ratcliffe, J. A. *Sun, Earth and Radio*. World University Library, London, 1970.

25. Ratcliffe, J. A. *An Introduction to the Ionosphere and Magnetosphere*. Cambridge University Press, 1972.

26. Rishbeth, H., and O. K. Garriott. *Introduction to Ionospheric Physics*. Academic Press, New York and London, 1969.

27. Simon, P. A., G. Heckman, and M. A. Shea (Eds.). *Solar-Terrestrial Predictions: Proceedings of a Workshop at Meudon, France, June 1984*. NOAA, Boulder, Col. and AFGL, Bedford, Mass., 1986.

28. Tascione, T. F. *Introduction ot the Space Environment*, 2nd ed. Krieger Publishing Company. Malabar, Fla., 1994.

29. Thompson, R. J., D. G. Cole, P. J. Wilkinson, M. A. Shea, D. Smart and G. Heckmann (Eds). *Solar-Terrestrial Predictions: Proceedings of a Workshop at Leura, Australia. October 16–20, 1989*. NOAA/ERL, Boulder, 1990.

## Specialized Ionospheric Physics

1. Kelley, M. C. *The Earth's Ionosphere*. International Geophysics Series, Volume 43, Academic Press, San Diego, 1989.

## CCIR Reports

1. CCIR Report 249. *The Use of Oblique Sounding for Propagation Analysis and Optimization*. Dubrovnik, 1986.

2. CCIR Report 252-2. *CCIR Interim Method for Estimating Sky-Wave Field Strength and Transmission Loss at Frequencies between the Approximate Limits of 2 and 30 MHz*. New Delhi, 1970.

3. CCIR Report 322-3. *Characteristics and Applications of Atmospheric Radio Noise Data*. Dubrovnik, 1986.

4. CCIR Report 727. *Short-Term Prediction of Solar-Induced Variations of Operational Parameters for Ionospheric Propagation*. Dubrovnik, 1986.

5. CCIR Report 886-1. *Special Properties of the High Latitude Ionosphere Affecting Radiocommunications*. Dubrovnik, 1986.

6. CCIR Report 888-1. *Short-Term Forecasting of Critical Frequencies, Operational Maximum Usable Frequencies and Total Electron Content*. Dubrovnik, 1986.

7. CCIR Report 889-1. *Real-Time Channel Evaluation of Ionospheric Radio Circuits*. Dubrovnik, 1986.

8. CCIR Report 890-1. *The Operational Use of Side-Scatter and Backscatter*. Dubrovnik, 1986.

9. CCIR Report 1012. *Operational Modeling of HF Radio Propagation Conditions at High Latitudes*. Dubrovnik, 1986.

## Applications

1. *The ARRL Antenna Book*. Published by The American Radio Relay League, Inc., Newington, Conn.

2. *Radio Communication Handbook*. Published by the Radio Society of Great Britain.

3. Braun, G. *Planning and Engineering of Shortwave Links*. Siemens Aktiengessellschaft, Heyden & Son Ltd., London, 1982.

4. Maslin, N. M. *HF Communications: A Systems Approach*. Plenum Press, New York and London, 1987.

# Index

absorption, 47, 86, 143, 146
absorption fading, 95
absorption limiting frequency, 79, 82, 136
absorption loss, 86, 145
acoustic gravity waves, 36
active regions, 8
Advanced SAPS, 79
afternoon-type TEP, 126
ALF, 79, 82, 136
amateur 6 m observations, 130
angle-of-arrival diversity, 96
angle of incidence, 39
angular plasma frequency, 20
anisotropic, 37
anomaly, 24, 30, 32, 48
antenna pattern, 45, 146
antenna selection, 45, 147
Appleton anomaly, 30
ARTIST, 57, 72
ASAPS, 79, 91, 135
atmospheric noise, 101, 142
attachment, 17
aurora, 98
auroral absorption zone, 100
auroral oval, 98
auroral zone, 37
automatic scaling, 72
autoscaling, 72

backscatter ionogram, 20, 61
ball and hill, 51
Bartels chart, 117
best usable frequency, 91, 92, 147
Brunt-Vaisala frequency, 36
bubbles, 129
BUF, 91, 92, 147
buoyancy frequency, 36

calibration curve, 27, 76
central meridian, 111
Chapman layer, 24
characteristic waves, 41
chordal mode, 129
chromosphere, 7
circularly polarized, 40
circular polarization, 41
collisional ionization, 98, 108
control point, 77
corona, 10, 114
coronal mass ejection, 105
correlation distance, 89
crests, 30
critical frequency, 19
cusp, 54

daylight fadeout, 108
day-to-day variation, 24, 37
deviative absorption, 47

Digisonde, 55, 61
dip angle, 32
dip equator, 32
dispersive, 37
dissociative recombination, 18
disturbed days, 26
diurnal variation, 22
diversity, 96
Doppler shift, 35
Doppler spread, 126, 128
drift, 30
DX, 152

effective sunspot number, 89
electric field, 30
electron concentration, 18
electron density, 18
electron density profile, 66
electron gyro-frequency, 31
elevation angle, 83, 149
elliptically polarized, 41
EMUF, 79, 136
equatorial anomaly, 30
equatorial ionosphere, 30
Es, 28, 97, 125
EUV light, 7, 17
evening-type TEP, 127
E × B drift, 30
excess system loss, 141
extraordinary wave, 41

faculae, 8
fading, 93
fading depth, 93, 96
fading margin, 93
fading of AM stations, 94
fading rate, 93
Faraday rotation, 94
fibril, 8
field-aligned irregularities, 34, 141
field-guided mode, 128
field strength, 86, 141
filament, 8
flares, 106
flutter fading, 94, 98
focussing, 63
FOT, 49
fountain effect, 30
free-space loss, 86, 145
frequency diversity, 96
frequency optimum travail, 49
frequency plan, 137
frequency spread, 34
Fresnel zone, 93

galactic noise, 101
G condition, 114
geomagnetic equator, 31

geomagnetic field, 31
geomagnetic latitude, 31
geomagnetic longitude, 31
geomagnetic storm, 117
glitches, 32
gradual commencement, 117
GRAFEX frequency predictions, 78, 135
GRAFEX symbols, 83
gray line, 138, 152
ground range, 39
ground reflection loss, 44, 86, 145, 148
ground wave, 39
group height, 53
group refractive index, 53
gyrofrequency, 31

$H_\alpha$, 6
$H_\alpha$ classification, 106
half-wave dipole, 147
high-latitude problems, 98
high ray, 40
high speed solar wind stream, 114
hop, 44
horizon focus gain, 87, 145
horizontally polarized, 40
HSSWS, 114
hysteresis, 150

$IF_2$, 151
IIP, 62
initial phase, 117
interference, 101, 142
interference fading, 94
interferometer, 11
internal gravity waves, 36
IONCAP, 135
IONCAP V. PC 2.2, 152
IONCAST, 152
ionogram, 20
ionosonde, 20
ionosonde network, 42
ionospheric illumination power, 62
ionospheric index, 28, 149
ionospheric modification, 33, 65
ionospheric storm, 111
irregularities, 33, 65
isotropic antenna, 146
IUWDS, 89

Jindalee, 61, 65

latitudinal variation, 24
leading edge, 62
local time map, 76
long path, 138
lower decile, 25
lowest usable frequency, 48, 77
low-latitude problems, 97

low ray, 40
LUF, 48, 77

magnetic index, 117
magnetic storm days, 117
main phase, 117
man-made noise, 101, 143
Maunder minimum, 11
maximum observed frequency, 85
maximum usable frequency, 42
median, 25
MF broadcast station, 94
$M(3000)F_2$, 43
microscopic level, 47
midlatitude seasonal anomaly, 24
mid-latitude trough, 99
MINIFTZ4, 152
MINIMUF, 135, 152
MINIPROP, 152
models of ionosphere, 75
MOF, 85
monthly median predictions, 75
MUF, 42, 77
multi-hop circuits, 137, 145
multipath fading, 93
multipath interference, 45
musical charts, 117

near vertical incidence, 146
noise, 101, 142
noise at 3 MHz, 143
noise field strength, 145, 155
noise power, 149, 155
noise/pathloss parameter, 88
non-deviative absorption, 47
non-great-circle, 129
now casting, 89
N/P, 88
NVIS, 146

oblique ionogram, 20, 57
obliquity factor, 42, 140
optimum working frequency, 49
ordinary wave, 41
OWF, 49, 136

pathloss, 86, 143
PCA, 109
PC programs, 152
PC-VOACAP, 152
phase refractive index, 54
photoionization, 17
photosphere, 6
pipelines, 117
plages, 7
plane polarized, 40
planetary waves, 36
planning predictions, 84
plasma, 17, 33
plasma cloud, 111

plasma frequency, 19
POLAN, 70
polar blackout, 109
polar cap absorption event, 109
polarization, 40
polarization coupling loss, 86, 145
polarization diversity, 96
polarization fading, 94
polarization mismatch, 100
power lines, 117
prominence, 8
propagation modes, 44
proton event, 108

radiative recombination, 18
radio amateurs, 130
radio interferometer, 11
radio telescope, 10
range spread, 34, 128
raypath, 39
real height analysis, 66
receiving antenna, 147
recombination, 17
recurrent storms, 117
refraction, 39
Regional Warning Centers, 153
relay station, 109
residential noise, 143
ring current, 105
Rossby waves, 36
rotation of E vector, 41
round-the-world propagation, 128
RTW propagation, 128
rural noise, 143

satellite communications, 34, 96, 110
satellite drag, 12
saturation, 150
scattering, 98
scintillations, 96
screening effect, 97
SDF, 116
seasonal variation, 24
secure communications, 43
seeing, 5
selective fading, 94
short path, 138
shortwave fadeout, 107
side scatter, 130
signal-to-noise ratio, 88, 141
single event upset, 32
skip fading, 95
skip focussing, 95
skip zone, 43
Skylab, 12
skywave, 39
S/N, 88, 141
solar cycle, 11
solar flares, 106
solar magnetic field, 8

solar observatories, 6
solar rotation period, 6
solar 10.7 cm flux, 11
solar wind, 10
solar wind stream, 114
South Atlantic anomaly, 32
space diversity, 96
SPARC, 153
spatial attenuation, 145
sporadic E, 28, 97, 125
sporadic-E obscuration loss, 86, 145
spread F, 34, 55
starting problem, 69
sudden commencement, 117
sudden disappearing filament, 116
sunspot groups, 6
sunspot number, 6
sunspots, 6
super mode, 126
susceptibility, 121
SWF, 108

TEP, 126
thermosphere, 37
thermospheric winds, 37
TID, 34
tides, 36
tilted bar magnet, 31
time diversity, 97
T index, 28, 151
topside ionograms, 51
trailing edge, 64
transequatorial propagation, 126
travelling ionospheric disturbances, 34
true height analysis, 66
two-hop circuit, 137

unusual propagation modes, 125
upper decile, 25, 137

valley, 69
van Allen belts, 105
vertical incidence ionogram, 51
vertically polarized, 40
virtual height, 53
VOACAP, 152

whip antenna, 147
whip tilt adapter, 147
white light image, 5
winds, 37
Winter anomaly, 48
world maps, 76

X rays, 9
X-ray classification, 106

zenith angle, 22